Springer Series in Synergetics   Editor: Hermann Haken

Synergetics, an interdisciplinary field of research, is concerned with the cooperation of individual parts of a system that produces macroscopic spatial, temporal or functional structures. It deals with deterministic as well as stochastic processes.

E. Schöll

Nonequilibrium Phase Transitions in Semiconductors

Self-Organization Induced by
Generation and Recombination Processes

With 165 Figures

Springer-Verlag Berlin Heidelberg New York
London Paris Tokyo

Priv.-Doz. Dr. rer. nat. Eckehard Schöll, Ph. D.

Institut für Theoretische Physik, Rheinisch-Westfälische Technische Hochschule, Templergraben, D-5100 Aachen, Fed. Rep. of Germany

Series Editor:

Professor Dr. Dr. h. c. Hermann Haken

Institut für Theoretische Physik der Universität Stuttgart, Pfaffenwaldring 57/IV, D-7000 Stuttgart 80, Fed. Rep. of Germany and
Center for Complex Systems, Florida Atlantic University, Boca Raton, FL 33431, USA

ISBN-13: 978-3-642-71929-5   e-ISBN-13: 978-3-642-71927-1
DOI: 10.1007/ 978-3-642-71927-1

Library of Congress Cataloging-in-Publication-Data. Schöll, E. (Eckehard), 1951- Nonequilibrium phase transitions in semiconductors. (Springer series in synergetics ; vol. 35) 1. Semiconductors. 2. Phase transformations (Statistical physics) 3. Phase rule and equilibrium. 4. Trapped-particle instabilities. I. Title. II. Title: Self-organization induced by generation and recombination processes. III. Series: Springer series in synergetics ; v. 35. QC611.S33   1987   537.6'22   87-4648

Typesetting: ASCO Trade Typesetting Limited, Hongkong

2153/3150-543210

To my parents

Preface

Semiconductors are complex dynamic systems which can in many cases exhibit electrical instabilities like current runaway, switching between a nonconducting and a conducting state, or spontaneous oscillations of current or voltage when they are driven far from thermodynamic equilibrium by strong external electric fields, irradiation, or current injection. There are a variety of different physical mechanisms that can give rise to such instabilities, but the observed phenomena are often similar, involving the spontaneous formation of spatial and temporal structures. Such *self-organizing* cooperative processes have been noted in a great number of different physical, chemical, and biological "synergetic" systems, when a state *far from thermodynamic equilibrium* is maintained by a continuous flux of energy or matter flowing through them. These instabilities bear a remarkable analogy with *phase transitions* of systems in thermal equilibrium, like ferromagnets or real gases.

Self-organization and nonequilibrium phase transitions in semiconductors are of considerable current interest for at least two reasons. First, these phenomena are the basis of a number of important semiconductor devices which are used in modern microelectronics and semiconductor technology. Second, semiconductors are particularly apt model systems for the study of complex nonlinear dynamics and self-organization, since recent advances in the technology of tailoring specific samples ("materials engineering"), and the direct observation via current and voltage measurements allowing for good reproducibility and high temporal and spatial resolution have opened up the possibility of a fruitful interaction between theory and experiment.

In this book we concentrate on those semiconductor instabilities whose physical mechanism is based upon nonlinear generation and recombination processes of the charge carriers. The aim of the book is to provide a coherent theoretical description of the spatial and temporal structures induced by simple generation-recombination mechanisms. It attempts to go beyond older work by taking into account in a systematic way the *nonlinearities* of these generation-recombination processes and all dynamic degrees of freedom of free *and trapped* carriers, and by introducing and developing the concepts of nonequilibrium phase transitions. A comparison with experiments is drawn where appropriate. The emphasis is on recent theoretical and experimental advances in this still-open field, and no comprehensive historic overview is attempted.

This work is intended to connect the field of semiconductor physics and the theory of nonlinear dynamic systems far from thermodynamic equilibrium. It might therefore be useful both to applied semiconductor physicists and to theoreticians by providing, on one hand, new concepts and viewpoints for the understanding of certain semiconductor instabilities, and, on the other hand, by presenting a new application of the theory of synergetic systems. The organization of the material is as follows. Chapter 1 contains the basic physical principles and some background of semiconductor instabilities and nonequilibrium phase transitions. In Chap. 2 a number of simple generation-recombination mechanisms for nonequilibrium phase transitions are presented, and potential applications to threshold switching in crystalline and amorphous semiconductors, nonlinear magneto-photoconductivity, and optical bistability are discussed. In Chaps. 3–6 the spatial and temporal structures resulting from the primary instability are developed. Chapter 3 discusses the linear modes describing the initial stages of the spatial and temporal instabilities. In Chap. 4 the fully developed stationary spatial structures (current filaments) are analyzed. Their stability, and transient phenomena like spinodal decomposition and nucleation processes are discussed in Chap. 5. Chapter 6 deals with temporal and spatio-temporal structures, in particular mechanisms for oscillations and for chaos are developed, and applied to explain recent experiments, in which intrinsic chaos was observed in semiconductors under a variety of experimental conditions.

I am very much indebted to Prof. P. T. Landsberg, Southampton/Gainesville, Fla., and Prof. F. Schlögl, Aachen, who have greatly influenced and stimulated the development of the ideas presented in this book. In the past I have benefited from valuable discussions with Prof. D. Bimberg, Berlin; Prof. V. Dohm, Aachen; Prof. B.C. Eu, Montreal; Dr. H.L. Grubin, Glastonbury, Conn.; Prof. R.P. Huebener, Tübingen; Prof. N. Klein, Haifa; Dr. V.V. Mitin, Kiev; Dr. J. Parisi and J. Peinke, Tübingen; Prof. W. Prettl, Regensburg; Prof. J. Schnakenberg, Aachen; Prof. M.P. Shaw, Detroit, Mich.; Prof. A. Stahl, Aachen; Prof. H. Thomas, Basel, and my colleagues at the Institute of Theoretical Physics of the RWTH Aachen, in particular Drs. C. Escher and W. Renz. I am obliged to Mrs. B. Schumacher for checking some of the formulas.

I would like to thank Prof. H. Haken for inviting me to write this monograph for the Springer Series in Synergetics. I thank Dr. H. Lotsch from Springer-Verlag for his excellent cooperation.

This book is partly based on a series of lectures given during the winter 1984 at the Institute for Amorphous Studies, Bloomfield Hills, Mich., USA, in conjunction with the Department of Electrical and Computer Engineering of Wayne State University Detroit, Mich. The hospitality of the Institute, and of Prof. M. P. Shaw from Wayne State University in particular is acknowledged gratefully. Part of this work was supported by the "Deutsche Forschungsgemeinschaft", and by the "Studienstiftung des deutschen Volkes".

It is a pleasure to thank Mrs. J. Elbert for her extremely fast and efficient typing of the manuscript.

Finally, I wish to offer sincere thanks to my wife and my children for their patience and encouragement during all stages of writing.

Aachen, April 1987   *Eckehard Schöll*

Contents

1. Introduction

This book deals with physical aspects of current instabilities in semiconductors, induced by generation and recombination (g-r) processes of the charge carriers. Although instabilities of semiconductors and insulators have been known for a very long time, for example in connection with dielectric breakdown in solids [1.1], the view of such an instability as a phase transition in a physical system far from equilibrium is a fairly recent development. The analogy of an overheating instability of the electron gas with an equilibrium phase transition was pointed out by *Volkov* and *Kogan* [1.2] in the late sixties, and *Pytte* and *Thomas* [1.3] drew this analogy in the case of the Gunn instability of the electron-drift velocity at about the same time. But generation-recombination (g-r) induced phase transitions in semiconductors were first noted by *Landsberg* and *Pimpale* [1.4] only a decade ago, stimulated by the similarity with *Schlögl*'s famous chemical reaction models for nonequilibrium phase transitions [1.5]. During these past ten years both the experimental observation and theoretical understanding of g-r induced phase transitions have made great progress, and have established a wealth of novel phenomena and models, thus giving birth to a new member of the growing family of physical and nonphysical systems that exhibit nonequilibrium phase transitions. The study of these systems has generated a new interdisciplinary field of science for which *Haken*, who pioneered these phenomena in the field of laser physics [1.6], has coined the name "synergetics" [1.7–9].

The particular fascination of the subject of this book is that it connects the discipline of semiconductor physics with the field of nonequilibrium thermodynamics. Both these fields are currently of great interest, and are experiencing a scientific "boom". Semiconductor physics is the basis of a vast variety of electronic devices [1.10, 11], ranging from VLSI (very large scale integrated) circuits, microprocessors, and high-frequency amplifiers and generators to semiconductor lasers, photodetectors, and other optoelectronic components. Many of these devices operate in the regime of controlled electrical instabilities which manifest themselves, for instance, by current or voltage oscillations, or by switching transitions between a conducting and a nonconducting state. They are thus connected with nonequilibrium phase transitions.

Nonequilibrium thermodynamics deals with open macroscopic systems with a large number of microscopic degrees of freedom driven by external fluxes and forces so far from thermal equilibrium that linear dynamic laws no longer hold. Due to the driving forces and the inherent nonlinearities of these systems, they may spontaneously evolve into a state of highly ordered spatial or temporal structures. These are called dissipative structures [1.12] since they can only be maintained by the continuous influx and dissipation of energy. Unlike an isolated, closed system, which after a perturbation always returns to a thermal equilibrium state characterized by maximum entropy, an open nonlinear system may exhibit a process of self-organization, in which the entropy is locally decreased. Such processes usually involve qualitative changes in the state of the system, similar to phase transitions in equilibrium systems. Progress has emerged in the discipline of nonequilibrium thermodynamics over the past years in a two-fold way: first, the mathematical tools to deal with such nonlinear differential systems have been developed and refined, including in particular the theory of bifurcation [1.13–16], the theory of dynamic systems [1.17–18], catastrophe theory [1.19–22], soliton theory [1.23], and the theory of chaos [1.24]. Second, some progress has been made in singling out relevant physical quantities and observables in nonequilibrium systems that are suitable to characterize and classify this novel behavior [1.25].

It is the aim of this book to apply the new methods developed in the field of nonequilibrium thermodynamics to semiconductors, thereby providing a better understanding of known, and a prediction of new, g-r induced instabilities in semiconductors. Emphasis will be on a coherent theoretical treatment of generation-recombination based mechanisms and simple models. The theory will be illustrated by experimental findings, in particular recent results, wherever possible. The selection of material reflects the author's personal interest and involvement, and does not attempt to give a comprehensive overview of all known current instabilities in semiconductors in which generation or recombination plays a role in some way. A number of reviews and monographs are available for that purpose [1.26–35]. Rather, the selection of material was guided by the consistency and coherence of the methods and models used. The material is organized in a progressive order of complexity of the dissipative structures, see Table 1.1.

In this chapter the basic phenomenology of instabilities in semiconductors, including a survey of different physical mechanisms, will be given, and the analogy with phase transitions in thermodynamic systems in equilibrium will be outlined. Chapter 2 contains a selection of g-r models which give rise to nonequilibrium phase transitions between different steady states. The analysis is on the simplest level of description, taking into account spatially homogeneous steady states only. It is shown that the g-r rate equations can result in bistability of these steady states, which leads to S-shaped static current density-versus-field characteristics. A number of g-r mechanisms involving a single type of charge carriers, two types of charge carriers (electrons and holes), and excitons, are presented.

In Chap. 3 the g-r mechanisms are coupled to charge-transport processes and to Maxwell's equations for the electromagnetic fields. The space- and time-dependent evolution is investigated from a linear version of these equations, linearized around the homogeneous nonequilibrium steady state. This allows one to draw conclusions about the local stability of this homogeneous steady state, i.e., its response to small space- and time-dependent fluctuations. As a result, a variety of filamentary, domain-type, and oscillatory instabilities are found for single-carrier g-r mechanisms.

In Chap. 4 the nonlinear transport equations are analyzed for stationary, but space-dependent solutions. Current filaments represent such solutions; they are calculated for different geometries and boundary conditions, and for both single-carrier and two-carrier g-r mechanisms. The stability of these stationary, space-

Table 1.1. Organization of the book (schematic)

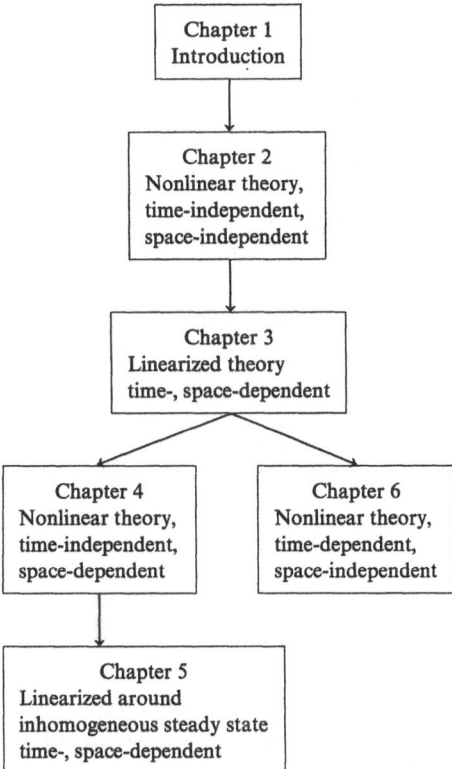

dependent solutions against small space- and time-dependent fluctuations is investigated in Chap. 5 by linearizing the transport equations around the spatially inhomogeneous steady state. This provides some insight into transient phenomena. In Chap. 6 time-dependent solutions of the nonlinear transport equations are studied; attention is mainly confined to spatial homogeneity. Self-sustained oscillations and chaos are obtained and compared with recent experimental findings.

The logic connection of the chapters is summarized schematically in Table 1.1. Particular emphasis is placed on the consistent and systematic treatment of the *nonlinear* generation-recombination processes, and of the dynamic degrees of freedom of free *and trapped* carriers throughout the book. This is not normally done in standard approximations of semiconductor transport theory or device modeling, but will in fact turn out to be essential for the description of nonequilibrium phase transitions and dissipative structures in semiconductors.

Throughout this book we restrict ourselves to a deterministic level of description, where charge transport and generation-recombination processes are described by a system of differential equations for a few macroscopic variables. On a mesoscopic level, a stochastic treatment of fluctuations would be necessary, either by adding random forces to the differential equations ("Langevin approach"), or by considering equations of motion for the distribution function of macroscopic variables ("Fokker-Planck equation", "Master equation", "Boltzmann equation"). The reader is referred to the vast literature on stochastic methods, e.g., [1.36–39].

1.1 Instabilities in Semiconductors

A semiconductor under sufficiently strong excitation conditions such as large electric or magnetic fields, strong optical irradiation, or high-current injection will in general exhibit transport properties which deviate substantially from linear ("Ohmic") current-voltage relations and lead in many cases to instabilities like current runaway, current oscillations, discontinuities in the current and/or voltage, switching, and hysteretic current-voltage characteristics. These instabilities are widespread in a variety of materials, temperature ranges, and excitation conditions [1.26–35]. Although such instabilities may often have a detrimental effect upon the performance of solid state devices, they have been used intentionally in a number of very important semiconductor devices for the generation of microwave power in the frequency range between 0.1 and 1000 GHz, for amplification in the GHz frequency range where normal transistors cannot be used, and for fast electronic switches.

1.1.1 Negative Differential Conductivity

The transport properties of a semiconductor show up most directly in its current-voltage ($I-V$) relation under time-independent (dc = direct current) conditions. It is determined in a complex way by the microscopic properties of the bulk semiconductor material, which gives the current density j as a function of the local electric field E, and by the contacts. A local, static, scalar $j(E)$ relation need not always exist, but in fact does in many cases.

If the $j-E$ characteristic has a regime of *negative differential conductivity*

$$\sigma_{\text{diff}} := \frac{dj}{dE} < 0 \ , \tag{1.1.1}$$

i.e., if the current density decreases with increasing electric field, or vice versa, then the corresponding time-independent states are generally unstable, and the actual electric response depends, for instance, upon the attached circuit which in general contains – even in the absence of external load resistors – unavoidable resistive and reactive components like lead resistances, lead inductances, package inductances, and package capacitances.

Negative differential conductivity (NDC) can be classified as NNDC or SNDC, depending upon the shape of the $j-E$ characteristic, which may resemble the letter N or S, respectively (Fig. 1.1). NNDC and SNDC are associated with voltage- or current-controlled instabilities, respectively. In the NNDC case the current density is a single-valued function of the field, but the field is multivalued: the $E(j)$ relation has three branches in a certain range of j. The SNDC case is complementary in the

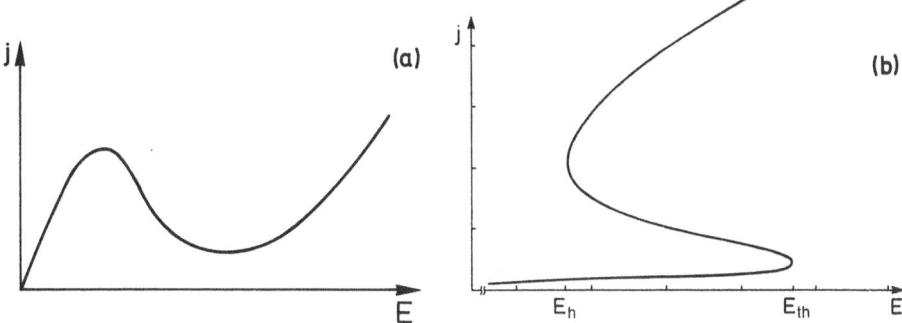

Fig. 1.1a, b. Current density j versus electric field E for two types of negative differential conductivity (NDC): (a) NNDC; (b) SNDC (schematic)

sense that E and j are interchanged. This duality is in fact far-reaching, and will be elaborated in later chapters (see, e.g., Sect. 6.2.1).

Combinations of SNDC and NNDC can also occur: regimes of SNDC and NNDC may follow one another on the static j–E characteristic as the electric field is increased [1.40]; the characteristic may evolve from SNDC to NNDC as time goes on [1.41]; or the static characteristic may have a more complicated shape than the simple N or S, being multivalued in both j and E [1.42, 43].

The current-voltage characteristic of a semiconductor can in principle be calculated from the j–E relation by integrating the current density j over the cross section of the current flow

$$I = \int j \, df \tag{1.1.2}$$

and the field E over the length L_z of the sample

$$V = \int_0^{L_z} E(z) \, dz \ . \tag{1.1.3}$$

Unlike the j–E relation, the I–V characteristic is not only a bulk property of the semiconductor material, but also depends on the geometry, the boundary conditions, and the contacts of the specific sample. If, however, the steady state is spatially homogeneous, and the contact resistance (which is in general – even for "Ohmic" contacts – strongly nonlinear) is negligible against the resistance of the bulk, then the j–E relation and the I–V relation are similar, i.e., up to scaling, identical.

It is, however, a general phenomenon that negative differential conductivity is associated with the instability of the homogeneous steady state against spatial fluctuations of the electric field and the carrier densities, and results in spatially inhomogeneous distributions of the current density or the field [1.44, 45, 26, 30]. A general thermodynamic argument for the formation of such spatial structures [1.44] has later been shown to be invalid [1.46, 47], since the invoked principle of minimum entropy production holds only in the linear, near-equilibirum regime of thermodynamics for systems with Onsager symmetry [1.48, 49], whereas NDC occurs far from thermal equilibrium. There are, however, numerous specific examples of

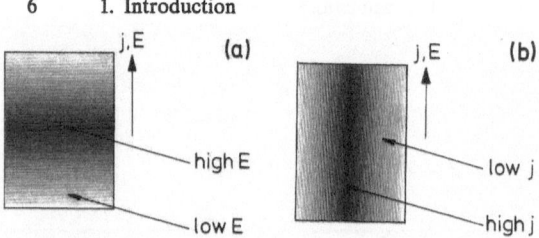

Fig. 1.2. Sketch of (**a**) a field domain, (**b**) a current filament, where j is the current density, E is the electric field (schematic)

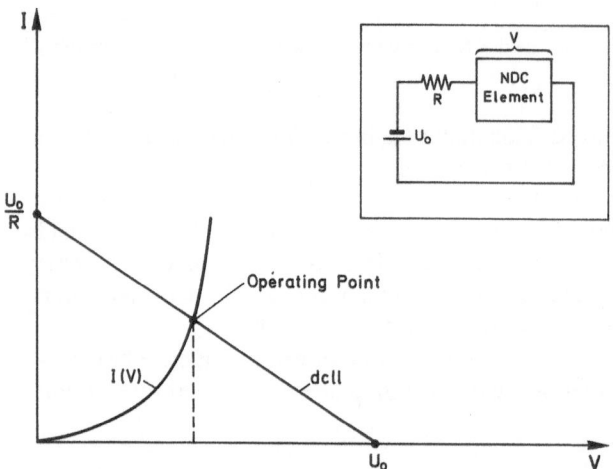

Fig. 1.3. Schematic behavior of an NDC element in a resistive circuit with an applied voltage U_0 and a load resistor R. The current I is plotted versus the voltage drop V across the NDC element. The steady state operating point is the intersection of the dc load line (dcll) and the $I(V)$ characteristic of the NDC element. The inset shows the circuit

NNDC, where the homogeneous steady state is unstable against the formation of moving or stationary inhomogeneous, layer-like field distributions with a hump-shaped field profile aross the sample ("high-field *domains*") (Fig. 1.2a). Similarly, in many examples of SNDC the homogeneous steady state is unstable against the formation of an inhomogeneous current-density distribution over the cross section of the current flow, with a cylindrical core of high-current density ("high-current *filament*") and a surrounding mantle of low-current density (Fig. 1.2b).

The simplest circuit in which an NDC element operates is shown in the inset of Fig. 1.3. Here it is in series with a load resistor R and a bias battery that provides a voltage U_0. If I is the current in the circuit and V the voltage drop across the NDC element, then

$$U_0 = IR + V \tag{1.1.4}$$

or $I = (U_0 - V)/R$ is the equation of the dc *load line* (dcll). This line is plotted in Fig. 1.3, its slope is $-1/R$, and its intersection with the device characteristic

$I(V)$ defines the steady state operating point. Operating points with negative differential conductance $dI/dV < 0$ (corresponding, e.g., to homogeneous steady states with negative differential conductivity dj/dE) are often, but not always, unstable both against the formation of inhomogeneous field or current-density distributions (space-charge nonuniformities) and/or circuit-controlled oscillatory effects. Although the detailed nature of these instabilities depends upon the particular NDC mechanism, some general results can be inferred from the shape of the $I(V)$ characteristic and the position of the load line. In case of N- or S-shaped $I(V)$ characteristics (Fig. 1.1), up to three intersection points of the $I(V)$ characteristic and the load line can exist. As the applied voltage is varied, the load line is shifted parallel, and the intersection points move along the $I(V)$ characteristic. When the load line becomes tangential to the characteristic (which may occur somewhere within the regime of negative dI/dV), two intersection points coalesce and, upon further variation of U_0, disappear: this is the simplest type of a bifurcation. Generally, bifurcations are intimately connected with the loss or exchange of stability of different solution branches. In our case this means that one of the two coalescing intersection points is stable, the other one is unstable – although both have negative differential conductance dI/dV. Thus the instability of certain branches of the $I(V)$ characteristic appears to be a generic property of their differential topology and their bifurcation behavior, rather than being simply determined by the sign of dI/dV. As explicit examples for the stabilization of points with $dI/dV < 0$ by a load line we refer to [1.50, 51].

1.1.2 Mechanisms for NDC

In this section we shall briefly review the major microscopic mechanisms that give rise to negative differential conductivity. They are effective in a variety of important semiconductor devices. Among those devices which are associated with NNDC are the tunnel diode [1.52, 53] and the Gunn diode [1.54]. The IMPATT diode [1.55, 56], multilayer devices like thyristors [1.57], pnpn diodes, and pin diodes [1.58], and thermal, electrothermal, or ovonic switches [1.59] operate in the SNDC regime. Tunnel, Gunn, and IMPATT (Impact ionization avalanche transit time) diodes are often used as microwave oscillators and generators, while the other devices represent electronic switches.

The NDC mechanisms can be dominated either by junction or bulk properties. The mechanisms of the tunnel diode and the pnpn diode are based upon junction effects. The *tunnel diode* (or Esaki diode) is a pn-junction device which operates by the quantum mechanical tunnelling of electrons through the potential barrier of the junction. The n- and the p-doped side are "degenerate", i.e., the Fermi-level in thermal equilibrium lies within the conduction band on the n-side (E_{Fn}), and the valence band on the p-side (E_{Fp}), such that the bottom of the conduction band on the n-side is filled with electrons, and the top of the valence band on the p-side is filled with holes, as shown schematically in Fig. 1.4a. When a small forward bias V is applied (i.e., the n-side is connected to a negative, the p-side to a positive voltage), E_{Fn} moves up in energy with respect to E_{Fp} by the amount eV, where e is the elementary charge (Fig. 1.4b). Thus, electrons below E_{Fn} on the n-side are placed

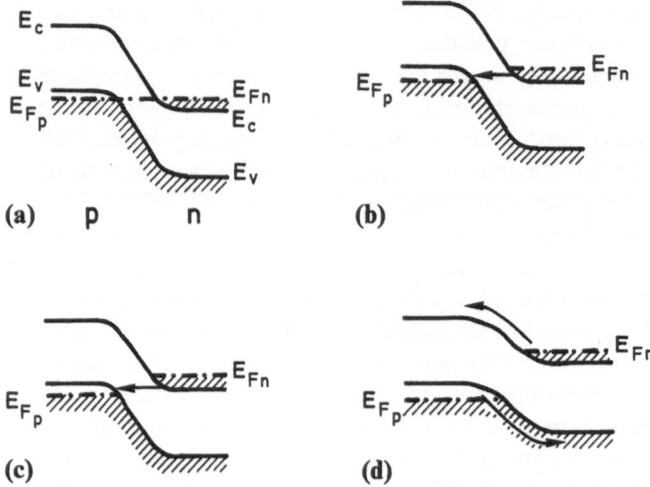

Fig. 1.4a–d. Schematic band diagrams of the tunnel diode: (**a**) thermal equilibrium (zero bias), (**b**) small forward bias: electron tunneling from n to p, (**c**) increased forward bias: electron tunneling from n to p decreases as bands pass by each other, (**d**) further increased bias: electrons and holes surmount the potential barrier. E_{Fn} and E_{Fp} are the (quasi-) Fermi levels on the n- and p-side, respectively; E_c and E_v denote the bottom of the conduction band and the top of the valence band, respectively

opposite empty states above E_{Fp} on the p-side. Electron tunneling occurs from n to p as shown. This forward tunneling current continues to increase with increased bias as more filled states are placed opposite empty states. However, as E_{Fn} continues to move up with respect to E_{Fp}, a point is reached at which the bands begin to pass each other. When this occurs, the number of filled states opposite empty states decreases (Fig. 1.4c). The resulting decrease in tunneling current produces a negative differential conductance $dI/dV < 0$. If the forward bias is further increased, the barrier between the p- and the n-region is lowered further, and the current begins to increase again due to the enhanced diffusion over the barrier (Fig. 1.4d). Altogether, an N-shaped current-voltage characteristic is produced. For a more detailed recent theoretical treatment see [1.53].

Another junction effect is used in the *pnpn diode*. This device is a four-layer (p-n-p-n) structure. There are many variations of the basic pnpn structure, including the thyristor or semiconductor controlled rectifier (SCR), which is used as power switch or light-dimmer switch: it effectively blocks current through two terminals until it is turned on by a small signal at a third terminal. The basic four-layer structure ("Shockley diode") consists of three consecutive junctions: j_1 (p-n), j_2 (n-p), and j_3 (p-n). If the p-side of j_1 is connected to a positive, and the n-side of j_3 to a negative voltage, i.e., the whole pnpn diode is forward biased, there are two possible states: a low-conductivity (small I) and a high-conductivity (large I) state. In the first case the junctions j_1 and j_3 are forward biased, but the junction j_2 is reverse biased, and hence blocking, while in the second case the potential distribution across the device is nonmonotonic such that all three junctions are forward biased, and hence

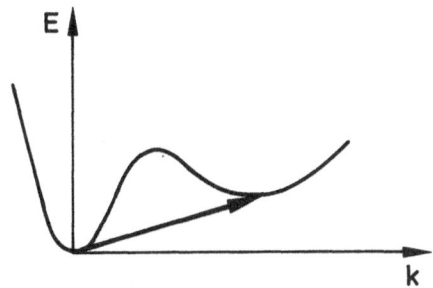

Fig. 1.5. Band structure (energy E versus wave vector k) of the conduction band of GaAs (schematic). The central valley at $k = 0$ has a small effective mass ($m^* \approx 0.07m_0$), while the satellite valley has a high effective mass ($m^* \approx 1.2m_0$)

conducting. This produces an S-shaped I–V characteristic. The mechanism by which the junction j_2 switches from reverse bias to forward bias can be understood by considering the pnpn structure as a combination of a pnp and an npn transistor [1.10], but this will not be discussed here.

Bulk-dominated NDC includes four major classes of mechanisms: the non-monotonic form of the current density j as a function of the field E, which leads to $dj/dE < 0$ in some range of E, can be due to a nonlinearity of the mobility (*drift instability*), of the carrier-density (*g-r instability*), of the electron temperature (*electron overheating instability*), or of the lattice temperature (*electrothermal instability*).

The best-known drift instability is the *Gunn effect* [1.31, 54]. It is used in Gunn diodes to generate and amplify microwaves at frequencies typically beyond 1 GHz. (These devices are called "diodes", since they are two-terminal devices, but no pn junction is involved!) The mechanism is based upon intervalley transfer of electrons from a state of high mobility to a state of low mobility by the influence of a strong electric field ($E > 3$ kV/cm). The band structure of GaAs and other III–V compound semiconductors is shown schematically in Fig. 1.5. At low electric fields the electrons are essentially in the minimum of the central valley, which has a low effective mass m^* and hence a high mobility. As the field E is increased, the electrons are heated up, and gain enough energy to be transferred to the satellite valley with a higher minimum energy, but larger effective mass, and hence lower mobility. As more and more electrons are transferred, the averaged mobility μ decreases strongly so that the current density $j = en\mu(E)E$ decreases with increasing field, as a result of negative differential mobility (NDM). When most electrons are in the upper valley, j increases again. Thus an NNDC characteristic is produced.

Other mechanisms for negative differential mobility are due to the anisotropy of equivalent satellite valleys [1.33], for instance the *Erlbach instability* in Ge [1.60, 61]. While the Gunn instability occurs for a current parallel to the applied field, the Erlbach instability involves off-diagonal elements of the differential conductivity tensor dj_α/dE_β. Consider in the simplest case two equivalent valleys 1 and 2 with anisotropic axes pointing in different directions. If an electric field E is applied in the symmetry direction between these two valleys, the current j is also in the x-direction. Now suppose that the field is slightly off the x-direction, such that the conductivity effective mass in valley 1 is greater than the mass in valley 2. The rate at which electrons absorb energy from the electric field, i.e., the rate at which they are heated, is inversely proportional to the conductivity effective mass, so the electrons in valley 2 are heated more than those in valley 1. Thus there will be a net transfer of electrons by intervalley scattering from the hotter valley to the cooler valley, and the electron population becomes greater in valley 1 than in valley 2. The decrease in the population of the higher-mobility valley 2 then leads to a negative contribution to the current in y-direction, which can result in transverse negative differential conductivity (NNDC). Another mechanisms for NDM and NNDC is based upon *Bragg scattering* at the boundaries of the Brillouin zone [1.62].

The large class of g-r instabilities is distinguished by a nonlinear dependence of the steady state carrier concentration n upon the field E, which yields a nonmonotonic current density-field relation $[j = e\mu n(E)E$ in the simplest case of unipolar electron currents] of either NNDC or SNDC type. This dependence is due to a redistribution of electrons between the conduction band and bound states (and possibly the valence band, in case of ambipolar currents) during the heating of the electron gas. The microscopic transition probabilities of the carriers between different states, and hence the g-r coefficients, generally depend upon the electric field. A particularly strong dependence is expected for the rate constants of the following g-r processes: *field-enhanced trapping*, which often leads to NNDC, and *impact ionization*, which is the key process for SNDC, as we shall show in Chap. 2.

Field-enhanced trapping occurs, e.g., in gold-doped n-Ge [1.30]. The gold atoms form deep impurity levels, corresponding to singly or doubly charged negative ions. Trapping of electrons into these levels requires the penetration of a Coulomb potential barrier. Therefore the trapping coefficient increases with field E while the (thermal) emission of electrons is practically field-independent, as long as E is well below the threshold for field ionization. Thus the free-carrier density decreases with rising field, $dn/dE < 0$, and the differential conductivity $dj/dE = e\mu(n + E\,dn/dE)$ can become negative. In larger fields the ionization coefficient increases and causes the carrier density to rise again with field, leading to a positive differential conductivity branch. Thus an N-shaped j–E characteristic is produced [1.63–67].

Impact ionization of carriers from impurity levels (shallow donors, acceptors, or deep traps) or across the bandgap (avalanching) is another process which can lead to NDC. If a free carrier has gained enough kinetic energy in the electric field, it can transfer this energy in a collision to a bound carrier, which is then released to the conduction band (or valence band, in case of a bound hole); thereby an additional free carrier is generated which may, in turn, impact ionize other carriers. Such a positive feedback ("autocatalysis") leads to a rapid increase of the free-carrier density. The impact-ionization coefficient strongly increases with E beyond a threshold field which is necessary to heat up the carriers so that they have enough energy for ionization. In Chap. 2 we will describe in detail some mechanisms based upon impact ionization in combination with other g-r processes, and show that they can under certain conditions lead to SNDC. Models of this type are relevant in a variety of materials and in different temperature ranges, for instance in low-temperature impurity breakdown [1.68, 69] or switching in amorphous thin films at room temperature [1.70, 71]. The general class of *g-r induced SNDC mechanisms* is the basis of most of the phenomena we shall deal with in this book.

Two important devices are also based upon g-r induced bulk negative differential conductance, but the coupling with junction effects is essential in these cases: pin diodes [1.72] and IMPATT diodes [1.73, 74]. The *pin diode* involves double injection of electrons and holes, and field-enhanced trapping. It consists of an intrinsic (undoped) layer adjacent to a p-doped and an n-doped region on either side. If the n-layer is connected to a negative and the p-layer to a positive voltage, electrons and holes are injected from the n-side and the p-side, respectively, into the central intrinsic (i) layer. We assume that the i-region contains deep acceptor-like recombination centers with a large cross section (or capture coefficient) for hole capture and a much smaller cross section for electron capture, and that these are completely occupied by electrons in thermal equilibrium. At low injection currents almost all of the injected holes will be captured by the recombination centers near the injecting *pi* junction, while the injected electrons freely traverse most of the i-layer. Thus there is a "recombination barrier" to the passage of holes. The resulting electron current is limited by the space charge that the injected electrons build up. At high injection currents, where the injected electron and hole concentrations exceed that of the recombination centers, all centers have trapped a hole, and the excess holes as well as the electrons traverse the i-layer. They are approximately equal in concentration, if the injection level is sufficiently high. The current is then carried by a quasi-neutral semiconductor plasma. Thus, at a given value of the applied voltage, there are two stable steady states: a low-current state, where the recombination centers are (except for a narrow region near the *pi* junction) occupied by electrons and the current is a single-carrier space-charge limited (SCL) current, and a high-current state, where the recombination centers are filled with holes, and the current is carried by an injected plasma. In between, there is an NDC state in which both the single-carrier SCL current and the semiconductor plasma extend some way into the i-region from both sides (from the *in* and the *pi* junction, respectively), and they are connected by regions of more complex carrier and field distributions [1.72].

The family of IMPATT (Impact ionization avalanche transit time) devices depends upon a combination of impact ionization and transit time effects. The IMPATT diodes can generate the highest cw (continuous wave) power output in the millimeter-wave regime (i.e., at frequencies > 30 GHz). The originally proposed device (Read diode) involves a reverse-biased n^+-p-i-p^+ structure, where n^+ and p^+ denote strongly n- or p-doped regions. In the n^+-p region (avalanche region), carriers are generated by impact ionization across the bandgap; the generated holes are swept through the i-region (drift region), and collected at the p^+ contact. When a periodic (ac) voltage is superimposed on the time-independent (dc) reverse bias, negative ac conductance can arise if the ac component of the carrier flow drifts opposite to the ac electric field. This corresponds to a phase lag of π of the current behind the voltage. This phase lag is due to the finite build-up time of the avalanche current ("avalanche delay") and the finite time it takes the carriers to cross the drift region ("transit-time delay"). If the sum of these delay times is approximately a one-half cycle of the operating frequency, negative conductance occurs. This can be achieved by properly matching the length of the drift region with the drift velocity and the frequency. Other transit-time devices are the BARITT (barrier injection and transit time) diode, the DOVETT (double velocity transit time) diode, and the TRAPATT (trapped plasma avalanche triggered transit) diode [1.11].

Negative differential conductivity can also arise from changes in the nature of the dissipation of the energy and momentum of the carriers as they are heated up by the electric field E [1.26, 30]. In such an *electron overheating instability* the energy and momentum relaxation times (τ_e and τ_m, respectively) depend in a nonlinear way upon the average energy per carrier which can often be related to an effective electron temperature T_e (Sect. 1.2.2). The mobility (in the simplest approximation: $\mu = e\tau_m/m^*$) is thus a nonlinear function of T_e which is specified by the particular scattering mechanisms, e.g., with acoustic phonons, optical phonons, or at impurities. The electron temperature T_e as a function of E follows in the steady state from a balance of the electrical power density $jE = en\mu(T_e)E^2$ with the power density $nP(T_e)$ dissipated by the electrons to the lattice. From this the differential conductivity dj/dE with $j = en\mu[T_e(E)]E$ can be calculated; both SNDC or NNDC are possible. For a recent example of such a mechanism involving optical phonon scattering in a two-dimensional system (quantum well) we refer to [1.75].

The preceding mechanisms were purely electronic, caused by the disturbance of the equilibrium of the carrier system through the electric field. Lattice heating due to the Joule effect can also give rise to NDC, sometimes in combination with electronic effects. Such thermal or *electrothermal* instabilities [1.34] have been known for a long time in connection with dielectric breakdown and switching [1.1, 76, 77]. If the Joule heating causes the lattice temperature T_L to rise locally above the ambient temperature, and if the conductivity strongly increases with T_L (e.g., through thermal generation of carriers), then an S-shaped current-voltage characteristic can be generated. A device in which NDC is induced in this way is called a thermistor. For a detailed theoretical description the heat-flow equation must be solved simultaneously with the electric-current equation under appropriate boundary conditions [1.34].

There are many other NDC mechanisms, like the acoustoelectric effect [1.30], or plasma instabilities involving a magnetic field, e.g., helicon waves [1.32], but these will not be treated here.

1.1.3 Semiconductor Transport

The theory of electrical transport in semiconductors [1.78–89] describes how charge carriers interact with electric and magnetic fields, and move under their influence. Thus it can be used for a quantitative explanation of negative differential conductivity. There are several different levels at which semiconductor transport can be modeled [1.78]:

(i)   Classical deterministic differential equations for the mean carrier densities and fields.
(ii) Semiclassical balance equations for the mean particle numbers, and the mean energy and momentum of the carriers.
(iii) A Boltzmann equation for the classical momentum- and position-dependent distribution function of the carriers.
(iv) Quantum transport theory based upon the von Neumann equation for the density matrix.
(v)   Monte Carlo simulations of the dynamics of the individual carriers.

The classical deterministic semiconductor equations (i) have been widely used in the modeling of all sorts of semiconductor devices [1.10, 11]. It is, however, important to realize the limits of validity with this approach which neglects statistical fluctuations as well as quantum effects. With the advent of very large scale integrated (VLSI) circuits it has often become necessary to use one of the more detailed theories, especially if short times (below picoseconds) and small lengths (below microns) are involved. Although our book is mainly based upon the approach (i), we will also survey the other approaches. All these theories are substantially nonlinear in the regime of high-electric fields where instabilities and negative differential conductivity occur.

(i) The Classical Deterministic Semiconductor Equations

These equations [1.11] are given by the continuity equations for the densities of electrons in the conduction band (n), of holes in the valence band (p), and of electrons trapped at various impurity levels $(n_{t_1}, n_{t_2}, \ldots, n_{t_M};$ denoted in a compact notation by a formal vector $n_t)$:

$$\dot{n} - \frac{1}{e}\mathbf{\nabla} \cdot \mathbf{j}_n = f_n(n, p, \mathbf{n}_t, \mathbf{E}) \tag{1.1.5}$$

$$\dot{p} + \frac{1}{e}\mathbf{\nabla} \cdot \mathbf{j}_p = f_p(n, p, \mathbf{n}_t, \mathbf{E}) \tag{1.1.6}$$

$$\dot{\mathbf{n}}_t = f_t(n, p, \mathbf{n}_t, \mathbf{E}) \ . \tag{1.1.7}$$

They are supplemented by Maxwell's equations (in Gauss units) for the electric field $\mathbf{E}$ and the magnetic field $\mathbf{H}$, assuming a homogeneous, isotropic, nonmagnetic material characterized by a static dielectric constant ϵ_s (at not too high frequencies):

$$\mathbf{\nabla} \cdot \mathbf{E} = \frac{4\pi e}{\epsilon_s}\left(N_D^* - n - \sum_{i=1}^{M} n_{t_i} + p\right) \tag{1.1.8}$$

$$\mathbf{\nabla} \times \mathbf{E} = -\frac{1}{c}\dot{\mathbf{H}} \tag{1.1.9}$$

$$\mathbf{\nabla} \cdot \mathbf{H} = 0 \tag{1.1.10}$$

$$\mathbf{\nabla} \times \mathbf{H} = \frac{\epsilon_s}{c}\dot{\mathbf{E}} + \frac{4\pi}{c}(\mathbf{j}_n + \mathbf{j}_p) \ . \tag{1.1.11}$$

The dot denotes the derivative with respect to time, $e > 0$ is the elementary charge, and $N_D^* := N_D - N_A$ is the "effective" donor concentration, where N_D and N_A are the donor and acceptor concentrations, respectively. The electron and hole current densities, $\mathbf{j}_n$ and $\mathbf{j}_p$, respectively, consist of drift and diffusion components, (we assume spatially homogeneous temperatures and neglect electrothermal currents, as well as magneto-transport currents):

$$j_n = e\mu_n n E + e D_n \nabla n \tag{1.1.12}$$

$$j_p = e\mu_p p E - e D_p \nabla p \ . \tag{1.1.13}$$

Here μ_n and μ_p are the (in general field-dependent) electron and hole mobilities, and D_n and D_p are the electron and hole diffusion constants. The negative sign in (1.1.5) results from the fact that the electrical current density j_n is opposite to the carrier flow because of the negative charge of the electrons.

The functions f_n, f_p, and $f_t := (f_{t_1}, f_{t_2}, \ldots, f_{t_M})$ are the generation-recombination rates; they depend nonlinearly upon the densities of the carriers involved in the respective g-r processes, and, through the g-r rate coefficients, upon the electric field. Since the g-r processes conserve the total number of carriers,

$$f_n - f_p + \sum_{i=1}^{M} f_{t_i} = 0 \tag{1.1.14}$$

always holds. Eqs. (1.1.5–13) represent nonlinear partial differential equations which have to be solved subject to suitable boundary and initial conditions. Only for simple device geometries and surface and contact properties can this be achieved analytically; otherwise numerical finite-difference or finite-element methods have to be used. Electromagnetic wave propagation and retardation effects can often be neglected, and therefore (1.1.5–8, 12, 13) represent the most important equations.

Appropriate boundary conditions have to be applied at surfaces and interfaces, in particular at junctions and contacts [1.90]. When a semiconductor is put into contact with a metal, the conduction band of the semiconductor is bent upward or downward near the contact, depending upon the metal workfunction, the electron affinity, and the bulk doping of the semiconductor, and possibly upon the charges trapped at interface states. Depletion or accumulation layers of charge carriers may thus be formed near the contacts. The case of a depletion layer of the majority carriers corresponds to the formation of a potential barrier ("Schottky barrier") between the semiconductor and the metal, and therefore the contact is blocking ("Schottky contact"), and has a high resistance. An accumulation layer of majority carriers corresponds to an injecting ("Ohmic") contact of low resistance. Note that here the word "Ohmic" does not mean that the contact resistance is linear (the contact resistance, is actually nonlinear, i.e., voltage dependent), but rather that it is small compared to the bulk resistance, and the contact can supply all the majority carriers that the bulk demands: it acts as a carrier reservoir. The effect of such a contact can often be simulated by imposing the boundary condition $E = 0$, which corresponds to an infinite reservoir of carriers [1.72]. Other, more realistic boundary conditions include a finite, fixed electric field E_c [1.31], or a fixed carrier density n_c at the contacts [1.67, 91, 92]. If interface recombination processes play a role, more complicated boundary conditions involving the carrier density and its spatial derivative at the interface are necessary [1.93]. The boundary conditions at free surfaces which are not traversed by the current flow will be treated in Sect. 4.3.

Equations (1.1.5–13) assume that the carrier-drift velocities respond instantaneously to changes in the electric field, that the mobility and diffusion coefficients are functions of the electric field alone, and that no differentiation is made between different valleys or other details of the band structure. These approximations require that the electric fields, carrier gradients, and current densities are not too large, and the dimensions are not too small. Since the equations contain only *mean* carrier numbers, neglecting phase relations between the wavefunctions of individual carriers, they cannot describe coherent excitations. Coherent macroscopic electronic excitations have been treated systematically elsewhere [1.94].

(ii) The Semiclassical Transport Equations

In a more detailed description [1.80, 83] the continuity equations for the particle densities (1.1.5, 6) are supplemented by balance equations for the mean electron energy $\langle E_e \rangle$ and for the mean electron momentum $\langle p_e \rangle$:

$$\langle \dot{E}_e \rangle + \langle v_e \rangle \cdot \nabla \langle E_e \rangle = -e \langle v_e \rangle \cdot E - \frac{1}{n} \nabla (n \langle v_e \rangle kT_e) - \frac{\langle E_e \rangle - E_e^0}{\tau_e(\langle E_e \rangle)} \qquad (1.1.15)$$

where

$$\langle E_e \rangle := \tfrac{1}{2} m_e^* \langle v_e^2 \rangle = \tfrac{1}{2} m_e^* \langle v_e \rangle^2 + \tfrac{3}{2} kT_e \qquad (1.1.16)$$

is the mean electron energy, $v_e(k) = \hbar^{-1} \nabla_k E_c(k)$ is the microscopic electron group velocity [$E_c(k)$ is the conduction-band structure, k is the wave-vector], and $T_e = \langle (v_e - \langle v_e \rangle)^2 \rangle m_e^*/(3k)$ is the electron temperature defined by the variance of the microscopic velocity. The term $E_e^0 \equiv \tfrac{3}{2} kT_L$ is the electron energy corresponding to thermal equilibrium with the lattice at temperature T_L, τ_e is the energy relaxation time, and m_e^* is the electron effective mass (for simplicity we assume isotropic parabolic bands). The balance equation for the mean electron momentum $\langle p_e \rangle = m_e^* \langle v_e \rangle = \langle \hbar k \rangle^1$ is:

$$\langle \dot{p}_e \rangle + (\langle v_e \rangle \cdot \nabla) \langle p_e \rangle = -eE - \frac{1}{n} \nabla (nkT_e) - \frac{\langle p_e \rangle}{\tau_m(\langle E_e \rangle)} \qquad (1.1.17)$$

where τ_m is the momentum relaxation time.

Equations (1.1.5, 15, 17) represent a closed system of equations for the variables n, $\langle E_e \rangle$, and $\langle p_e \rangle$, since T_e can be eliminated by (1.1.16) and the electron current density is given by

$$j_n = -en \langle v_e \rangle = -en \langle p_e \rangle / m_e^* \;,$$

which includes the diffusion component. Similar equations can be obtained for the hole energy and momentum. These balance equations hold if the electron distribution function $f(k)$ is spherically symmetric around the mean wavevector $\langle k \rangle$, such that a scalar electron temperature T_e can be defined, and that no heat flow term occurs in (1.1.15). In particular, this is satisfied if $f(k)$ is a displaced Maxwellian in k-space so that moments of higher than second order of the distribution function vanish:

---

[1] The second equality holds for isotropic parabolic bands only.

$$f(k) = \exp\left\{-\left[\frac{1}{2m_e^*}\hbar^2(k - \langle k\rangle)^2 - E_F\right]\Big/(kT_e)\right\} . \tag{1.1.18}$$

Here E_F is the (quasi-) Fermi energy of the electrons. Equation (1.1.18) holds, e.g., if the crystal momenta $\hbar k$ of all carriers are increased by $\Delta p = e\tau E$ in an electric field E.

The balance equations (1.1.15–17) can describe substantial carrier heating causing transient velocity overshoot of the drift velocity over its steady state value, which occurs if the momentum relaxation time is smaller than the energy relaxation time.

(iii) The Classical Boltzmann Equation

The classical carrier distribution function $f(k, r, t)$ for carriers of charge q ($= -e$ for electrons) is governed by the Boltzmann equation [1.85]:

$$\frac{\partial}{\partial t}f + \frac{q}{\hbar}E \cdot \nabla_k f + v_e \cdot \nabla_r f = \left(\frac{\partial f}{\partial t}\right)_{\text{scattering}} \tag{1.1.19}$$

where

$$\left(\frac{\partial f}{\partial t}\right)_{\text{scattering}} := -\frac{1}{(2\pi)^3}\int [W(k, k')f(k, r, t) - W(k', k)f(k', r, t)]\, d^3k'$$

is the collision integral which contains all scattering processes which occur, e.g., optical and acoustic phonon scattering, impurity scattering, electron-electron scattering, etc. The probability of transition per unit time from a state k to k' is denoted by $W(k, k')$. The basic assumptions of the classical Boltzmann equation (1.1.19) are:

(a) A unique carrier distribution function exists that depends on the position r and the wavevector k, which is meaningful if the distribution function varies spatially only over distances much larger than the de Broglie wavelength of the carriers.
(b) The carrier density is low enough so that only binary collisions occur.
(c) The time between successive collisions is much longer than the duration of a collision.
(d) The density gradients are small over the range of the interparticle potentials.

The Boltzmann equation (1.1.19) can be solved analytically by expanding f in spherical harmonics [1.85], or making a relaxation time approximation for the collision term, or by iterative procedures [1.82]. Alternatively, solutions can be obtained from moment or cumulant expansions [1.95], or numerically by Monte Carlo techniques [1.79]. In the first case [1.95], a hierarchy of equations for the mean values $\langle k^\alpha\rangle$ ($\alpha = 0, 1, 2, 3, \ldots$) of the powers of the wavevector k is derived; the hierarchy is truncated at some order α_{max}. In the latter case the motion of individual carriers is simulated using a random generator to model the microscopic scattering processes.

(iv) Quantum-Transport Theory

In very small-scale devices (<0.1 μm), dense systems, or with strong nonlocal scattering, quantum effects become dominant [1.80, 81], and the semiconductor should be described by a statistical density matrix ρ, governed by the von Neumann equation

$$i\hbar \frac{\partial}{\partial t}\rho = [H,\rho] \ , \tag{1.1.20}$$

where H is the Hamilton operator, and the square brackets denote the commutator. From (1.1.20) Pauli's Master equation for the probability of finding the semiconductor in a quantum mechanical state $|i\rangle$ can be derived by a coarse-graining process [1.87]. Alternatively, Wigner distribution functions $f_\sigma(k,r,t)$ can be used to directly calculate expectation values of physical observables, e.g., current density [1.81]. These Wigner functions are the quantum-mechanical analogs of classical distribution functions; however, they are not necessarily positive definite, which complicates their physical interpretation.

The classical "hydrodynamic" deterministic semiconductor equations (1.1.5–13) will be the basis of the following chapters. The continuity equations (1.1.5–7) and the energy and momentum balance equations (1.1.15, 17) can be derived from the Boltzmann equation (1.1.19) in the standard way [1.96] by forming the moments

$$n := \int f \, d^3 k$$

$$\langle p_e \rangle := \frac{1}{n}\int \hbar k f \, d^3 k = m_e^* \langle v_e \rangle = -m_e^* j_n/(en)$$

$$\langle E_e \rangle := \frac{1}{n}\int \frac{\hbar^2 k^2}{2 m_e^*} f \, d^3 k$$

assuming a distribution function f which is spherically symmetric around $\langle k \rangle$, and making a relaxation-time approximation for the collision integrals. The equation for the current density j_n (1.1.12) can be obtained from the momentum balance equation (1.1.17) by further neglecting the inertia terms $\langle \dot{p}_e \rangle + (\langle v_e \rangle \cdot \nabla)\langle p_e \rangle$ and the space- dependence of T_e, and introducing the mobility $\mu_n := (e/m_e^*)\tau_m$ and the diffusion constant $D_n := \mu_n k T_e/e$. Alternatively, (1.1.5–7) can be systematically derived from first principles starting from a quantum-mechanical von Neumann equation [1.87].

1.2 Phase Transition Analogies

Instabilities in semiconductors bear a remarkable similarity to phase transitions of systems in thermal equilibrium. This observation provides a deeper insight into the physical nature of these instabilities, and allows one to use methods and concepts developed for the theoretical description of equilibrium phase transitions. Such phase-transition-like behavior is a widespread phenomenon in nonlinear systems driven far from equilibrium by external forces or fluxes. We shall show in this section that highly excited semiconductors represent indeed such nonequilibrium systems, and outline some mathematical tools and physical notions for their description.

1.2.1 Equilibrium and Nonequilibrium Phase Transitions

Thermodynamic systems can exist in qualitatively different stable states ("phases"). Transitions between these phases are characterized by singular or discontinuous behavior of certain macroscopic observables. Standard examples of equilibrium phase transitions are the gas-liquid transition of a Van der Waals' gas, and the para-ferromagnetic transition of a magnetic system in a mean field description.

The Van der Waals' equation for one mole of a real gas is

$$P = \frac{RT}{V-b} - \frac{a}{V^2} \; , \tag{1.2.1}$$

where P is the pressure, V the volume, T the temperature, R the universal gas constant, and $a, b > 0$ are material-dependent constants. In terms of the reduced variables $v := V_c/V, \tau := T/T_c, \pi := P/P_c$, where the subscript c denotes values at the critical point, (1.2.1) can be written as

$$\pi = v^3 - 3v^2 + \tfrac{1}{3}(8\tau + \pi)v = 0 \; . \tag{1.2.2}$$

In Fig. 1.6 the isotherms are plotted in the $P-V$ diagram. Unstable parts of the isotherms, where $(\partial P/\partial V)_T > 0$, are dashed. Phase coexistence between the vapor phase (volume V_g) and the liquid phase (V_l) is possible if the vapor pressure $P_{co}(T)$ at a given temperature T satisfies Maxwell's rule

$$\int_{V_l}^{V_g} [P(V, T) - P_{co}(T)] \, dV = 0 \; . \tag{1.2.3}$$

Phase transitions from the liquid to the gas phase along the coexistence line $P = P_{co}$ are marked by double arrows. The metastable states $(\partial P/\partial V)_T < 0$ of the isotherms

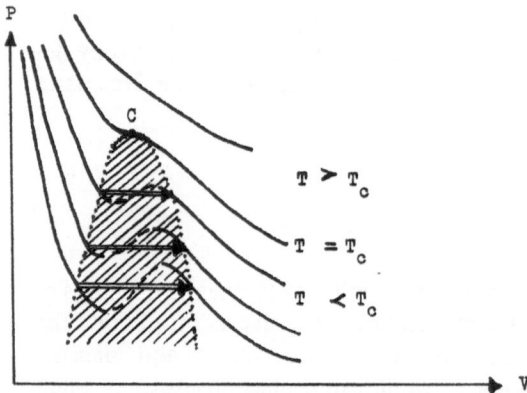

Fig. 1.6. Isotherms in the pressure (P)-volume (V) diagram for the Van der Waals' equation (schematic). Broken lines represent unstable states, the region of phase coexistence is hatched. The arrows symbolize first-order phase transitions along the coexistence line given by Maxwell's construction of equal areas (1.2.3). The critical point is denoted by C

on the left-hand side of the coexistence region correspond to an overheated liquid, those on the right side correspond to an oversaturated vapor. At the critical point C the two phases become indistinguishable. According to Ehrenfest, a phase transition can be classified as nth order if all derivatives of the Gibbs free energy G of order lower than n are continuous, whereas nth derivatives have a finite discontinuity.

The gas-liquid phase transition (arrows in Fig. 1.6) is of first order since $V = (\partial G/\partial P)_T$ has a finite jump, and so has the entropy $S = -(\partial G/\partial T)_P$; the specific heat $c_P = T(\partial S/\partial T)_P$ diverges. This reflects the latent heat that is always connected with a first-order phase transition. If the critical point C is reached along a path in the $P-V$ diagram, coming from outside the coexistence region, a second-order phase transition occurs. In a second-order transition all extensive variables, such as V, are continuous, but have discontinuous derivatives with respect to intensive variables, such as the specific heat. There is no latent heat, and no phase coexistence.

Another standard mean field example is the equation of state for a Weiss ferromagnet:

$$M = M_0 \tanh\left[\frac{m}{kT}(H + wM)\right] \tag{1.2.4}$$

with macroscopic magnetization M, magnetic field H, saturation magnetization M_0, molecular magnetic moment m, and the Weiss factor of the mean field $w > 0$. It can be derived from the Ising model in mean-field approximation. An $M-T$ plot of (1.2.4) is shown in Fig. 1.7 for fixed values of H ($0 < H_1 < H_2$). In the absence of an external magnetic field, the magnetization M which represents an order parameter,

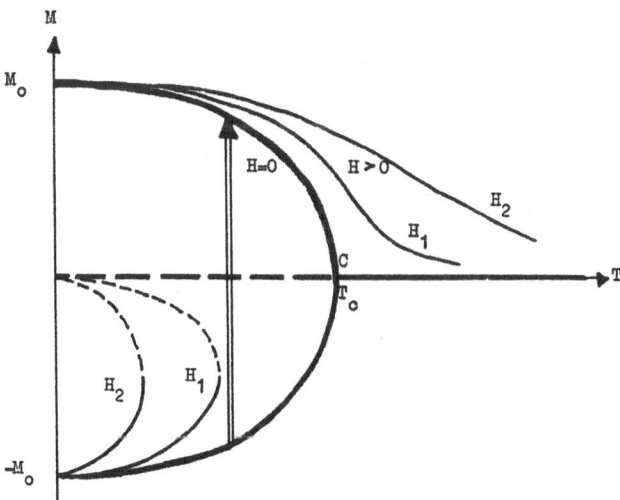

Fig. 1.7. Magnetization M versus temperature T for a Weiss ferromagnet (schematic). The parameter of different curves is the magnetic field H ($0 < H_1 < H_2$). C denotes the critical point (Curie point), below which spontaneous magnetization occurs

is zero for $T > T_c$ (paramagnetic phase) and nonzero for $T < T_c$ (ferromagnetic phase). At the critical point (Curie temperature T_c, $H = 0$) the disordered state $M = 0$ turns unstable, and the spatial symmetry is broken by a spontaneous magnetization $M > 0$ or $M < 0$. This a second-order phase transition since M changes continuously but with a discontinuous derivative with respect to T. First-order phase transitions between the two ferromagnetic phases $M < 0$ and $M > 0$ are also possible (arrow in Fig. 1.7). A magnetic field $H > 0$ smoothes out the symmetry-breaking second-order phase transitions, and singles out the stable state $M > 0$. The unstable (dashed) and metastable states (full lines) corresponding to $M < 0$ are also shown in Fig. 1.7 for H_1 and $H_2 > 0$.

An open thermodynamic system that is coupled to strong external fluxes of energy or matter can be driven into a state far from thermal equilibrium and maintained in a regime where the linear thermodynamic laws of irreversible processes are no longer valid, but where the dynamics must be described by nonlinear equations for the variables characterizing the macroscopic state. At certain threshold values of the *control parameter* (driving forces or fluxes, pump strength, external fields) the state of the system may change in a singular way. Such phenomena are called *nonequilibrium phase transitions* because of their similarity to equilibrium phase transitions. The transition is of *first order* if the state variables have a discontinuity at the transition point. It is of *second order* if the state variables change continuously but have a discontinuous derivative with respect to the control parameter. Contrary to the equilibrium case, where the state variables and control parameters are well defined as extensive observables and adjoint intensive parameters, the choice of the variables and control parameters in nonequilibrium systems is not a priori unambiguously determined but is guided more or less by empirical expediency. Nor is the stable steady state of a nonequilibrium system singled out by well-defined extremal conditions of thermodynamic potentials, as in equilibrium. [Note that the minimum entropy production principle holds close to equilibrium only [1.49]].

The phase transition analogies include the similarity of appropriate state diagrams with those shown in Figs. 1.6, 7, the spontaneous symmetry breaking of an order parameter, critical and tricritical points, critical slowing down of relaxation times, the buildup of long-range spatial correlations, spatial coexistence of phases described by "equal-areas rules" similar to Maxwell's construction (1.2.3), the growth of long-wavelength fluctuations, spinodal decomposition, and nucleation phenomena similar to the condensation of droplets. As the story of this book unfolds, we shall indeed encounter all these phenomena in nonequilibrium semiconductor systems.

An important ingredient of the phase transition analogy is the critical behavior [1.97]. We shall first discuss equilibrium systems. In the framework of the present treatment we will restrict ourselves to a mean-field theory, although more sophisticated approaches should include the modern techniques of renormalization group theory [1.98]. Close to the critical point, thermodynamic systems of quite different physical nature and of different materials show a universal asymptotic power-law behavior, characterized by critical exponents. In the magnetic system (1.2.4), for instance, the critical exponents α, β, γ, δ are defined by

$$c_H := T \left(\frac{\partial S}{\partial T} \right)_H \sim (T - T_c)^{-\alpha} \quad T > T_c, H = 0 \tag{1.2.5a}$$

$$M \sim (T_c - T)^\beta \qquad\qquad T < T_c, H = 0 \tag{1.2.5b}$$

$$\chi := \left(\frac{\partial M}{\partial H} \right)_T \sim (T - T_c)^{-\gamma} \quad T > T_c, H = 0 \tag{1.2.5c}$$

$$H \sim |M|^\delta \qquad\qquad T = T_c \tag{1.2.5d}$$

and describe the singularities of the specific heat, the order parameter, the susceptibility, and the ordering field, respectively, as the critical point is approached. The
mean field exponents following from the thermal equation of state (1.2.4), and – in
case of α – the caloric equation of state, are $\alpha = 0$, $\beta = \frac{1}{2}$, $\gamma = 1$, $\delta = 3$; they are the
same for the Van der Waals' gas (1.2.1) and other mean-field models. Much better
agreement with the experimental values of the critical exponents is obtained from
renormalization group theory, which has been extensively applied to static and
dynamic critical phenomena. Recent theoretical developments include the non-
asymptotic critical regime [1.99, 100], and the critical behavior in finite-size systems
and at surfaces [1.101–103].

At the critical point spatial correlation lengths and relaxation times also diverge
with universal critical exponents. The divergence of the relaxation times is known
as "critical slowing down". Moreover, the fluctuations of thermodynamic observables become critical and diverge. This is a consequence of the static fluctuation-
dissipation theorem, which generally relates the variance of the fluctuation $\Delta M :=
M - \langle M \rangle$ of a macroscopic extensive variable M to the susceptibility of its mean
value $\langle M \rangle$ with respect to its thermal adjoint ζ:

$$\langle (\Delta M)^2 \rangle = kT \frac{\partial}{\partial \zeta} \langle M \rangle , \tag{1.2.6}$$

where ζ is defined via the entropy S by

$$\zeta := -T \frac{\partial S}{\partial \langle M \rangle} . \tag{1.2.7}$$

If, in case of a ferromagnet, M is chosen as the magnetization, then ζ is the magnetic
field H. Equation (1.2.6) gives a remarkable connection between equilibrium fluctuations and the macroscopic equation of state which describes the mean value $\langle M \rangle$
as a function of ζ. The relation has been generalized [1.104] to higher-order
cumulants of M

$$\langle M^l \rangle_c = (kT)^{l-1} \left(\frac{\partial^{l-1}}{\partial \zeta^{l-1}} \langle M \rangle \right)_{\zeta=0} \tag{1.2.8}$$

where the cumulant $\langle M^l \rangle_c$ of order l of the random quantity M is defined as the
lth expansion coefficient of the power expansion of the generating function

$$\ln \langle e^{sM} \rangle = \sum_{l=0}^{\infty} \frac{s^l}{l!} \langle M^l \rangle_c .$$

(1.2.9)

Here the thermal mean values $\langle \ \rangle$ are formed with the canonical distribution

$$\rho = Z^{-1} \exp[-\mathcal{H}/(kT)]$$

with the Hamiltonian $\mathcal{H}$ and partition function Z. Note that $\langle M^2 \rangle_c = \langle (\Delta M)^2 \rangle$, $\langle M^3 \rangle_c = \langle (\Delta M)^3 \rangle$, but $\langle M^4 \rangle_c = \langle (\Delta M)^4 \rangle - 3 \langle (\Delta M)^2 \rangle^2$. Far from the critical point the higher cumulants become smaller with increasing order. Yet near the critical point the cumulants increase and diverge the more strongly the higher their order is, thereby indicating strong deviations from a normal (Gaussian) distribution near the critical point. For the Weiss ferromagnet, e.g., $\langle M^2 \rangle_c \sim |T - T_c|^{-1}$, $\langle M^3 \rangle_c \sim |T - T_c|^{-5/2}$, $\langle M^4 \rangle_c \sim |T - T_c|^{-4}$.

In nonequilibrium systems the homogeneous steady state can often be described by mean-field equations similar to the thermal equations of state (1.2.1, 4), whence the dynamic variable can be identified with the order parameter, say M, and the control parameters can be considered analogous to the temperature T and the ordering field H (thermal adjoint of M). Thus the critical exponents β, γ, δ may be defined in analogy with (1.2.5b–d) [1.105]. The exponent α can not, however, be defined in this way, since an analog of the caloric equation of state is, in general, not available. To overcome this difficulty, *Schlögl* [1.106, 107] considered the bit-number cumulants of a generalized probability distribution ρ:

$$\Gamma_l := \langle b^l \rangle_c$$

(1.2.10)

where $b := -\ln \rho$ is the bit number, and pointed out that for equilibrium systems described by a generalized canonical distribution $\rho = Z^{-1} \exp[(-\mathcal{H} + \zeta M)/(kT)]$ the cumulant of order $l = 2$, i.e., the *bit-number variance* Γ_2, reduces to the specific heat $c_\zeta = T(\partial S/\partial T)_\zeta$, while the first cumulant ($l = 1$) becomes the equilibrium entropy S. He suggested the use of the bit-number variance as a characteristic critical quantity that generalizes the specific heat for nonequilibrium systems, and calculated the critical behavior of this quantity for various equilibrium and nonequilibrium phase transitions. These systems fall into two classes A and B, where Γ_2 has a divergence $\Gamma_2 \sim |\tau|^{-1}$ or a finite discontinuity, respectively. Here τ is the control parameter corresponding to $T - T_c$, which vanishes at the critical point. The merit of Γ_2 is that it represents an additional, independent quantity for the characterization of the phase transition, and that it is particularly sensitive to the onset of correlations, in fact more sensitive than Γ_1 which is the (more familiar) information entropy. This sensitivity is the reason for the critical behavior of Γ_2, which lends itself to a natural generalization of c_H in (1.2.5a).

The critical behavior of semiconductor phase transitions will be investigated in Chap. 2 for specific g-r models. We will see that, in addition to simple critical points, tricritical points [1.108] occur as well, and the critical and tricritical exponents can be classified into three different classes. Similar tricritical phenomena were found in a laser phase transition [1.109].

To conclude this section, we point out that a large variety of physical and even nonphysical systems, ranging from lasers, semiconductors, chemical reaction systems, hydrodynamic, biological, and neurophysical systems, to models in sociology, economics, and even simulations of traffic flow on motorways, exhibit nonequilibrium phase transitions and similar cooperative phenomena. The unifying viewpoint of synergetics has stimulated a vast amount of research papers, conferences, and monographs, of which we can only cite a few [1.2–9, 12, 37, 38, 110–117]. Further references can be found therein.

In semiconductors, apart from g-r induced nonequilibrium phase transitions, there is another class of phase transitions, in which the coupling of the charge carriers to photons is the essential feature: the semiconductor laser transition [1.115]. The coupled nonlinear g-r kinetics of electrons and photons gives rise to a second-order phase transition from a nonlasing state (no photons, if spontaneous emission is neglected) to a coherent lasing state when the pump strength given by the injected current across a p-n-junction exceeds a certain threshold. At still higher injection currents, further instabilities lead to relaxation oscillations of the optical intensity. This can be used as a simple and efficient method of picosecond optical-pulse generation.

1.2.2 Semiconductors as Nonequilibrium Systems

Let us now explore the conditions under which a semiconductor represents a system far from thermodynamic equilibrium.

In global thermodynamic equilibrium, i.e., in the absence of externally applied electric or magnetic fields, current injection, or optical excitation, the semiconductor is characterized by a spatially uniform, common temperature T of the charge carriers (electrons, holes) and the lattice (phonons) – *thermal equilibrium* – and a uniform electrochemical potential of the carriers, the Fermi level E_F – *chemical equilibrium*. The distribution of carriers into different energy states E_i of the bands and of localized levels can then be described by a grand canonical ensemble

$$\rho = \Xi^{-1} \exp[(-E_i + E_F N)/(kT)] , \qquad (1.2.11)$$

where Ξ is the grand partition function, and N is the total number of carriers. It then follows from standard statistical mechanics that the mean number of electrons in a conduction band state E, i.e., the occupation probability, is given by the Fermi-Dirac distribution

$$f(E) = \{1 + \exp[(E - E_F)/(kT)]\}^{-1} , \qquad (1.2.12a)$$

and the number of holes in a valence band state E is

$$f_h(E) = 1 - f(E) . \qquad (1.2.12b)$$

The number of electrons in localized states is given by slightly modified occupation probabilities taking into account spin degeneracy, excited states, and the possibility of binding more than one electron (multivalent levels) [Ref. 1.123, p. 380]. For nondegenerate semiconductors with parabolic bands, (1.2.12) can be approximated by a Maxwell-Boltzmann distribution, and the equilibrium density of electrons in the conduction band n_0 and of holes in the valence band p_0 is

$$n_0 = N_c \exp[(E_F - E_c)/(kT)] \qquad\qquad (1.2.13a)$$

$$p_0 = N_v \exp[(E_v - E_F)/(kT)] \ , \qquad\qquad (1.2.13b)$$

where $N_c := 2(2\pi m_e^* kT/h^2)^{3/2}$ and $N_v := 2(2\pi m_h^* kT/h^2)^{3/2}$ are the effective densities of state of the conduction band and the valence band with effective masses m_e^* and m_h^*, respectively, and E_c, E_v are the edges of the conduction and valence band, respectively. The Fermi level is determined from the condition of quasi neutrality, and depends upon temperature and doping. From (1.2.13a, b) it follows that the product

$$n_0 p_0 = N_c N_v \exp[-(E_c - E_v)/(kT)] =: n_i^2 \qquad\qquad (1.2.14)$$

is independent of the Fermi level, and hence of doping. In (1.2.14) we have introduced the intrinsic equilibrium carrier density n_i.

The transport coefficients $\mu_{n,p}$ and $D_{n,p}$ of the carriers at or close to thermal equilibrium are not independent but are related by the following fluctuation-dissipation theorem, see (1.2.6),

$$eD_{n,p} = \mu_{n,p} kT \qquad \text{(Einstein relation)} \ . \qquad\qquad (1.2.15)$$

[Note that the diffusion constant can be defined on a microscopic level via the second cumulant of the particle coordinate $x(t)$:

$$D := \lim_{t \to \infty} \frac{1}{2} \frac{d}{dt} \langle (x - \langle x \rangle)^2 \rangle] \ .$$

In going away from global thermodynamic equilibrium, several different paths may be taken:

(i)   The common temperature T may vary in space.
(ii) The Fermi level may vary in space.
(iii) The temperature of the carriers T_e and of the lattice T_L may become different.
(iv) The Fermi level may split into two distinct quasi-Fermi levels $E_{Fn} \neq E_{Fp}$ for electrons and holes, respectively.

Although often a combination of several of these effects will occur, it is important to separate these conceptually in order to assign a precise, quantitative meaning to the notion "*far from equilibrium*".

If (i) or (ii) alone occur, and the gradients are not too large, one may still have local equilibrium. Large gradients of the temperature, the Fermi level, or the electrostatic potential, however, lead to deviations from the linear flux (current density) – force ($\nabla T, \nabla n, \nabla \phi \equiv -E$) relations [1.95]. Case (i) applies, e.g., to local Joule heating effects, and may lead to thermal instabilities if the temperature gradients are large. Case (ii) is associated with carrier injection, e.g., from contacts or surfaces, and results in carrier density gradients and diffusion currents.

Case (iii) occurs if the carriers are not in equilibrium with the lattice. This happens when they gain kinetic energy from an electric field or by optical excitation faster than they can transfer this energy to the phonon system, i.e., if the excitation time constant is shorter than the energy relaxation time. The mean energy of the electrons $\langle E_e \rangle$ (or of the holes) is then larger than the equilibrium value $\frac{3}{2}kT_L$, and one can define an electron temperature $T_e > T_L$ by setting

$$\langle E_e \rangle := \tfrac{3}{2}kT_e \ . \tag{1.2.16}$$

This definition makes physical sense only if the electron distribution in the conduction band obeys a Maxwell-Boltzmann law with an effective temperature T_e, which requires a strong carrier-carrier scattering and elastic impurity scattering in order to randomize the direction of the k-values and produce an isotropic distribution in k-space. A generalization of the definition (1.2.16) to a *displaced* Maxwellian was given in (1.1.16). Hot-electron phenomena in high-electric fields are the subject of extensive research, especially in submicron structures where large local fields occur [1.79–84, 1.118]. The relation between T_e and the field E in a spatially homogeneous steady state follows from the energy balance equation, cf., (1.1.15),

$$e\mu_n E^2 = \tfrac{3}{2}k(T_e - T_L)/\tau_e \ . \tag{1.2.17}$$

T_e and T_L can actually be measured independently, for instance, T_e follows from the optical lineshape and T_L from the occupation ratio of two impurity levels [1.119].

If the electron distribution in the conduction band is strongly perturbed by external fields or laser irradiation, the concept of an electron temperature breaks down. The distribution function f can no longer be parametrized by an effective electron temperature, but has to be determined from the Boltzmann equation (1.1.19), taking into account various scattering mechanisms. An example of such a non-Maxwellian distribution, based upon infrared absorption, optical phonon scattering, and impact ionization, was treated in [1.120], see also the extensive bibliographies given in [1.79–82].

Case (iv) occurs if the electrons and holes are not in equilibrium with each other, but the carriers in each band (electrons or holes, respectively) are in quasi-equilibrium, described by two quasi-Fermi distributions of the form (1.2.12) with different quasi-Fermi levels $E_{Fn} \neq E_{Fp}$ [1.121]. (In addition, one may have two different effective temperatures T_e, T_h.) The total electron and hole concentrations n, p can then be parametrized by E_{Fn}, E_{Fp}, respectively, in analogy with (1.2.13):

$$n = N_c \exp[(E_{Fn} - E_c)/(kT)] \ , \tag{1.2.18a}$$

$$p = N_v \exp[(E_v - E_{Fp})/(kT)] \ . \tag{1.2.18b}$$

The equilibrium condition (1.2.14) is modified as follows:

$$np = n_i^2 \exp[(E_{Fn} - E_{Fp})/(kT)] \ . \tag{1.2.19}$$

The concentrations n and p are no longer fixed by the thermal equilibrium condition, but are determined by *nonequilibrium generation-recombination* processes. Examples

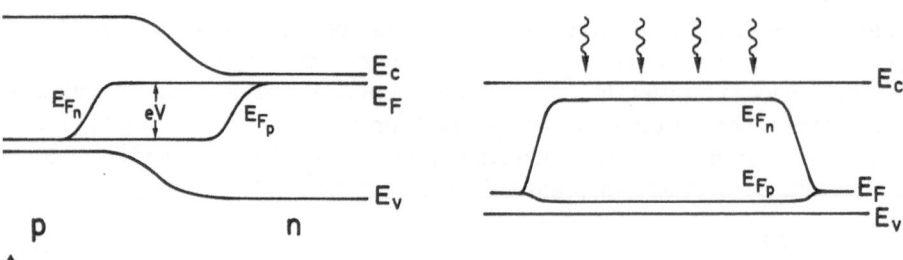

Fig. 1.8. Nonequilibrium band diagram of a forward-biased pn junction (schematic). Within a distance of the order of a diffusion length from either side of the junction the Fermi level E_F is split into two quasi-Fermi levels of the electrons E_{Fn} and the holes E_{Fp}, separated by eV where V is the applied voltage. This reflects the increased electron concentration on the p-side and the increased hole concentration on the n-side

Fig. 1.9. Nonequilibrium band diagram of a p-photoconductor under steady state illumination, which generates excess electron-hole pairs (schematic). The strong increase in the minority carrier (electron) concentration leads to a strong rise of the electron quasi-Fermi level E_{Fn}, while the smaller *relative* increase of the majority carrier (hole) concentration leads to a smaller shift of the hole quasi-Fermi level E_{Fp}

are provided by a forward-biased pn-junction (Fig. 1.8) and by a photoconductor (Fig. 1.9). In a pn-junction excess electrons are injected from the n-side into the p-region, and excess holes are injected from the p-side into the n-region, resulting in a separation of the quasi-Fermi levels in the transition region by an amount eV, where V is the applied voltage. In a photoconductor, excess electron-hole pairs are optically generated, and $E_{Fn} - E_{Fp}$ increases with the optical-excitation intensity. Generally, the separation of the quasi-Fermi levels provides a quantitative measure of the distance from equilibrium, by (1.2.19):

$$E_{Fn} - E_{Fp} = kT \ln \frac{np}{n_i^2} \ . \tag{1.2.20}$$

This may be illustrated by the threshold condition of a semiconductor injection laser:

$$E_{Fn} - E_{Fp} > E_g \ ,$$

where E_g is the bandgap. It shows that the laser phase transition occurs indeed in a rigorous sense "far from equilibrium".

The case (iv) is the essential nonequilibrium condition for the g-r instabilities which we shall treat in this book, though it is in many cases coupled with carrier heating (iii) and spatial inhomogeneities (ii). If impurity states are involved in the g-r kinetics, a separate quasi-Fermi level should in general be assigned also to each group of localized levels.

The most common recombination processes of carriers involving the conduction band, the valence band, and localized impurity levels (donors, acceptors, traps, or recombination centers) are shown in Fig. 1.10. The arrows represent the transitions of electrons; in those cases in which holes are involved ($T_2^S, B_2, T_2, T_3, T_4$) the

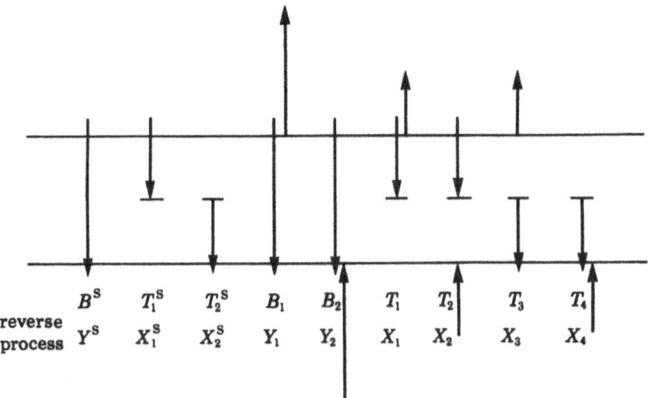

Fig. 1.10. Recombination processes in semiconductors. The arrows represent electron transitions between the conduction band, the valence band and impurity levels. The symbols denote the recombination coefficients (B^S, T_1^S, T_2^S: single-electron processes; B_1, B_2, $T_1 - T_4$: two-electron processes) and the corresponding generation coefficients of the reverse processes

transition of the holes is opposite to the arrows. The reverse processes representing generation are also possible. The first three processes are single-electron processes: radiative or nonradiative band-band recombination (B^S), and band-trap capture (T_1^S, T_2^S). The reverse processes are optical, thermal, or field-enhanced band-band generation (Y^S), and emission of trapped electrons (X_1^S) or holes (X_2^S). The other processes are two-electron processes: Auger recombination involving two bands (B_1, B_2), or one band and a trap level (T_1, T_2, T_3, T_4), and their reverse, namely impact ionization (Y_1, Y_2; X_1, X_2, X_3, X_4). A survey of these and of other g-r processes, including transitions between different impurities (e.g., donor-acceptor transitions), and involving excitons, and including a compilation of experimental and theoretical values of the various g-r coefficients, has previously been given [1.125]. Other comprehensive reviews on nonequilibrium g-r kinetics in semiconductors exist [1.121, 127]. Three-particle Auger processes, e.g. $2e + 2h \rightarrow e + h$ or $3e + h \rightarrow 2e$, $3h + e \rightarrow 2h$, have also been suggested theoretically [1.128].

The g-r rate, i.e., the number of carriers undergoing a transition per unit time and unit volume, can be calculated as follows. Consider e.g., band-to-band recombination and let the quantum-mechanical transition probability of an electron in a conduction-band state of energy E_e into an empty valence band state of energy E_h per unit time be denoted by $W(E_e, E_h)$. The total recombination rate between the two bands is then obtained by integrating over all initial and final states:

$$r = \int_{E_c}^{\infty} dE_e \int_{-\infty}^{E_v} dE_h f(E_e, E_{Fn}) D_c(E_e)[1 - f(E_h, E_{Fp})] D_v(E_h) W(E_e, E_h) \qquad (1.2.21)$$

where $D_c(E_e)$, $D_v(E_h)$ are the respective densities of states in the two bands, and the integrals over the bands have been extended to $+\infty$ and $-\infty$, respectively. Under nondegenerate conditions ($n \ll N_c$, $p \ll N_v$) the quasi-Fermi distributions $f(E_{e,h}, E_{Fn,Fp})$ can be approximated by

$$f(E_e, E_{Fn}) \approx \exp\left(\frac{E_{Fn} - E_e}{kT}\right) \approx \frac{n}{N_c}\exp\left(\frac{E_c - E_e}{kT}\right)$$

$$1 - f(E_h, E_{Fp}) \approx \exp\left(\frac{E_h - E_{Fp}}{kT}\right) \approx \frac{p}{N_v}\exp\left(\frac{E_h - E_v}{kT}\right) \tag{1.2.22}$$

using (1.2.12, 18). Thus the dependence upon n and p can be extracted from the integral (1.2.21):

$$r = B^S np , \tag{1.2.23}$$

where B^S is the appropriate band-band recombination constant defined by (1.2.21–23). Equation (1.2.23) reflects the mass-action kinetics which is familiar from dilute chemical reaction systems. In fact, the radiative recombination process can be written as a "bimolecular reaction":

$$e + h \overset{B^S}{\to} \gamma , \tag{1.2.24}$$

where e, h, γ represent the electron, hole, and photon, respectively.

Similarly, the g-r rates of the other processes in Fig. 1.10 can be inferred from the form of the "reaction" equation, assuming nondegenerate semiconductors; the results agree with the expressions derived from first principles [1.87, 123]:

$$(T_1^S)\, e + h_t \rightleftharpoons e_t \qquad : \; T_1^S np_t - X_1^S n_t , \tag{1.2.25a}$$

$$(T_2^S)\, h + e_t \rightleftharpoons h_t \qquad : \; T_2^S pn_t - X_2^S p_t , \tag{1.2.25b}$$

$$(B_1)\, 2e + h \rightleftharpoons e \qquad : \; B_1 n^2 p - Y_1 n , \tag{1.2.25c}$$

$$(B_2)\, 2h + e \rightleftharpoons h \qquad : \; B_2 p^2 n - Y_2 p , \tag{1.2.25d}$$

$$(T_1)\, 2e + h_t \rightleftharpoons e + e_t \quad : \; T_1 n^2 p_t - X_1 nn_t , \tag{1.2.25e}$$

$$(T_2)\, e + h + h_t \rightleftharpoons h + e_t : \; T_2 npp_t - X_2 pn_t , \tag{1.2.25f}$$

$$(T_3)\, e + h + e_t \rightleftharpoons e + h_t : \; T_3 npn_t - X_3 np_t , \tag{1.2.25g}$$

$$(T_4)\, 2h + e_t \rightleftharpoons h + h_t \qquad : \; T_4 p^2 n_t - X_4 pp_t . \tag{1.2.25h}$$

Here the subscript t denotes trapped carriers. Note that the only "autocatalytic reactions" (positive feedback) are the impact-ionization processes Y_1, Y_2, X_1, X_4. The g-r rates f_n, f_p, f_t introduced in the continuity equations (1.1.5-7) include the various terms of (1.2.23, 25). They are nonlinear in the carrier concentrations. Only if the various carrier concentrations are close to their equilibrium values n_0, p_0, n_{t0}, p_{t0}, can the rate equations be linearized around equilibrium, and a *constant* electron (τ_n) or hole (τ_p) lifetime can be defined by

$$\delta\dot{n} = -\delta n/\tau_n , \qquad \delta\dot{p} = -\delta p/\tau_p , \tag{1.2.26}$$

where $\delta n := n - n_0$, $\delta p := p - p_0$. If, on the other hand, the semiconductor is far from equilibrium in the sense of case (iv) above, then the nonlinearities of the g-r rates are essential, and the concept of a (carrier concentration independent) lifetime is no longer useful. This case applies to the g-r instabilities which will be the subject of this book.

In addition, the g-r coefficients in general depend upon an applied electric field E. This dependence is particularly pronounced in case of the impact ionization coefficients Y_1, Y_2, X_1, X_2, X_3, X_4, since the ionizing carrier must have a certain threshold energy before it can ionize, cf., Fig. 1.10. The simplest field dependence is given by Shockley's Lucky Drift Model [1.131]:

$$Y_1, Y_2, X_1, X_2, X_3, X_4 \sim \exp[-E_i/(e\lambda E)]^{\cdot},$$

where E_i is a characteristic threshold ionization energy, and λ is the mean free path of the ionizing carrier. More sophisticated expressions for band-band impact-ionization coefficients [1.129, 130, 132] and band-trap impact-ionization coefficients [1.129, 133–138] have been derived.

In equilibrium the principle of detailed balance holds. It follows from the invariance of the microscopic equations of motion against time reversal (micro-reversibility), and states that the forward and reverse rate of each g-r process balance individually ("in detail"). For the bimolecular band-band recombination, for instance, this implies, with (1.2.14),

$$B^S n_0 p_0 \equiv B^S n_i^2 = Y^S . \tag{1.2.27}$$

Similarly, one can express the generation coefficient of every reverse rate in terms of the recombination coefficient of the forward rate and the equilibrium carrier densities. If the g-r coefficients do not change (though the carrier densities do change!) when the system is driven away from equilibrium, then (1.2.27) and the analogous detailed balance relations for other processes can still be used for non-equilibrium g-r kinetics to eliminate the rate coefficients of all reverse processes $(Y^S, X_1^S, X_2^S; Y_1, Y_2, X_1, X_2, X_3, X_4)$ in (1.2.25):

$$B^S (np - n_i^2)$$

$$T_1^S (np_t - n_1 n_t)$$

$$T_2^S (pn_t - p_1 p_t)$$

$$B_1 n (np - n_i^2)$$

$$B_2 p (np - n_i^2)$$

$$T_1 n (np_t - n_1 n_t)$$

$$T_2 p (np_t - n_1 n_t)$$

$$T_3 n (pn_t - p_1 p_t)$$

$$T_4 p (pn_t - p_1 p_t) , \tag{1.2.28}$$

where $n_1 := n_0 p_{t0}/n_{t0}, p_1 := p_0 n_{t0}/p_{t0}$. This is the basic assumption of the Shockley-Read-Hall kinetics [1.122]; it leads to a unique, stable steady state for any values of the recombination coefficients, and thus excludes g-r induced nonequilibrium phase transitions.

If, however, some g-r coefficients are externally driven, e.g., by electric fields, and thereby depend upon the distance from thermodynamic equilibrium, the effective nonequilibrium g-r rates given in (1.2.28) are no longer valid, but instead the generation coefficients must be retained as independent control parameters as in (1.2.25). This applies in particular to the strongly field-dependent impact-ionization coefficients $(Y_1, Y_2, X_1 - X_4)$, and is a precondition for g-r induced bistability of the homogeneous steady state, as treated in Chap. 2. It is a subtle point that it is not sufficient for this kind of g-r instability that the semiconductor is far from equilibrium in the sense of strong deviations of the carrier *densities* from their equilibrium values, leading to nonlinear g-r rate equations, as implied by case (iv) of our above discussion, but rather that the g-r *coefficients* must meet additional far from equilibrium conditions.

In summary, a semiconductor driven far from thermodynamic equilibrium by field-assisted generation-recombination processes represents a physical system which can potentially display a variety of nonequilibrium phase transitions. This is important in two respects. First, by a careful treatment of g-r induced nonequilibrium phase transitions, a thorough understanding of the physical mechanisms of semiconductor instabilities and of the influence of geometrical and material parameters can be achieved. This is a necessary basis for the material engineering of novel semiconductor devices operating in the regime of g-r instabilities, and giving possibly better efficiencies, greater reliability, and lower cost of production. Second, a nonequilibrium semiconductor represents a particularly apt model system for the study of nonlinear synergetic effects, like nonequilibrium phase transitions and self-organization, since in semiconductor physics extremely sophisticated techniques of measurement are available, offering better reproducibility and better spatial and temporal resolution (covering the complete range from minutes to picoseconds) than in many other, e.g., hydrodynamic, systems. Nevertheless, g-r instabilities in semiconductors have in the past not been studied under the aspect of nonequilibrium phase transitions as intensively as, for example, the laser transition, or hydrodynamic (Bénard, Taylor) instabilities.

1.2.3 Bifurcation Phenomena

In many cases nonequilibrium systems can be theoretically modeled as *dynamic systems* [1.17]

$$\dot{q} = F(q, k) , \qquad\qquad (1.2.29)$$

where $q := (q_1, q_2, \ldots, q_n)$ is a set of dynamic variables, $k := (k_1, \ldots, k_m)$ is a set of control parameters, and $F := (F_1, F_2, \ldots, F_n)$ is a set of nonlinear functions of q, which may contain spatial derivatives of q as well. In particular, the semiconductor transport equations (1.1.5-13) can be cast into the general form (1.2.29), where q

represents the carrier densities and internal fields and k stands for external parameters like applied currents or fields, or optical irradiation intensities.

Nonequilibrium phase trasitions show up as bifurcations of this nonlinear
dynamic system. In the following we shall review some basic aspects of stability and
bifurcations of dynamic systems which will be needed in later chapters. Except for
section (E) on continuous bifurcation, we will assume that the function F does not
contain spatial derivatives.

A special case of (1.2.29) arises if the functions F do not depend explicitly upon time. In this case (1.2.29) is called an *autonomous* dynamic system, and all the
time dependence is through $q(t)$. Examples are semiconductors under steady state
irradiation or dc bias. Such a situation will be prevailing in this book. The semiconductor equations can often be reduced to a two-variable autonomous system
(Chaps. 2, 4):

$$\dot{q}_1 = F_1(q_1, q_2; k) \;,$$

$$\dot{q}_2 = F_2(q_1, q_2; k) \;. \tag{1.2.30}$$

Such systems are conveniently discussed in the (q_1, q_2) *phase plane* of dependent
variables. For every phase point (q_1, q_2), except for *singular* points (q_1^0, q_2^0) where

$$F_1(q_1^0, q_2^0; k) = F_2(q_1^0, q_2^0; k) = 0 \;, \tag{1.2.31}$$

the differential equations (1.2.30) give a unique direction of motion (F_1, F_2) in the
phase plane. (Note that the uniqueness property is lost for nonautonomous systems,
since F_1 and F_2 may have different values for the same values of q_1 and q_2 at different
times, due to the explicit dependence on t). By plotting the vector field (F_1, F_2) in
the whole phase plane, one can construct a field of directions which is tangent to
the curve or trajectory along which the system moves in the phase plane. For a
given initial phase point $(q_1(0), q_2(0))$ a unique trajectory can be graphically constructed from this field of directions. By choosing other initial points, other trajectories are obtained and a *phase portrait* emerges, similar to the streamlines of a fluid.
Due to the uniqueness of the field of directions, a trajectory can never cross itself
or other trajectories. (This is not true for nonautonomous systems!)

A simple way of obtaining a qualitative phase portrait is by drawing the two
curves $F_1(q_1, q_2) = 0$ and $F_2(q_1, q_2) = 0$, the so-called null-isoclines. On the $F_1 = 0$
curve dq_1/dt vanishes and $dq_1/dq_2 = 0$, hence the phase-plane trajectories cross this
curve parallel to the q_2-axis. Similarly, on the $F_2 = 0$ curve dq_2/dt and dq_2/dq_1
vanish, hence the trajectories cross parallel to q_1-axis. This, together with the
determination of the sign of dq_1/dt on either side of the $F_1 = 0$ curve, and of dq_2/dt
on either side of the $F_2 = 0$ curve, yields a qualitative idea of the phase portrait. The
intersections of the two null-isoclines are, by (1.2.31), the singular points, and
represent time-independent *steady states*. They are sometimes also called equilibrium states, which is, however, misleading since they are usually states far from
thermodynamic equilibrium.

An analytical expression for the geometrical form of the trajectories can be
obtained by eliminating time from (1.2.30), assuming $F_1 \neq 0$:

$$\frac{dq_2}{dq_1} = \frac{F_2(q_1, q_2; k)}{F_1(q_1, q_2; k)} .$$

(1.2.32)

The solution of the differential equation (1.2.32) is the phase-plane trajectory. In order to obtain the actual time dependence $(q_1(t), q_2(t))$ as the system moves along a trajectory, one must perform one further integration.

Stability

Any solution $q = u_{q_0}(t)$, which may be regarded as the path of a particle in q-space is uniquely determined by its initial value q_0 at the initial time t_0. Since the initial values are normally subject to perturbations, the question of interest becomes: what happens to the path of the particle if the initial condition is slightly different than the initial choice? A trajectory is said to be *locally stable* if other trajectories, obtained from initial conditions, different from, but close to q_0, remain close to the original trajectory for all later times as both phase points move about in the phase plane. If a neighborhood of q_0 cannot be found such that all initial departures within this neighborhood satisfy this criterion, the trajectory u_{q_0} is unstable. It is *locally asymptotically stable* if for all trajectories $v_{q_0'}$, close to u whose phase points fulfill the stability criterion, the condition

$$|u_{q_0}(t) - v_{q_0'}(t)| \to 0 \quad \text{as } t \to \infty$$

also holds. These definitions include the special, but very important, case that u consists of a single (singular!) point, i.e., a steady state.

A weaker form of stability, which is important when limit cycles (defined below) are considered, is *orbital stability*. It involves only the geometric form of the trajectories in phase space, and not the distance of neighboring phase points on these trajectories at fixed times. A trajectory u is orbitally stable if all phase points sufficiently close to u at some time t_0 remain within a certain given distance from the curve u for all later times. Orbital stability allows for a phase lag between phase points on the original and the neighboring trajectories. *Asymptotic orbital stability* is then defined by requiring additionally that the distance between u and an arbitrary phase point $q(t)$ on a neighboring trajectory tends to zero as $t \to \infty$.

Global stability requires that not only neighboring phase points but *all* points of the relevant phase plane meet the above conditions of local stability. Thus there cannot be more than one globally stable trajectory. Note, however, that a phase plane may contain several locally stable trajectories or singular points. In Hamiltonian and in gradient systems, global stability can readily be determined by looking at the form of the potential curves. In many systems which do not have a potential a "Lyapunov function" (a generalization of a potential) can be found [1.8].

Yet another notion is that of structural stability [1.19]. It does not involve the evolution of an individual phase point as time increases, but rather the change of the whole field of trajectories as the control parameters are varied. A dynamic system is *structurally stable* if the topological structure of the phase portrait is not affected by small changes in the differential equations.

Finally, as a special case, which shall be needed in later chapters, we will examine the local asymptotic stability of a singular point of a two-variable autonomous system in some more detail. In the vicinity of a singular point (q_1^0, q_2^0) the dynamic system given by (1.2.30) may be linearized around (q_1^0, q_2^0). In this approximation the trajectories near a singular point can be computed explicitly. Setting

$$\delta q_1 := q_1 - q_1^0 , \qquad \delta q_2 := q_2 - q_2^0 , \tag{1.2.33}$$

we obtain from (1.2.30)

$$\delta \dot{q} = A(q_1^0, q_2^0) \delta q , \tag{1.2.34}$$

where the Jacobian matrix $A(q_1^0, q_2^0)$ is defined by its matrix elements

$$A_{ij} := \frac{\partial F_i}{\partial q_j} , \qquad i, j = 1, 2 . \tag{1.2.35}$$

(Note that this definition can easily be generalized to arbitrary dimensions). The general solution of the system of linear differential equations (1.2.34) is in the generic case $(\lambda_1 \neq \lambda_2)$

$$\delta q(t) = c_1 \eta_1 e^{\lambda_1 t} + c_2 \eta_2 e^{\lambda_2 t} , \tag{1.2.36}$$

where c_1, c_2 are constants, η_1, η_2 are eigenvectors of A, and λ_1, λ_2 are the eigenvalues of A, given by

$$\lambda^2 - T\lambda + D = 0 \qquad \text{or} \tag{1.2.37}$$

$$\lambda = \tfrac{1}{2}[T \pm (T^2 - 4D)^{1/2}] . \tag{1.2.38}$$

Here

$$T := A_{11} + A_{22} \qquad \text{and}$$

$$D := A_{11} A_{22} - A_{12} A_{21} \tag{1.2.39}$$

are the trace and the determinant of the matrix A, respectively. For an asymptotically stable singular point ("sink") all trajectories (1.2.36) are required to approach it as $t \to \infty$, hence the real parts of both λ_1 and λ_2 must be negative.

The five qualitatively different regimes of solutions (1.2.36) are shown in Fig. 1.11.

(a) *Stable focus*: $D > 0$, $T < 0$, $T^2 < 4D$. The eigenvalues are conjugate complex and have negative real parts. The solutions are damped and complex, causing δq to execute damped oscillations in the phase plane with angular frequency $\omega = (D - \tfrac{1}{4}T^2)^{1/2}$ and damping constant $\tfrac{1}{2}T$. The general form of the trajectory is an elliptical spiral converging on the origin.

(b) *Unstable focus*: $D > 0$, $T > 0$, $T^2 < 4D$. The eigenvalues are complex and have positive real parts; here the elliptical spiral diverges.

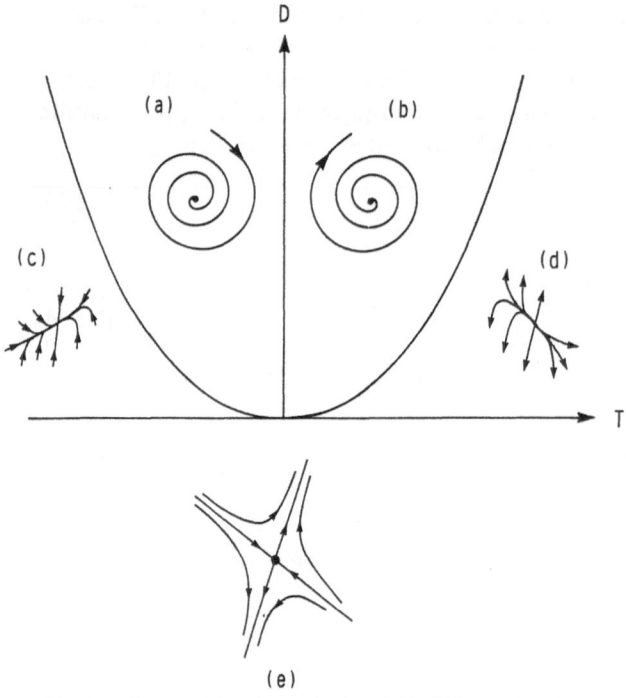

Fig. 1.11. Regimes of stability of the singular points of a two-variable autonomous dynamic system. T and D denote the trace and the determinant, respectively, of the Jacobian matrix defined in (1.2.35). The insets show the qualitative nature of the phase portraits near the singular point for the five generic cases (a)–(e), corresponding to (a) stable focus, (b) unstable focus, (c) stable node, (d) unstable node, (e) saddle point

(c) *Stable node*: $D > 0$, $T < 0$, $T^2 > 4D$. Both eigenvalues are negative. The solutions are real and the oscillations are replaced by an exponential decay.

(d) *Unstable node*: $D > 0$, $T > 0$, $T^2 > 4D$. Both eigenvalues are positive; the trajectories diverge from the origin.

(e) *Saddle point*: $D < 0$. The eigenvalues are real, with one value positive and the other negative. The positive value causes an instability in the direction of the corresponding eigenvector in phase space.

The borderlines between these five regimes obviously represent structurally unstable cases. They require more detailed investigations, involving higher orders in the Taylor expansion of the system around the singular point [1.17]. In particular, for $T = 0$ and $D > 0$ the trajectories near the singular point are either ellipses, or slowly converging or diverging spirals (depending on the nonlinear terms of the expansion). In the first case the singular point is called a center, in the second case a weakly stable/unstable focus. The borderlines between regime (a) and (b), and between (c) and (e) are associated with various bifurcations which we shall study below.

Bifurcation

Bifurcation is a phenomenon peculiar to nonlinear dynamic systems and is closely related to the loss of stability [1.13]. It describes the branching of solutions (steady states, oscillatory or spatially nonuniform solutions) as a control parameter k of the system described by (1.2.29) is varied. When different solution branches intersect or coalesce, they usually change their stability character. For steady states this means that at least one of the eigenvalues of the Jacobian A (1.2.35) has a vanishing real part. Also, at the bifurcation point k_0 the whole system is structurally unstable.

In the following, a topological classification of the most important types of bifurcations is given. Although most of these bifurcations are more general, restriction here is to autonomous two-variable systems [1.17], whose behavior can readily be visualized in the corresponding phase portraits.

(A) *Zero-Eigenvalue Bifurcations*

Local bifurcations [1.18] can be completely characterized by the phase portrait near the bifurcating singular point. Relevant information about the bifurcation is contained in the linearized system given by the matrix A (1.2.35). The simplest bifurcation occurs if one simple real eigenvalue λ of A turns from negative to positive values upon variation of the control parameter k. At the bifurcation value, k_0, the eigenvalue is zero, and the singular point does not fall in any one of the simple categories (a–e) of Fig. 1.11. Rather, it is a *multiple singular point*, which is generated by the coalescence of several simple singular points. From (1.2.37) it follows that $\lambda = 0$ results in $D = 0$, while $T \neq 0$. Thus, the bifurcation is described by a simple analytical condition for the Jacobian matrix A of a two-variable system. In the physical literature this type of bifurcation is often called a *soft-mode instability* because $\lambda \rightarrow 0$ implies that the corresponding linear mode softens, i.e., slows down critically in its temporal evolution.

The following four subclasses may occur (Fig. 1.12):

(A1) *Saddle-node bifurcation*: a saddle point (sa) and a stable node (sn) coalesce and disappear. At the bifurcation point they form a so-called saddle node. This is typical of threshold switching and SNDC in semiconductors. It is associated with first-order phase transitions involving other solution branches (see, e.g., Fig. 2.8 below).

(A2) *Transcritical bifurcation*: A saddle point and a stable node coalesce and separate again. Often the unstable branch for $k < k_0$ turns back again in a saddle-node bifurcation of type (A1), or it lies in a nonphysical regime of q-space. The exchange of stability of the two crossing solution branches corresponds to a second-order phase transition.

(A3) *Supercritical pitchfork bifurcation*: A saddle point and two stable nodes coalesce. This is typical of a symmetry-breaking bifurcation, as it occurs at a critical point.

(A4) *Subcritical pitchfork bifurcation*: Two saddle points and a stable node coalesce. Often the two unstable branches for $k < k_0$ turn back again in a saddle-node bifurcation.

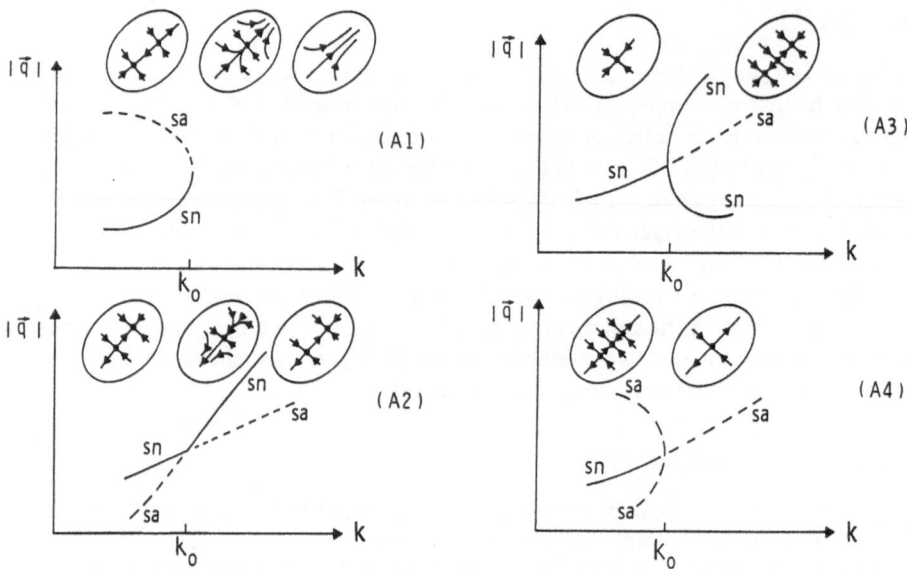

Fig. 1.12. Bifurcation diagrams for a soft-mode instability (zero eigenvalue bifurcation). The steady state coordinate $|q|$ is plotted versus the control parameter k. Full lines represent stable steady states, broken lines are unstable steady states. Saddle points and stable nodes are denoted by sa and sn, respectively. The insets represent schematic phase portraits corresponding to values of $k < k_0$, $k = k_0$, and $k > k_0$ where k_0 is the bifurcation point

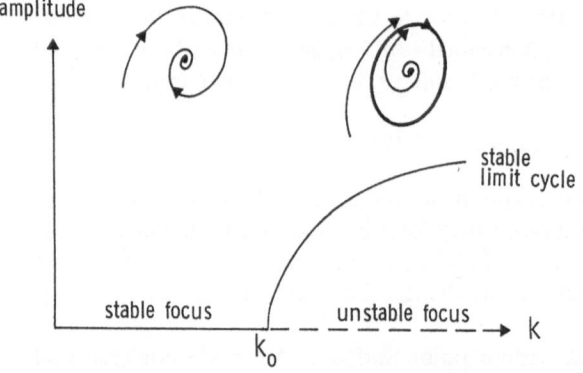

Fig. 1.13. Bifurcation diagram for a Hopf bifurcation of a limit cycle. The amplitude of the limit cycle is plotted versus the control parameter k. The insets show schematic phase portraits for $k < k_0$ (stable focus) and $k > k_0$ (unstable focus + stable limit cycle)

(B) *Hopf Bifurcation of Limit Cycles*

If a pair of complex conjugate eigenvalues of A cross the imaginary axis, the singular point changes from a stable to an unstable focus, and a time-dependent periodic solution bifurcates from the steady state solution branch with zero amplitude, as shown in Fig. 1.13. This periodic solution is asymptotically orbitally stable, i.e., all trajectories starting in a neighborhood (inside and outside) of this closed trajectory in the phase plane will asymptotically approach it. It is therefore called a stable *limit cycle* [1.139–142]. The inverted configuration with an unstable limit cycle and a stable focus for $k_0 > 0$, and an unstable focus for $k_0 < 0$, also exists; it can be obtained from Fig. 1.13 by time reversal. A system which possesses a limit cycle is structurally stable (except at the bifurcation point). These two properties – orbital and structural stability – distinguish a limit cycle from conservative oscillations around a center. The latter are also self-sustained oscillations, but not asymptotically orbitally stable, since there is an infinite continuum of neighboring oscillations in the phase plane. Also, conservative oscillations are not structurally stable since a damping term – however small it may be – destroys the closed trajectories and converts the center into a stable focus. For these two stability reasons, limit cycles are extremely important, as we shall see in Chap. 6.

The Hopf bifurcation may be supercritical with the bifurcation scheme:

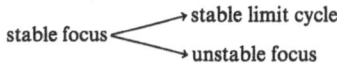

$$\text{stable focus} \Big\langle \begin{array}{l} \rightarrow \text{stable limit cycle} \\ \rightarrow \text{unstable focus} \end{array}$$

or subcritical with the bifurcation scheme

$$\text{unstable focus} \Big\langle \begin{array}{l} \rightarrow \text{unstable limit cycle} \\ \rightarrow \text{stable focus} \end{array}$$

The bifurcation point is given, cf., (1.2.37) by the condition $T = 0, D > 0$; it corresponds to a weakly stable or unstable focus, from which the limit cycle bifurcates at frequency $\omega = \sqrt{D} \neq 0$. Since the frequency does not tend to zero, i.e., does not "soften", the bifurcation is sometimes called a *hard-mode instability*. We shall encounter examples of Hopf bifurcations in Sect. 6.2.

(C) *Limit-Cycle Bifurcations by Condensation of Paths*

If a stable and an unstable limit cycle are around the same singular point (e.g., a stable focus), they can coalesce and disappear (Fig. 1.14). At the bifurcation both

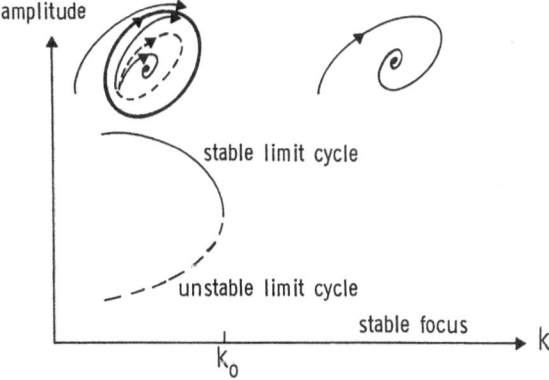

Fig. 1.14. Bifurcation diagram for a limit-cycle bifurcation by condensation of paths. The insets show a stable (——) and an unstable limit cycle (– – –) around a stable focus for $k < k_0$, and a stable focus for $k > k_0$

the amplitude and the frequency are nonzero, as opposed to the Hopf bifurcation
(B). Such bifurcations, and also type (D) below, are called *global* bifurcations, since
they do not merely involve the neighborhood of a singular point, but a larger region
of phase space.

(D) *Bifurcation of Limit Cycles from a Separatrix*

Two possible topologies are shown in Fig. 1.15. In Fig. 1.15a a saddle point and a
stable node, which lie on a closed curve formed by two trajectories ("separatrices"),
coalesce and disappear, whereby a stable limit-cycle bifurcates from the separatrix.
In Fig. 1.15b a saddle-to-saddle loop is formed from which the limit cycle bifurcates.
Here the saddle does not disappear. In both cases the limit cycle appears immediately with a nonzero amplitude, but with zero frequency. Thus the period of the

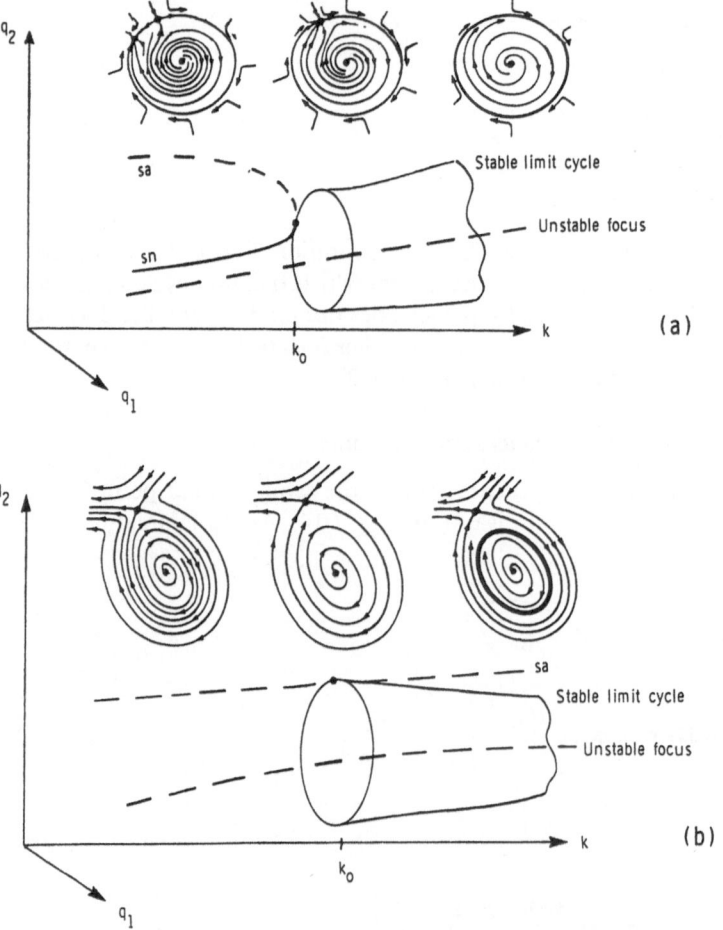

Fig. 1.15a, b. Bifurcation diagrams for limit-cycle bifurcations from a separatrix. The limit cycle is
represented by a cylindrical surface in the three-dimensional space of dynamic variables q_1, q_2 and
control parameter k

oscillation tends to infinity at the bifurcation point. The unstable focus in the interior of the limit cycle does not change its stability character. Examples of such global bifurcations have been described in chemical reaction systems [1.143].

(E) *Continuous Bifurcation*

In dynamic systems allowing for diffusion and/or drift, i.e., containing spatial derivatives, there is yet another type of bifurcation: the bifurcation of nonuniform spatial structures from a uniform steady state. The condition for bifurcations is obtained in a similar way as for uniform structures by linearizing the dynamic system around the uniform state and using the ansatz

$$\delta q \sim \exp(\lambda t + i\boldsymbol{K}x) \ , \tag{1.2.40}$$

where $\boldsymbol{K}$ is a wave vector and x the spatial coordinate. From this a dispersion relation $\lambda(\boldsymbol{K})$ is obtained. It describes the damping-out of small periodic spatial fluctuations if the real part of $\lambda(\boldsymbol{K})$ is negative. If $\mathrm{Re}\{\lambda(\boldsymbol{K})\}$ changes sign at some value of $\boldsymbol{K}$, a spatially periodic solution with the wavevector $\boldsymbol{K}$ bifurcates from the uniform steady state. Upon variation of the control parameters the dispersion relation $\lambda(\boldsymbol{K})$ is changed, and a whole family of spatially nonuniform solutions bifurcates. The family is specified by appropriate boundary conditions and may include a "solitary solution" of infinite period as a limiting case.

If $\lambda(\boldsymbol{K})$ is zero at the bifurcation point, the bifurcating solutions are time independent. An important example is the current filamentation associated with SNDC. We shall treat this case in Sect. 3.2 and Chap. 4.

If $\lambda(\boldsymbol{K})$ is purely imaginary at the bifurcation point, a standing wave (see Sect. 3.5) or a traveling wave (see Sect. 3.3) bifurcates. Such nonlinear propagating waves may be periodic, or solitary, like the high-electric field domains occurring in the Gunn-Hilsum effect. This can lead to current oscillations governed by the transit time of a domain between the cathode and anode, as we shall see in Sect. 6.1.2.

Sometimes the solitary waves have an additional property, besides retaining their shape while propagating: they emerge unperturbed in shape and velocity after a collision with another solitary wave. In this case they are called *solitons*. Solitons are special solutions of nonlinear dispersive wave equations, and occur in a variety of areas, ranging from shallow water waves to Josephson junctions [1.23]. The first extensive theoretical treatment of soliton motion was given in connection with the propagation of dislocations in solids and their interaction with phonons [1.144].

More complicated sequences of bifurcations leading eventually to completely irregular, chaotic states are possible in autonomous systems with more than two variables. This will be elaborated upon in Sect. 6.3.

In conclusion, we have seen that bifurcations of nonlinear dynamic systems can produce a great wealth of dissipative structures, like bistable homogeneous steady states, stationary spatial patterns, solitary waves, limit-cycle oscillations, and chaos. These bifurcations are connected with various first- and second-order nonequilibrium phase transitions. In the course of this book we will show that all these phenomena can be induced by nonlinear generation-recombination processes in semiconductors. This might help to understand old and predict new instabilities in semiconductors.

2. Bistability of Homogeneous Steady States

In this chapter we will elaborate a variety of g-r mechanisms that give rise to three spatially homogeneous steady states (two of which are stable) in a certain range of applied electric fields and material parameters. This results in S-shaped current density-field relations (SNDC), and in various nonequilibrium phase transitions of first and of second order between the different steady states. Impact ionization turns out to be a key process for these phenomena. The connection with threshold switching and – in the case of exciton g-r kinetics – optical bistability is pointed out.

Throughout the present chapter, attention is confined to spatially homogeneous states.

2.1 One-Carrier Models

First, let us consider models in which only one species of mobile carriers (electrons or holes) is involved [2.1–13]. For convenience all formulas will be given for the case of electrons, corresponding to an n-type semiconductor, but all models can readily be applied to p-type material by replacing the concentrations of electrons and of donors with those of holes and of acceptors, respectively, and replacing the appropriate g-r constants.

We assume an n-semiconductor with N_D donors, partially compensated by $N_A < N_D$ acceptors. At sufficiently low temperatures g-r processes involving the valence band may be neglected. The acceptors are assumed to be fully occupied so that they do not enter the g-r kinetics explicitly. The electrons may be trapped at an impurity (a donor atom, or – at higher temperatures where all donors are thermally ionized – a deep trap). We include the possibility that the trapped electron is either in the ground state or in one of the excited states, which are similiar to the energy spectrum of an H atom in the case of shallow (i.e., of energy only slightly below the conduction-band edge) impurities.

The spatially homogeneous solutions of the transport equations (1.1.5–13) are characterized by position-independent n, n_{t_i}, and $E = E_0 e_z = (V/L)e_z$, where E_0 is the static electric field applied along the z-axis, V is the voltage across the sample, L the sample length, and e_z the unit vector in the direction of the applied field. Then the basic equations (1.1.5, 7, 8, 12) reduce to the following: from (1.1.12) the conduction current density as a function of the applied field is given by

$$j_0 = e\mu_n n(E_0)E_0 \ . \tag{2.1.1}$$

We assume that the field dependence of the mobility μ_n can be neglected in the field range of interest in the following. Maxwell's equation (1.1.8) reduces to the condition of local neutrality

$$N_D^* - n - \sum_{i=1}^{M} n_{t_i} = 0 \; , \tag{2.1.2}$$

where $N_D^* := N_D - N_A$ is the effective donor density. From (2.1.2) one of the trapped electron densities, say n_{t_M}, can be eliminated in (1.1.5, 7). This gives the g-r rate equations:

$$\dot{n} = g_0(n, n_{t_1}, \dots, n_{t_{M-1}}, E_0) \tag{2.1.3a}$$

$$\dot{n}_{t_i} = g_i(n, n_{t_1}, \dots, n_{t_{M-1}}, E_0) \quad i = 1, \dots, M-1 \tag{2.1.3b}$$

where

$$g_\lambda(n, n_{t_1}, \dots, n_{t_{M-1}}, E_0)$$

$$:= f_{t_\lambda}\left(n, n_{t_1}, \dots, n_{t_{M-1}}, n_{t_M} = N_D^* - n - \sum_{i=1}^{M-1} n_{t_i}, E_0\right)$$

$$\lambda = 0, 1, \dots, M-1; t_0 \equiv n \; . \tag{2.1.4}$$

Assuming nondegenerate statistics, we can employ for g_λ the usual mass-action-type polynomials discussed in Sect. 1.2.2. The homogeneous steady states $n(E_0)$ are given by the simultaneous solution of the system of M nonlinear algebraic equations

$$0 = g_\lambda(n, n_{t_1}, \dots, n_{t_{M-1}}, E_0) \quad \lambda = 0, 1, \dots, M-1 \; . \tag{2.1.5}$$

With this, (2.1.1) represents the static current density-field (or, under spatially uniform conditions, current-voltage) characteristic.

We shall now illustrate this general procedure by specific examples of g-r kinetics.

2.1.1 Second-Order Phase Transitions

The simplest model arises if one trap level only – the ground state – is taken into account [2.7, 8]. It involves a single independent concentration variable, n, since the trapped electron density n_t can be eliminated by (2.1.2):

$$n_t = N_D^* - n \; . \tag{2.1.6}$$

We assume that the donors are fully ionized, and the trap density is $N_t > N_D^*$. The number of unoccupied traps is then given by

$$p_t = N_t - N_D^* + n \; . \tag{2.1.7a}$$

An important special case arises if the traps are provided by the donors. In this case the donors enter the g-r kinetics explicitly, and the acceptors only are assumed to be completely ionized. In all formulas $N_t = N_D$ must be set. Thus (2.1.7a) becomes:

Table 2.1. Equivalence of donor g-r kinetics and trap g-r kinetics

	Shallow donors	Deep traps
Concentration of impurities involved in g-r kinetics	N_D	N_t
Compensating impurity density	N_A (completely ionized acceptors)	$N_t - N_D^*$ ($N_D^* \equiv N_D - N_A$: completely ionized effective donor density)
Density of trapped electrons in the ground state	n_D	n_{t_1}
Density in the excited state	n_D^*	n_{t_2}
Density of unoccupied impurities	p_D	p_t
Necessary condition for phase transitions	$N_A < N_D$	$N_A < N_D < N_A + N_t$

$$p_t = N_A + n \; . \tag{2.1.7b}$$

The replacements to be made are listed in Table 2.1.

The respective rate equation (2.1.3a) is explicitly given by, see (1.2.25a, e),

$$\dot{n} = X_1^S n_t + X_1 n n_t - T_1^S n p_t - T_1 n^2 p_t \tag{2.1.8}$$

or, by (2.1.6, 7),

$$\dot{n} = X_1^S N_D^* + n[X_1 N_D^* - T_1^S(N_t - N_D^*) - X_1^S]$$
$$- n^2[X_1 + T_1^S + T_1(N_t - N_D^*)] - T_1 n^3 \; . \tag{2.1.9}$$

If $X_1^S > 0$, there is always a unique steady state solution $n > 0$ of (2.1.9), since the sequence of the coefficients of the cubic polynomial in n changes sign once.

We shall now consider the case $X_1^S = 0$, which corresponds to the neglect of thermal or optical ionization of the traps. The g-r processes are shown schematically in Fig. 2.1. Additionally, we shall neglect Auger recombination ($T_1 = 0$). These conditions can be realized at sufficiently low temperatures. The rate equation (2.1.9) is then

$$\dot{n} = n[X_1 N_D^* - T_1^S(N_t - N_D^*) - n(X_1 + T_1^S)] \; . \tag{2.1.10}$$

It always admits the steady state solution $n = n_1 \equiv 0$, and an additional steady state

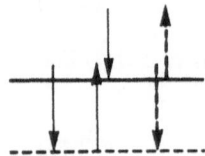

$T_1^S \quad X_1 \quad T_1$

Fig. 2.1. Simplest model for g-r induced second-order phase transitions, involving band-trap recombination (with rate constant T_1^S), impact ionization (X_1), and Auger recombination (T_1). The dashed process (T_1) is not necessary for the phase transition

solution $n = n_2 > 0$ if

$$X_1 \geq X_{1c} := (N_t/N_D^* - 1)T_1^S . \tag{2.1.11}$$

The steady state is stable if any fluctuation δn regresses in time. This is tested by linearizing (2.1.10) around the steady state:

$$\delta\dot{n} = [X_1 N_D^* - T_1^S(N_t - N_D^*)]\delta n = (X_1 - X_{1c})N_D^*\delta n$$

$$\text{for } \delta n := n - n_1 , \tag{2.1.12}$$

and

$$\delta\dot{n} = -2n_2(X_1 + T_1^S)\delta n = -2(X_1 - X_{1c})N_D^*\delta n \quad \text{for } \delta n := n - n_2 .$$

Thus the stable steady states are

$$n = \begin{cases} 0 & \text{for } X_1 \leq X_{1c} \\ \dfrac{X_1 N_D^* - T_1^S(N_t - N_D^*)}{X_1 + T_1^S} & \text{for } X_1 \geq X_{1c} . \end{cases} \tag{2.1.13}$$

Regarding the impact-ionization coefficient X_1 as a control parameter which can be increased by increasing the electric field E_0, one finds that the steady state solution $n(X_1)$, and hence also $n(E_0)$, has a discontinuous derivative $dn/dX_1 (dn/dE_0)$ at the critical value X_{1c} (corresponding to a critical field E_c). This is a second-order nonequilibrium phase transition from a nonconducting ($n = 0$) to a conducting state ($n > 0$). It corresponds to a transcritical bifurcation as shown in Fig. 1.12 (A2), where the unstable branch for $k < k_0$ lies in the negative, i.e., nonphysical regime of n-values.

The threshold condition (2.1.11) of the second-order phase transition is reached most easily if the ratio $C := N_t/N_D^*$ is close to unity, and if the band-trap recombination coefficient T_1^S is small, e.g., for repulsive traps. For $C \leq 1$ the phase transition disappears altogether, and the conducting state $n_2 > 0$ exists and is stable throughout the entire range of the impact-ionization coefficient X_1, including arbitrarily small X_1.

Physically, the second-order phase transition is connected with the well-known phenomenon of impurity breakdown in semiconductors. The usual breakdown condition for impact ionization [2.14–16] agrees with the critical point X_{1c} in (2.1.11). Experimentally, the following typical numerical values were obtained (for gold in silicon, with trap depth 0.54 eV below the conduction-band edge):

breakdown fields $\approx 10^3 - 10^4$ V/cm ,

$X_1 \approx 10^{-8} - 10^{-7}$ cm^3 s^{-1} ,

$N_D^*, N_t \approx 10^{15} - 10^{16}$ cm^{-3} ,

$T_1^S \approx 10^{-8} - 10^{-9}$ cm^3 s^{-1} (at $77 - 300$ K) .

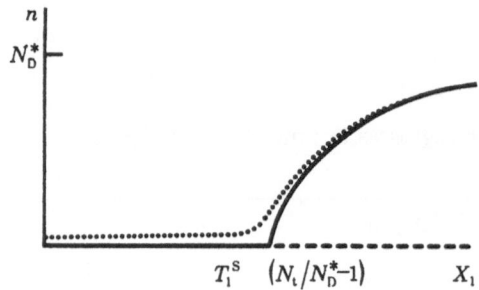

Fig. 2.2. Steady state electron concentration versus impact-ionization coefficient X_1 for the simple model of Fig. 2.1 with $T_1 = 0$, $X_1^s = 0$ (——) and $T_1 = 0$, $X_1^s \neq 0$ (–––) (schematic). This diagram is topologically similar to the current-field characteristic

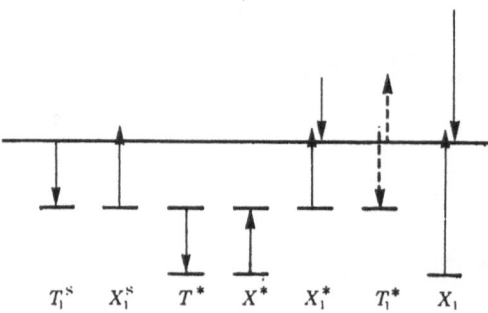

Fig. 2.3. Elaboration of the simple g-r model of Fig. 2.1, by including the first excited state of the traps. These processes give rise to second-order phase transitions similar to Fig. 2.2 when X^* or X_1^s are zero

A nonzero value of the thermal and optical ionization coefficient X_1^s, however small, destroys the second-order phase transition, by smoothing out the discontinuity of dn/dX_1 at the critical point, just as a magnetic field does in the case of a second-order ferromagnetic equilibrium phase transition (cf., Sect. 1.2.1). But for small X_1^s the physical phenomenon of a sharp rise in the carrier concentration [and, by (2.1.1), the current density] by several orders of magnitude is still retained, see Fig. 2.2. Thus in the idealized limit $X_1^s \to 0$ (which is a very good approximation at low temperatures and with shielding against external radiation) impurity breakdown in semiconductors represents a second-order nonequilibrium phase transition.

Let us now discuss some elaborations of the simplest g-r mechanism for a second-order transition. The effect of the inclusion of Auger recombination (T_1) is slight in that it affects n_2, but not the critical point X_{1c} which is still given by (2.1.11). The stable steady state solutions are

$$n = \begin{cases} 0 & \text{for } X_1 \leq X_{1c} \\ \dfrac{b}{2T_1}(\{1 + 4T_1[X_1 N_D^* - T_1^s(N_t - N_D^*)]/b^2\}^{1/2} - 1) & \text{for } X_1 \geq X_{1c} \end{cases}$$

$$b := X_1 + T_1^s + T_1(N_t - N_D^*) \ . \tag{2.1.14}$$

By involving also the first excited state of the trap, a further elaboration becomes possible (Fig. 2.3). From the conservation of impurities and charge neutrality we have respectively

$$p_t + n_{t_1} + n_{t_2} = N_t \ , \qquad n_{t_1} + n_{t_2} = N_D^* - n \ . \tag{2.1.15}$$

The condition $\dot{n}_{t_1} = 0$ for a steady state then yields

$$n_{t_2} = \omega(N_D^* - n)$$

$$n_{t_1} = (1 - \omega)(N_D^* - n) \, , \tag{2.1.16}$$

where

$$\omega := \frac{X^* + X_1 n}{T^* + X^* + X_1 n} \, . \tag{2.1.17}$$

Substituting in the rate equation for n, one finds in the steady state

$$0 = \dot{n} = X_1^S \omega N_D^* + [X_1(1 - \omega)N_D^* + X_1^* \omega N_D^* - T_1^S(N_t - N_D^*) - X_1^S \omega]n$$

$$- [T_1^S + X_1^* \omega + T_1^*(N_t - N_D^*) + X_1(1 - \omega)]n^2 - T_1^* n^3 \, . \tag{2.1.18}$$

As before, the Auger recombination (T_1^*) does not affect the qualitative nature of the phase transition and will therefore be neglected. In this section we consider only three special cases in which second-order phase transitions similar to those of the basic model occur, when X^* or X_1^S tend to zero. (In Sect. 2.1.2 we shall show that the general model can also lead to first-order phase transitions.) The critical point at which the transition from a nonconducting to a conducting state occurs, is different in each case.

The first case uses impact ionization of the excited-trap level as the key reaction, and neglects the following:

(i) $T_1^* = X_1 = X_1^S = 0 \, . \tag{2.1.19}$

Here $\omega \equiv X^*/(T^* + X^*)$ is independent of n and represents, from (2.1.16), the fraction of electrons which occupy the excited-trap level in the nonconducting state. Then (2.1.18) reduces to (2.1.10), provided that the control parameter X_1 in (2.1.10) is replaced by ωX_1^*. The critical point is given by

$$\omega X_{1c}^* := (N_t/N_D^* - 1)T_1^S \, , \tag{2.1.20}$$

and stability arguments go through as before.

The other two cases use impact ionization of the ground-trap level as key reactions and are distinguished by the neglect of X_1^S or X^*, respectively. First consider

(ii) $T_1^* = X_1^* = X_1^S = 0 \, , \tag{2.1.21}$

when ω depends on n by (2.1.17). For the steady state (2.1.18) becomes

$$0 = [\omega' X_1 N_D^* - T_1^S(N_t - N_D^*)]n$$

$$- [\omega' X_1 + T_1^S + X_1(N_t - N_D^*)T_1^S/(T^* + X^*)]n^2$$

$$- [X_1 T_1^S/(T^* + X^*)]n^3 \, , \tag{2.1.22}$$

where $\omega' := T^*/(T^* + X^*)$ is the fraction of electrons which occupy the ground-trap level in the nonconducting state. The critical point is given by

$$\omega' X_{1c} := (N_t/N_D^* - 1)T_1^S .\tag{2.1.23}$$

Next consider

(iii) $T_1^* = X_1^* = X^* = 0 .$ \hfill (2.1.24)

Here the steady state equation (2.1.18) for n reads

$$0 = [X_1 N_D^* - \zeta T_1^S(N_t - N_D^*)]n$$
$$- [X_1 + \zeta T_1^S + X_1(N_t - N_D^*)T_1^S/(T^* + X_1^S)]n^2$$
$$- [X_1 T_1^S/(T^* + X_1^S)]n^3 .\tag{2.1.25}$$

The critical point is given by

$$X_{1c} := (N_t/N_D^* - 1)\zeta T_1^S ,\tag{2.1.26}$$

where $\zeta := T^*/(T^* + X_1^S)$ is the probability for relaxation of the excited trap level, and therefore ζT_1^S represents the effective recombination coefficient for electron capture via an excited level. In contrast to model (i) and (ii), all electrons occupy the trap ground level in the nonconducting state. From $0 \leq \omega, \omega', \zeta \leq 1$ it is evident that in cases (i) and (ii) the critical threshold of the impact-ionization coefficient is raised or remains the same, whereas in (iii) it is lowered or remains the same, compared with the basic model (2.1.11).

A system of practical interest involving excited trap levels is furnished by an alkali-halide crystal with F-centers. The energy gap is then sufficiently large to justify the neglect of processes involving the valence band, and the first excited state F* is usually considered in the reaction kinetics [2.17–19]. The defects consist of anion vacancies ('α-centers') which can capture electrons to form F-centers. It has been deduced from photoconductivity experiments [2.18] that besides these F-centers (of concentration N_D^*) alkali-halide crystals contain additional empty vacancies of concentration $N_t - N_D^*$ which are frozen in at high temperatures when the crystal samples are quenched to room temperature; the charge of these is compensated by other charged defects. The processes of Fig. 2.3 can be realized as electron capture by vacancies (T_1^S), thermalization of F*-centers (X_1^S), radiative or nonradiative relaxation into the ground state (T^*), optical excitation by irradiation into the F-band (X^*), impact ionization of F* or F-centers (X_1^*, X_1), and Auger recombination (T_1^*). Second-order transitions of type (i) and (ii) are to be expected at low temperatures where X_1^S is negligible, when the impact ionization of F*-centers (i) or of F-centers (ii) reaches the threshold given by (2.1.20) or (2.1.23) respectively. The threshold for type (i) is lowered by simultaneous strong irradiation into the F-band; for type (ii) the converse holds. Type (iii) corresponds to the absence of F-band irradiation ($X^* = 0$) and high temperatures ($X_1^S \neq 0$).

In all three cases the phase transition can occur only if initially not all vacancies are occupied by electrons, i.e., if additional frozen-in α-centers are present, otherwise there is a unique stable steady state with $n > 0$ for all choices of the rate constants. The required impact-ionization coefficient can be estimated from the following typical experimental data for KCl:

$$(N_t/N_D^* - 1) \approx 10^{-7} \text{ to } 10^{-4}$$

$$T_1^S \approx 5 \times 10^{-3} \text{ to } 10^{-5} \text{ cm}^3 \text{ s}^{-1} \; ; \qquad \left. \right\} \qquad \qquad \text{[2.18]}$$

$$T^* \approx 10^6 \text{ s}^{-1} \; ; \qquad\qquad\qquad\qquad\qquad\qquad \text{[2.20]}$$

$$X^* \begin{cases} \approx 10^{-6} \text{ to } 10^{-2} \text{ s}^{-1}, \text{ for normal irradiation ;} & \text{[2.18]} \\ \text{or up to } 10^8 \text{ s}^{-1}, \text{ for laser irradiation ;} & \text{[2.21]} \end{cases}$$

$$X_1^S \approx 10^{12} \exp(-0.15 \text{ eV}/kT) \text{ s}^{-1} \; . \qquad\qquad\qquad \text{[2.20]}$$

From this it can be seen that under appropriate conditions, impact ionization coefficients in the range 10^{-8} to 10^{-12} cm^3 s^{-1} could induce a second-order transition to a conducting state. It can be shown by explicit calculation that the addition of single electron processes involving F' centers (i.e., anion vacancies with two bound electrons) does not change the critical point of the second-order transition.

Note that the present models furnish remarkably simple examples of the much discussed phenomenon that new stable solutions emerge from the thermodynamic branch if a physical, chemical, or biological system is driven far from thermal equilibrium [1.12]. This can be achieved in a semiconductor for example by sufficiently strong electric fields and/or irradiation. As a measure of the 'distance from equilibrium' one may take the ratio of the critical value X_{1c} and the equilibrium value X_{10} of the impact-ionization coefficient, cf., the discussion in Sect. 1.2.2. One finds, using detailed balance to determine X_{10} that

$$X_{10} n_0 n_{t0} = T_{10} n_0^2 p_{t0} \; , \qquad\qquad\qquad\qquad\qquad (2.1.27)$$

or, with (2.1.6, 7),

$$X_{10} = T_{10} n_0 \frac{N_t - N_D^* + n_0}{N_D^* - n_0} \qquad\qquad\qquad\qquad (2.1.28)$$

where the subscript 0 denotes equilibrium values. Combination of (2.1.27, 11) yields

$$\frac{X_{1c}}{X_{10}} = \frac{T_1^S}{T_1 n_0} \frac{(N_t - N_D^*)(N_D^* - n_0)}{(N_t - N_D^* + n_0)N_D^*} \approx \frac{T_1^S}{T_1 n_0} \; , \qquad\qquad (2.1.29)$$

if the assumption that

$$n_0 \ll N_D^* \; , N_t - N_D^* \qquad\qquad\qquad\qquad\qquad (2.1.30)$$

is made.

Typical values for Ge at room temperature [1.121, 125] are $T_1^S \approx 10^{-9}$ cm^3 s^{-1}, $T_1 \approx 10^{-26}$ cm^6 s^{-1}, $n_0 \approx 10^{15}$ cm^{-3}, and thus $X_{1c}/X_{10} \approx 10^2$. This shows that the second-order phase transition does indeed occur far from thermal equilibrium (in particular with deep traps at low temperature, when n_0 is very small).

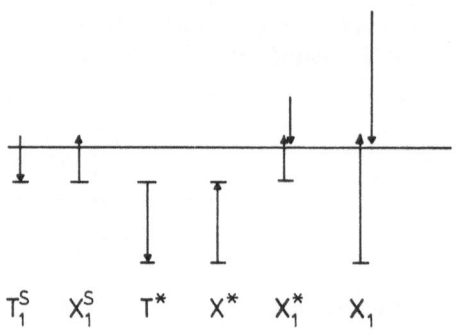

Fig. 2.4. Generation-recombination model for first-order phase transitions, involving the conduction band, the trap (or donor) ground state, and its first excited state

$$T_1^S \quad X_1^S \quad T^* \quad X^* \quad X_1^* \quad X_1$$

2.1.2 First-Order Phase Transitions

Next, we shall analyse the g-r model with two trap levels (the ground state and the first excited state) in detail, and show that this model – as opposed to the single-level model – can give bistability of the homogeneous steady states, and first-order nonequilibrium phase transitions [2.9, 10]. The g-r processes considered are shown in Fig. 2.4. The rate equations (2.1.3) are given by

$$\dot{n} = X_1^S n_{t_2} - T_1^S n p_t + X_1^* n n_{t_2} + X_1 n n_{t_1} \, , \tag{2.1.31a}$$

$$\dot{n}_{t_1} = -X^* n_{t_1} + T^* n_{t_2} - X_1 n n_{t_1} \, , \tag{2.1.31b}$$

$$\dot{n}_{t_2} = -\dot{n} - \dot{n}_{t_1} \, . \tag{2.1.31c}$$

Eliminating p_t and n_{t_2} by the conditions of conservation of impurities and of charge neutrality (2.1.15), we obtain two rate equations for the two independent variables n and n_{t_1}:

$$\dot{n} = a_0 + a_1 n + a_2 n_{t_1} + a_3 n n_{t_1} + a_4 n^2 \tag{2.1.32}$$

$$\dot{n}_{t_1} = b_0 + b_1 n + b_2 n_{t_1} + b_3 n n_{t_1} \tag{2.1.33}$$

with

$$a_0 := X_1^S N_D^* \qquad\qquad b_0 := T^* N_D^*$$

$$a_1 := X_1^* N_D^* - X_1^S - T_1^S (N_t - N_D^*) \qquad b_1 := -T^*$$

$$a_2 := -X_1^S \qquad\qquad b_2 := -(T^* + X^*)$$

$$a_3 := X_1 - X_1^* \qquad\qquad b_3 := -X_1 \, .$$

$$a_4 := -(T_1^S + X_1^*) \tag{2.1.34}$$

In the steady state we obtain from (2.1.33)

$$n_{t_1} = \frac{T^*(N_D^* - n)}{T^* + X^* + X_1 n} \, , \tag{2.1.35}$$

and n is determined by (2.1.32):

$$0 = P_3(n) := an^3 + bn^2 + cn + d \tag{2.1.36}$$

with

$$a := -(T_1^S + X_1^*)X_1 < 0$$

$$b := X_1^* X_1 N_D^* - [X_1^S + T_1^S(N_t - N_D^*) + T^*]X_1 - X^* X_{1^*} - T_1^S(T^* + X^*)$$

$$c := (X_1^S + T^*)N_D^* X_1 + X^* N_D^* X_1^* - X_1^S X^* - T_1^S(N_t - N_D^*)(T^* + X^*)$$

$$d := X_1^S X^* N_D^* \geq 0 . \tag{2.1.37}$$

Depending upon the rate constants, there are up to three physical solutions $0 \leq n \leq N_D^*$. Necessary and sufficient conditions for three positive solutions are

$$b > 0 , \qquad c < 0$$

$$\left(\frac{3ac - b^2}{9a^2}\right)^3 + \left(\frac{b^3}{27a^3} - \frac{bc}{6a^2} + \frac{d}{2a}\right)^2 < 0 . \tag{2.1.38}$$

The local stability of the steady state (n, n_{t_1}) against spatially homogeneous perturbations δn, δn_{t_1} follows from the linearized rate equations:

$$\begin{pmatrix} \delta \dot{n} \\ \delta \dot{n}_{t_1} \end{pmatrix} = A \begin{pmatrix} \delta n \\ \delta n_{t_1} \end{pmatrix} \tag{2.1.39}$$

with

$$A := \begin{pmatrix} a_1 + a_3 n_{t_1} + 2a_4 n & a_2 + a_3 n \\ b_1 + b_3 n_{t_1} & b_2 + b_3 n \end{pmatrix} \tag{2.1.40}$$

evaluated in the steady state.

The steady state (n, n_{t_1}) is locally asymptotically stable if A has two eigenvalues λ_1, λ_2 with negative real parts. This leads to the stability conditions

$$\text{tr}\{A\} \equiv \lambda_1 + \lambda_2 < 0$$

$$\det\{A\} \equiv \lambda_1 \lambda_2 > 0 . \tag{2.1.41}$$

From (2.1.40, 34, 32, 15) it follows that

$$\text{tr}\{A\} = -X_1^S n_{t_2}/n - X_1^S - (T_1^S + X_1^*)n - (T^* + X^*) - X_1 n , \tag{2.1.42}$$

which is always negative for n, $n_{t_2} \geq 0$. From (2.1.40, 35, 36) one finds

$$\det\{A\} = -\frac{d}{dn}P_3(n) . \tag{2.1.43}$$

From (2.1.37) it follows that $P_3(0) > 0$ and $P_3(\infty) \to -\infty$. Hence in case of a single, simple solution $n \geq 0$ of (2.1.36), $\det\{A\} > 0$ holds, and the steady state is therefore a stable node or focus ("antisaddle"), cf., Fig. 1.11. In case of three distinct solutions $0 \leq n_1 < n_2 < n_3$, we have a saddle ($\det\{A\} < 0$) located between two stable anti-saddles ($\det A > 0$) in the (n, n_{t_1}) phase plane. If two of these solutions n_i ($i = 1, 2, 3$) coincide, a saddle and a stable node merge into a saddle node, i.e., a multiple singular point ($\det\{A\} = 0$) in the phase plane, and one eigenvalue of A vanishes.

The negativeness of $\mathrm{tr}\{A\}$ (2.1.42) is a general property of mass-action systems without second-order autocatalysis. It has been used by *Hanusse* [2.22], and *Tyson* and *Light* [2.23] to exclude stable limit cycles around unstable foci in a large class of chemical mass-action reaction systems with two independent variables. Although (2.1.32, 33) are not of simple mass-action type in terms of the two variables n, n_{t_1} (due to the negativeness of a_2), the nonexistence theorem carries over, since the generation-recombination kinetics is of mass-action form in terms of the four (dependent) variables n, n_{t_1}, n_{t_2}, p_t. We shall, however, see in Chap. 6 that limit-cycle oscillations become possible if the temporal variation of the electric field (corresponding to a displacement current) is also taken into account.

So far we have only considered stability against homogeneous perturbations conserving local neutrality. A normal mode analysis for more general perturbations $\delta E(x, t)$, $\delta n(x, t)$, $\delta n_{t_1}(x, t)$, $\delta n_{t_2}(x, t)$ will be performed in Chap. 3.

Control Parameter Plane

The rate constants depend upon the applied electric field. We assume that in the bistability domain the field dependence of the impact-ionization coefficients X_1, X_1^* dominates that of all other rate constants and of the mobility. Therefore X_1 and X_1^* may be considered as external control parameters. The domain of bistability is most conveniently discussed in the (X_1, X_1^*) plane of control parameters, see Fig. 2.5. As

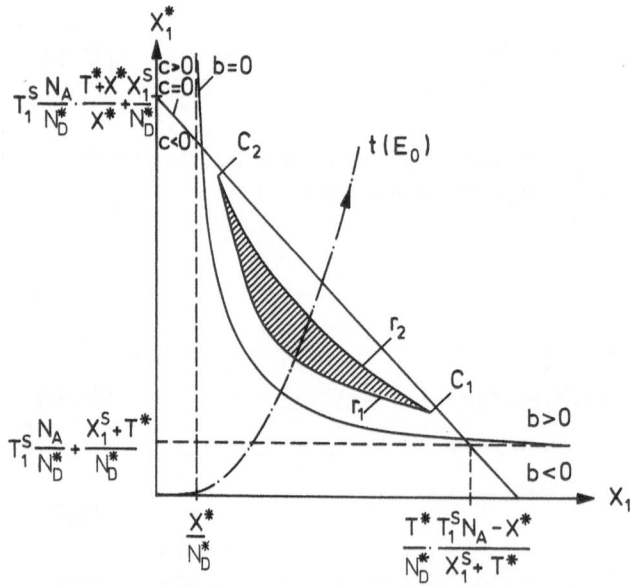

Fig. 2.5. Control parameter plane of the impact-ionization coefficients X_1, X_1^* (schematic). The hatched area represents the bistability domain. The field dependence of X_1, X_1^* defines a trajectory $t(E_0)$. The notation corresponds to donor kinetics; for trap kinetics replace N_A by $N_t - N_D^*$, cf., Table 2.1

additional control parameters one might choose X_1^S and X^*, since these can be varied externally by changing the temperature or optical irradiation.

For values of X_1 and X_1^* taken from the hatched region in Fig. 2.5, there are three spatially homogeneous steady states with $n_1 < n_2 < n_3$, two of which, corresponding to n_1 and n_3, are stable. At the boundaries r_1 and r_2, the unstable solution (n_2) and one stable solution (n_3 and n_1, respectively) merge. This is a saddle-node bifurcation as shown in Fig. 1.12 (A1). Outside the hatched region there is one homogeneous steady state only. When the boundaries r_1 and r_2 are crossed, first-order, i.e., discontinuous nonequilibrium phase transitions between the steady states n_1 and n_3 are induced.

At the points C_1 and C_2 two stable states (n_1, n_3) and one unstable state merge in a supercritical pitchfork bifurcation (Fig. 1.12 (A3); i.e., the two phases n_1 and n_3 become identical. Therefore C_1 and C_2 may be called critical points in analogy with equilibrium phase transitions. Along the lines r_1 and r_2, including C_1 and C_2, $\det\{A\} = -dP_3/dn = 0$ (Sect. 1.2.3), and therefore, because of $\lambda_1\lambda_2 = \det A$, the relaxation time of the steady state is slowed down critically.

For typical numerical parameters (Table 2.2) the control parameter plane is shown in Fig. 2.6 and the steady electron concentration $n(X_1, X_1^*)$ is represented as a surface in (n, X_1, X_1^*) space in Fig. 2.7. The singularities corresponding to C_1 and C_2 are of the "cusp" type, well known from gradient systems in catastrophe theory.

Current-Field Characteristic

The electric-field dependence of the impact-ionization coefficients $X_1(E_0)$, $X_1^*(E_0)$ defines a trajectory $t(E_0)$ in the (X_1, X_1^*) plane along which the system travels with increasing applied field E_0. The corresponding steady state concentration $n(E_0)$ yields, by (2.1.1), the static current density-field characteristic $j_0(E_0)$. It is S-shaped if the trajectory intersects the bistability domain in Fig. 2.5 (Figs. 2.8a–c). Threshold switching from the high-(n_1) to the low-(n_3) resistivity branch of the current-field characteristic occurs when the threshold field E_{th} defined by the intersection of $t(E_0)$ and r_2 is reached and exceeded. With decreasing field the system switches back to the high-resistivity ("off") state at a lower holding field $E_h < E_{th}$, defined by the intersection of $t(E_0)$ and r_1. Thus the typical hysteresis behavior observed in threshold switching is produced. Note that for spatially homogeneous steady states the current density-field characteristic is, up to rescaling of the axes, identical with the current-voltage characteristic.

In general the external circuit coupled to the sample has a nonzero resistance R, so that the sample voltage $V = E_0 L$ is not controlled directly, but is fixed by the intersection of the load line $U_0 = RI + V$ with the current-voltage characteristic $I(V)$, where U_0 is the external voltage and I the current through the sample. This includes voltage-controlled ($R \to 0$) and current-controlled ($R \to \infty$) conditions as limiting cases.

If the generation constants X^* or X_1^S are increased by suitable irradiation or heating of the sample, the bistability region in the (X_1, X_1^*) plane becomes smaller. If

$$T_1^S(N_t - N_D^*) < X^*(2 + X_1^S/T^*) , \tag{2.1.44}$$

Table 2.2. Numerical parameters used in Figs. 2.6, 7, 8, 11, 13, 15

Fig.	$T_1^S N_D^*$ [μs^{-1}]	T^* [μs^{-1}]	X_1^S [μs^{-1}]	X^* [μs^{-1}]	$(N_t - N_D^*)/N_D^*$	$X_1 N_D^*$ [μs^{-1}]	$X_1^* N_D^*$ [μs^{-1}]
2.6a	10^3	1	10^{-4}	1	1	—	—
2.6b	10^3	1	1	1	1	—	—
2.7							
2.8a	10^3	1	0.5–10	1	0.5	$50\exp(-6/E_0)$	$10^3\exp(-1.5/E_0)$
2.8b	10^3	1	0.5	1–2	0.5	$50\exp(-6/E_0)$	$10^3\exp(-1.5/E_0)$
2.8c	10^3	1	0.5	0.5	0.2–0.4	$50\exp(-6/E_0)$	$10^3\exp(-1.5/E_0)$
2.11	10^3	1	0	1–10	0.5	$5\times10^2\exp(-6/E_0)$	$10^3\exp(-1.5/E_0)$
2.13	10^3	1	2–50	0	0.5	$5\times10^2\exp(-6/E_0)$	$10^3\exp(-1.5/E_0)$
2.15	10^3	1	0.5	1	0.5	$50\exp(-6/E_0)$	$10^3\exp(-1.5/E_0)$

The effective donor concentration N_D^* enters as a scaling factor only.
A typical value is $N_D^* = 10^{15}$ cm^{-3}.

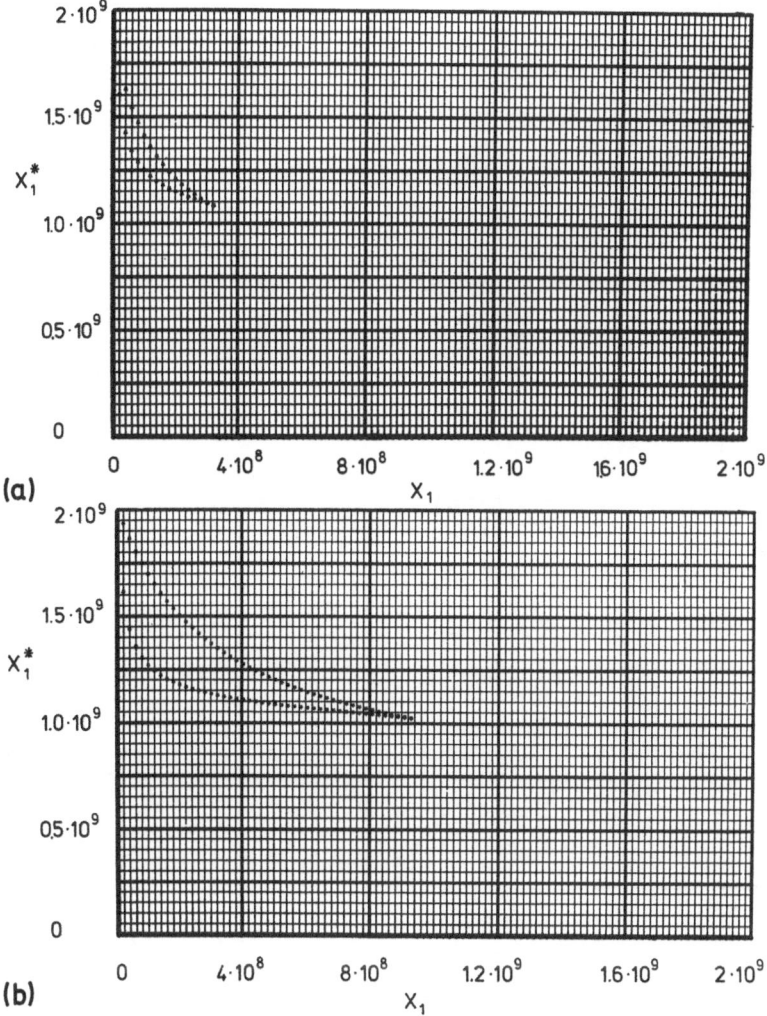

Fig. 2.6a, b. Control parameter plane (X_1, X_1^*) for the numerical parameters given in Table 2.2. The bistability regime is indicated by asterisks

the bistability conditions (2.1.38) cannot be satisfied. This is sufficient but not necessary for the disappearance of SNDC. It occurs at sufficiently large X^* or X_1^S, or at sufficiently small compensation ratio $(N_t - N_D^*)/N_D^*$ (Fig. 2.8a–c), which agrees well with experiments (Fig. 2.9).

The occurrence of S-type current-voltage characteristics is not bound to any particular field dependence of the impact-ionization coefficients, as long as $X_1(E_0)$ and $X_1^*(E_0)$ are increasing functions of E_0. In Figs 2.8a–c, 11, 13 a field dependence derived from Shockley's Lucky Electron Model [1.131] has been used:

$$X_1 = X_1^0 \exp\left(-\frac{E_D}{e\lambda E_0}\right), \qquad X_1^* = X_1^{*0} \exp\left(-\frac{E_D^*}{e\lambda E_0}\right), \qquad (2.1.45)$$

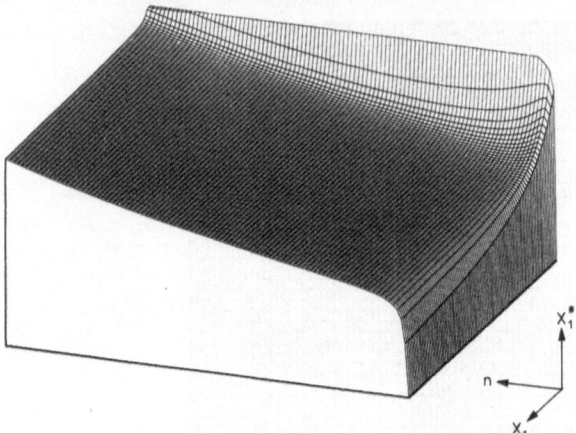

Fig. 2.7. Steady state electron concentration $n(X_1, X_1^*)$ for the numerical parameters given in Table 2.2

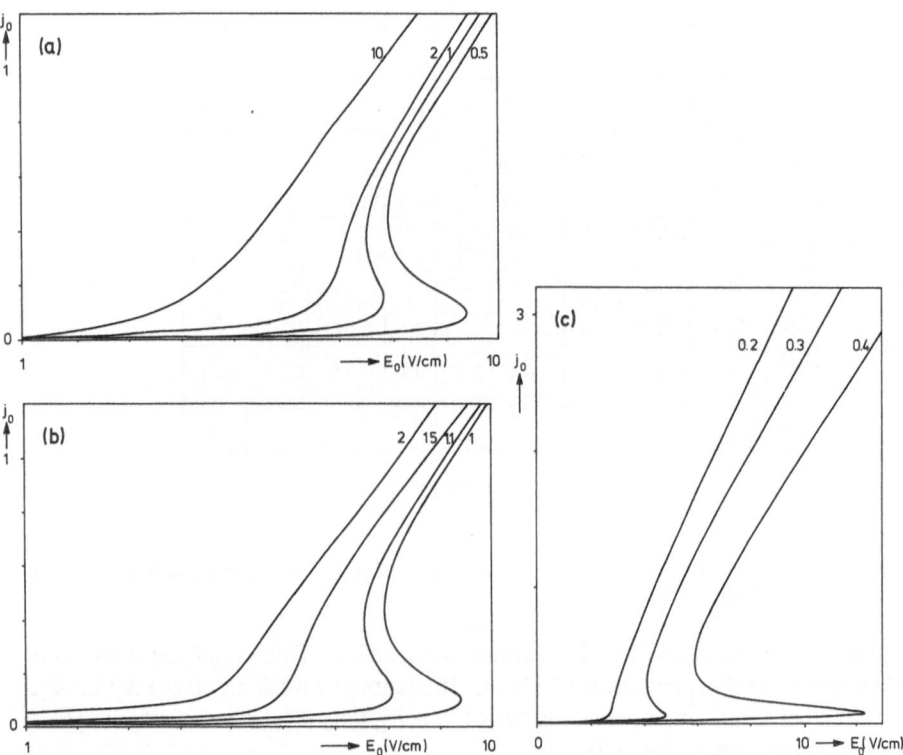

Fig. 2.8a–c. Current density-field characteristics, calculated with the numerical parameters of Table 2.2. Here and in Figs. 2.11, 13 j_0 is plotted in units $e\mu_n N_D^*$. The parameters of the different curves are (a) X_1^s in μs^{-1}, (b) X^* in μs^{-1}, (c) $(N_t - N_D^*)/N_D^*$

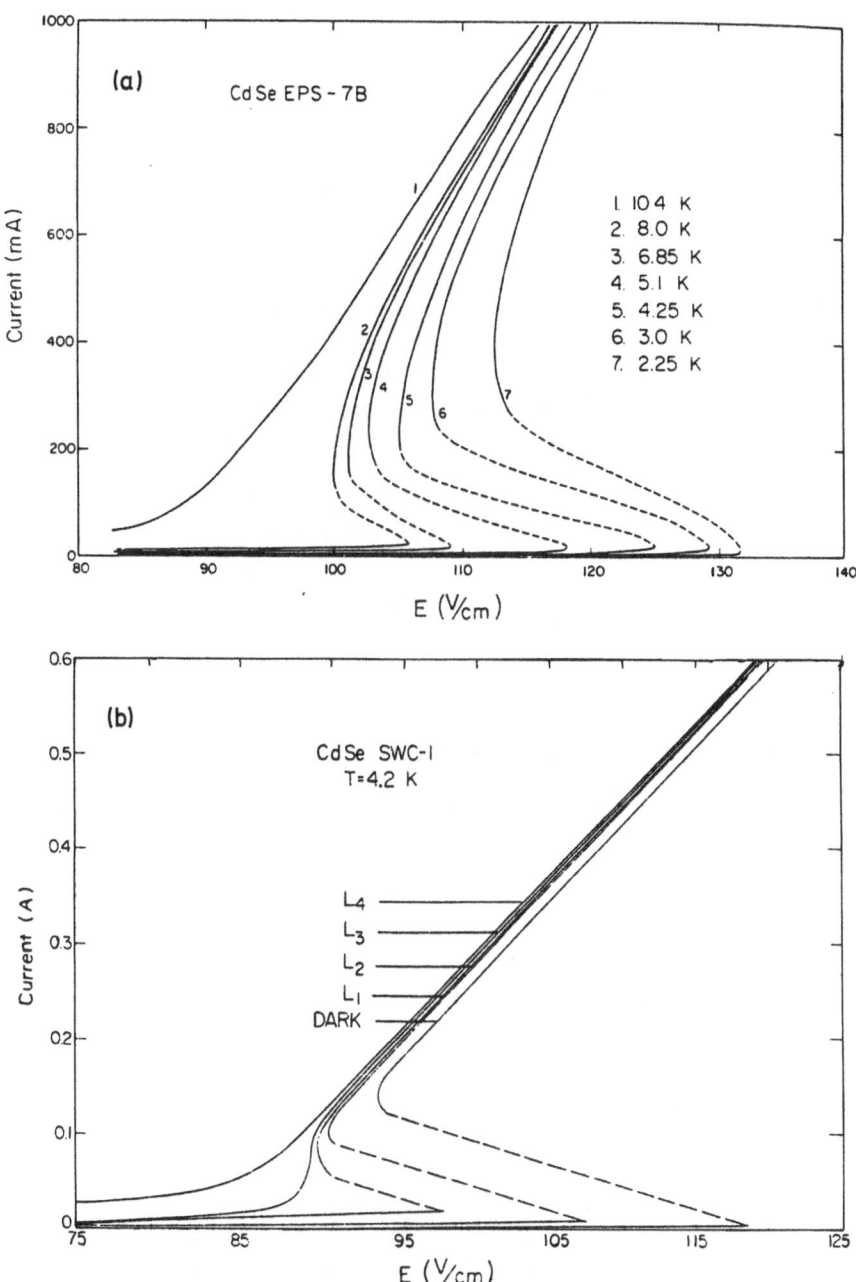

Fig. 2.9a, b. Measured current-field characteristics for CdSe. In (**a**) the temperature is increased from 1. to 7., in (**b**) the optical irradiation intensity is increased from dark to L_4 [2.34]

where E_D, E_D^* are the ground state and excited donor energy levels, λ is the mean free path of the electrons, and X_1^0, X_1^{*0} are approximately constants.

The numerical parameters of Figs 2.8, 11, 13 (Table 2.2) correspond to typical experimental data for n-type GaAs. Impact ionization of impurities associated with SNDC has been observed at low temperatures in Ge [2.24–28], GaAs [2.29–33], CdSe [2.34], CdTe [2.35], and InSb [2.36]. At low temperatures (e.g., 4.2K) GaAs shows SNDC induced by impact ionization of shallow donors (1s-ground state at 5.9 meV, 2p-excited state at 1.5 meV below the conduction band) at threshold fields of 6–8 V/cm, corresponding to impact-ionization coefficients $X_1 \approx 2 \times 10^{-8}$ cm^3 s^{-1}, $X_1^* \approx 6 \times 10^{-7}$ cm^3 s^{-1} [2.31]. The other g-r coefficients are $T_1^S \approx 10^{-6}$ cm^3 s^{-1} [2.31], corresponding to a trapping cross section of $\sigma \approx 10^{-12}$ cm^2 at 4.2K, and X_1^S, $X^* \approx 1$ μs^{-1} corresponding to optical absorption coefficients $\alpha \approx 25$ cm^{-1} [2.32] and photon flux densities $\approx 10^{19}$ cm^{-2} s^{-1}. For materials with deeper donor or trap states the threshold fields can be scaled to higher values without changing the control parameter space shown in Fig. 2.5 by simply redefining the field dependence of X_1, X_1^*. Experimental evidence for the validity of our two-level model in Sb-doped n-Ge at 4K has been presented by *Brown* et al. [2.28], who observed avalanche photomultiplication due to impact ionization, both from the ground level (1s) and an excited level (2p±), and optical excitation from 1s to 2p± in FIR photodetectors at $\lambda = 151$ μm.

Special Cases and Second-Order Phase Transitions

We shall now study special cases in which more explicit analytical results can be obtained. Consider a sample at sufficiently low temperature and weak irradiation, such that at least one of the generation processes X_1^S, X^* is negligible. From (2.1.36) we then obtain with $d = 0$:

$$n_1 = 0$$

$$n_{2,3} = \frac{-b \mp (b^2 - 4ac)^{1/2}}{2a} \quad \text{for } b^2 > 4ac \ . \tag{2.1.46}$$

For $c > 0$ we have $n_2 < 0$ and $n_3 > 0$, noting that always $a < 0$. The stability matrix A of (2.1.40) associated with the solution $n_1 = 0$ satisfies

$$\det\{A\} = -c \tag{2.1.47}$$

by (2.1.43, 36), hence the steady state $n_1 = 0$ is unstable for $c > 0$.

For $c < 0$, $b > 0$, $b^2 > 4ac$ there are three nonnegative solutions n, two of which (n_1, n_3) are stable.

Case (i): $X_1^S = 0$

The steady state $n_1 = 0$, $n_{t_1} = \omega' N_D^*$, $n_{t_2} = \omega N_D^*$, where $\omega' = 1 - \omega := T^*/(T^* + X^*)$, is associated with the eigenvalues

$$\lambda_1 = (X_1\omega' + X_1^*\omega)N_D^* - T_1^S(N_t - N_D^*) = c/(T^* + X^*)$$

$$\lambda_2 = -(T^* + X^*) .$$ (2.1.48)

Hence it is stable for $c < 0$.

In Fig. 2.10 the (X_1, X_1^*) control parameter plane is represented schematically. At the boundaries r_1, r_2 of the bistability region first-order phase transitions between n_1 and n_3 take place. When the lines l_1, l_2 defined by $c = 0, b < 0$ are crossed, phase transitions of second order occur such that the carrier concentration rises from $n_1 = 0$ to $n_3 > 0$ continuously, but with a discontinuous derivative with respect to the control parameters. As $c = 0$ is approached, the carrier relaxation time $1/\lambda_1$ is slowed down critically and becomes infinite. This is a general feature of a second-order phase transition.

The points C_{t_1}, C_{t_2} with

$$X_1^{t_{1,2}} = \frac{T_1^S(N_t - N_D^*)}{2N_D^*}\left[1 \pm \left(1 - \frac{4X^*N_t(T^* + X^*)}{[T_1^S T^*(N_t - N_D^*)^2]}\right)^{1/2}\right]$$ (2.1.49)

are given by $c = 0, b = 0$. Here the second-order transition changes into a first-order one, hence these points may be called *tricritical* points in analogy with equilibrium phase transitions.

From (2.1.49) one obtains a necessary and sufficient condition for the existence of a bistability domain in the (X_1, X_1^*) plane:

$$T_1^S(N_t - N_D^*)^2 > 4X^*N_t(T^* + X^*)/T^* .$$ (2.1.50)

Hence it is evident that SNDC vanishes with increasing generation X^* and decreasing compensation $N_t - N_D^*$. Typical current-field characteristics with X^* as a parameter are plotted in Fig. 2.11. For generation coefficients X^* smaller than the "tricritical" value one obtains first-order phase transitions associated with threshold switching and hysteresis, whereas for larger values only second-order transitions are possible, corresponding to simple breakdown. These second- and first-order phase transitions are associated with normal and inverted transcritical bifurcations of the steady state, respectively (Fig. 1.12 (A2)).

Case (ii): $X^* = 0$

The steady state $n_1 = 0$, $n_{t_1} = N_D^*$, $n_{t_2} = 0$ is stable for $c = X_1 N_D^*(T^* + X_1^S) - T_1^S T^*(N_t - N_D^*) < 0$. Fig. 2.12 respresents the (X_1, X_1^*) control parameter plane. As in case (i), first- and second-order phase transitions are possible. There is again a tricritical point, given by

$$X_1^t = T_1^S\frac{N_t - N_D^*}{N_D^*}\frac{T^*}{T^* + X_1^S} ,$$

$$X_1^{*t} = T_1^S\frac{N_t - N_D^*}{N_D^*} + (T^* + X_1^S)\frac{N_t}{(N_t - N_D^*)N_D^*} .$$ (2.1.51)

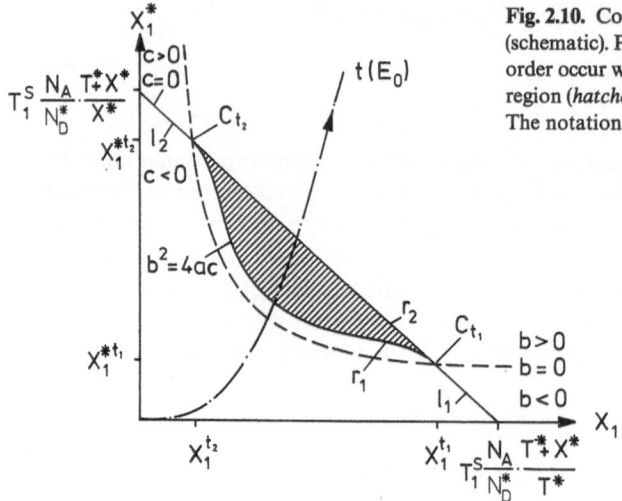

Fig. 2.10. Control parameter plane for $X_1^S = 0$ (schematic). Phase transitions of first or second order occur when $t(E_0)$ intersects the bistability region (*hatched*) or the lines l_1, l_2, respectively. The notation corresponds to donor kinetics

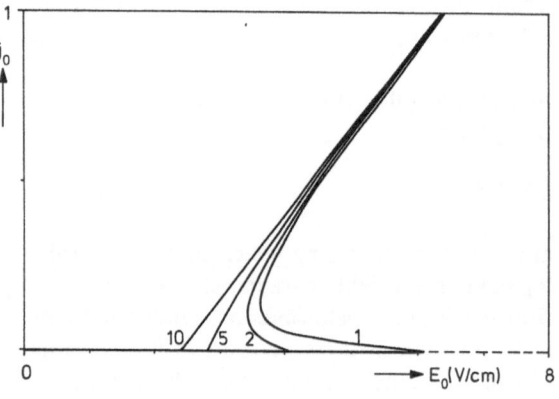

Fig. 2.11. Current-field characteristics for $X_1^S = 0$ with X^* as a parameter (in μs^{-1}), calculated with the numerical parameters of Table 2.2

The bistability region exists for any nontrivial choice of the rate constants, but becomes smaller with increasing X_1^S. The threshold field E_{th} for switching from the nonconducting (n_1) to the conducting (n_3) state is defined implicitly by

$$X_1(E_{th}) = T_1^S \frac{N_t - N_D^*}{N_D^*} \frac{T^*}{T^* + X_1^S} . \tag{2.1.52}$$

For a given field dependence of the impact-ionization coefficients the trajectory $X_1(E_0)$, $X_1^*(E_0)$ does *not* cross the bistability domain if and only if

$$X_1^*(E_{th}) < X_1^{*t} . \tag{2.1.53}$$

For the particular field dependence (2.1.45) an explicit condition for the disappearance of SNDC is readily deduced from (2.1.52, 53) by setting

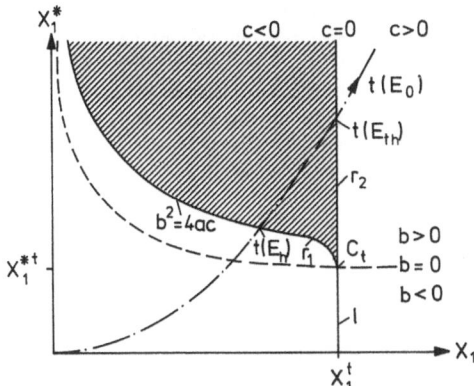

Fig. 2.12. Control parameter plane for $X^* = 0$ (schematic). First-(*hatched region*) or second-(line *l*) order phase transitions are possible. The notation corresponds to donor kinetics

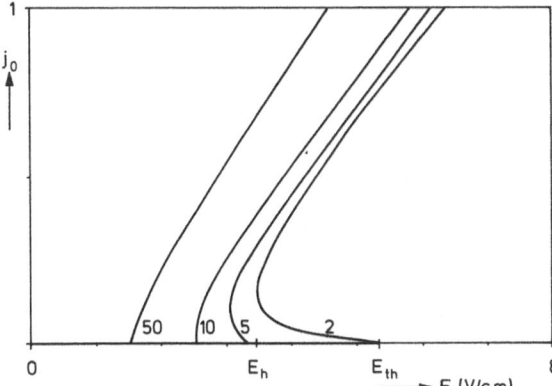

Fig. 2.13. Current-field characteristics for $X^* = 0$ with X_1^S as a parameter (in μs^{-1}), calculated with the numerical parameters of Table 2.2

$$X_1^*(E_{\rm th}) = X_1^{*0}\left(\frac{1}{X_1^0}\,T_1^S\,\frac{N_t - N_{\rm D}^*}{N_{\rm D}^*}\,\frac{T^*}{T^* + X_1^S}\right)^{E_{\rm D}*/E_{\rm D}} \tag{2.1.54}$$

see Fig. 2.13.

Physical Interpretation of the SNDC Mechanism

The physical mechanism leading to SNDC can be interpreted as a coupled two-step impact-ionization process: first, the impurity (donor) ground state is depleted by the process X_1, so that the steady state population ratio $n_{t_2}/n_{t_1} = (X_1 n + X^*)/T^*$ increases with increasing $X_1 n$. Although the total concentration of occupied impurities $n_{t_1} + n_{t_2} = N_{\rm D}^* - n$ decreases with increasing n due to charge neutrality, the concentration of excited impurities n_{t_2} grows for $X_1 N_{\rm D}^* > X^*(T^* + X^*)/T^*$. Next, the enhanced excited impurity population is impact ionized with a rate $X_1^* n_{t_2} n$ increasing hence stronger than linearly in n. This two-step impact-ionization process is similar to a "second-order autocatalysis" in chemical reactions. In the bistability range, i.e., in an intermediate range of X_1, X_1^*, impact ionization is negligible at low n, and the recombination rate r is essentially balanced by the single-electron generation X_1^S. Due to the superlinear increase of the impact-ionization rate another stable

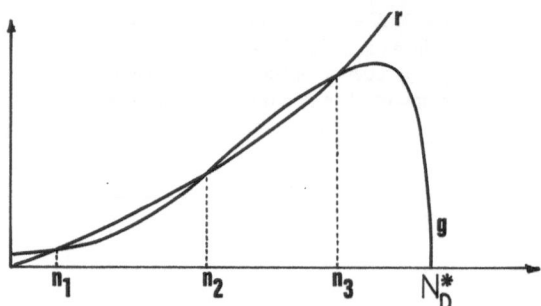

steady state becomes possible at high n, recombination being balanced by impact ionization as shown in Fig. 2.14. With a single impact-ionization process (either X_1 or X_1^*) the generation rate g would increase only linearly in n and could not compensate the recombination rate for a second time.

2.1.3 General M-Level Mechanisms

The two-level SNDC mechanism can easily be generalized to M-level g-r models involving the ground state and $M - 1$ excited states of the trapped electrons [2.13]. For this purpose it is appropriate to use a slightly different elimination procedure in computing the homogeneous steady state than that outlined in (2.1.2–5). Incidentally, this approach shows the profound analogies between a drift nonlinearity leading to NNDC, and a g-r nonlinearity leading to SNDC. The idea is to eliminate the n_{t_i}'s from the steady state rate equations *first*, and *afterwards* use the condition of charge neutrality.

The basic constitutive equations for homogeneous steady states are given by the current density – field relation $j_0(n, E_0)$ (2.1.1), and the condition of vanishing charge density ρ (2.1.2):

$$0 = \rho := e[N_D^* - n_{\text{tot}}(n, E_0)] \quad \text{where} \tag{2.1.55}$$

$$n_{\text{tot}}(n, E_0) := n + \sum_{i=1}^{M} n_{t_i}(n, E_0) \tag{2.1.56}$$

is the total density of negatively charged carriers. For standard g-r kinetics the g-r rates $f_{t_i}(n, n_{t_1}, \ldots, n_{t_M}, E_0)$ are linear in n_{t_i}; hence in the steady state the equations $0 = f_{t_i}$ can be solved to give $n_{t_i}(n, E_0)$ in (2.1.56). The general explicit formulas will be given in Chap. 3, see (3.1.13) and Table 3.1, but here we only remark that the function $n_{\text{tot}}(n, E_0)$ thereby obtained is in general strongly nonlinear in n. The explicit dependence of n_{tot} on E_0 occurs through the g-r coefficients, and we shall assume in the following that this dependence is on the *magnitude* of E_0 only, and is monotonically decreasing, since an increase in field will in most cases decrease the number of trapped electrons through field-enhanced generation or impact ionization. Bistability of the homogeneous steady state occurs if the dependence of n_{tot} upon n is *nonmonotonic* (N-shaped) in some field range such that (2.1.55) has *three solutions*.

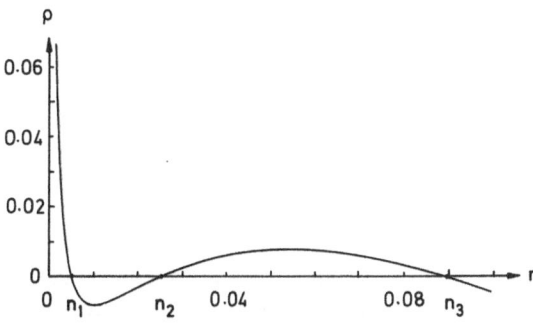

It should be stressed that the nonmonotonicity of the constitutive relation $n_{tot}(n, E_0)$ in the case of g-r induced SNDC is dual to the nonmonotonicity of the constitutive $v(E_0)$ relation, where v is the drift velocity, in the case of a drift instability leading to NNDC (like intervalley transfer in the Gunn effect, see Sect. 1.1.2). The SNDC and the NNDC mechanisms essentially influence the charge density (2.1.55) or the current density (2.1.1), respectively.

A typical plot of ρ versus n for such a nonmonotonic $n_{tot}(n, E_0)$ is shown in Fig. 2.15. It is representative; and can be profitably used for a unified treatment, of a variety of different bulk single-carrier g-r mechanisms which yield SNDC [2.3–13]. Physically, when ρ versus n at fixed E_0 is increasing, i.e., when n_{tot} versus n displays a falling region in some intermediate field range, this corresponds to a state where a *decrease* in the total carrier concentration n_{tot} leads to an increase in the density of free carriers n. It is easy to see that this cannot happen if the distribution of carriers between free and trapped states is governed by a thermal equilibrium distribution, or more generally, a quasi-Fermi distribution

$$n_{t_i} = N_t[1 + \exp(E_i - E_F)/(kT)]^{-1} \, , \tag{2.1.57}$$

where the quasi-Fermi level E_F is given by $n = N_c \exp[(E_F - E_c)/(kT)]$, see (1.2.18a), and E_i is the ith trap level. From (2.1.57) it follows by eliminating E_F that

$$n_{t_i} = N_t n[n + N_c \exp(E_i - E_c)/(kT)]^{-1} \tag{2.1.58}$$

and hence also n_{tot}, is a *monotonically* increasing function of n.

A nonmonotonic $n_{tot}(n)$ relation can occur if the rates which describe the balance of generation and recombination depend strongly upon n, such that the presence of a few additional free carriers strongly depletes the trapped state. This can be achieved by impact ionization. It may be illustrated by the two-level model shown in Fig. 2.4. From (2.1.31b, c) it follows in the steady state with $p_t = N_t - n_{t_1} - n_{t_2}$ that

$$n_{t_1}(n, E_0) = N_t T^* T_1^S n/\Delta \tag{2.1.59a}$$

$$n_{t_2}(n, E_0) = N_t(X^* + X_1 n) T_1^S n/\Delta \tag{2.1.59b}$$

with

$$\Delta := (X_1^S + T_1^S n + X_1^* n)(X^* + X_1 n) + T^*(T_1^S + X_1)n \, .$$

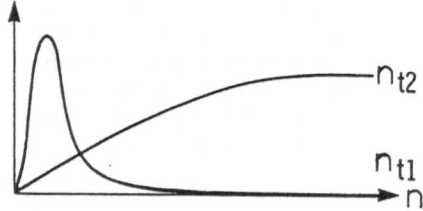

Fig. 2.16. Trapped electron concentration in the ground state (n_{t_1}) and in the excited state (n_{t_2}) versus free-electron concentration, without assumption of charge neutrality (schematic). The sharp decrease in n_{t_1} is due to impact ionization

The occupancies of the two levels are plotted schematically for an intermediate field range in Fig. 2.16. Note that the explicit E_0-dependence of n_{t_1} and n_{t_2} occurs through the impact-ionization coefficients X_1, X_1^*, which are increasing functions of E_0. For very low fields the depletion of the ground state n_{t_1} is diminished, while for very high fields it is enhanced; in both cases the resulting $n_{tot}(n)$ relation is no longer nonmonotonic. The underlying physical mechanism is the following: for small n_{tot}, impact ionization is negligible, and the distribution between free and trapped states is governed by the balance of capture and thermal ionization. For large n_{tot}, capture is balanced by impact ionization. In an intermediate range of n_{tot}, both states are possible, and additionally there is an NDC state, where n_{tot} is decreasing with n, since impact ionization depletes the ground state and forms a highly nonthermal distribution of free and trapped carriers.

Other mechanisms which lead to a qualitatively similar form of ρ as a function of n include the screening of electron-phonon and electron-impurity scattering through an increased electron concentration, which heats up the electrons and leads to enhanced impact ionization with a complicated, non-mass-action rate that has to be evaluated numerically [2.3]. Two-level mechanisms similar to the one of Fig. 2.4 have been proposed earlier, but no stability analysis was performed [2.4, 5]. A general M-level model was evaluated numerically [2.6]. A three-level model taking into account the splitting of the $2p$ donor state in a magnetic field was also treated, and was used to explain impact-ionization induced negative far-infrared photoconductivity in n-GaAs [2.11].

Effect of Contacts

The assumption of spatially homogeneous steady states in the preceding analysis is clearly an idealization. Near the contacts, the electric field and the carrier concentrations are expected to be a function of the coordinate z in the direction of the current flow. In the simplified theory of an injecting ("Ohmic") contact one may assume that diffusion is negligible and that the boundary condition for the electric field $E(z)$ at the contact $z = 0$ is $E(0) = 0$ [2.37]. In the steady state the transport equations (1.1.5, 7, 8, 12) for free and trapped electrons are then reduced to

$$j_0 = e\mu_n nE = \text{const} \tag{2.1.60}$$

$$\frac{dE}{dz} = \frac{4\pi}{\epsilon_s}\rho(n, E) \tag{2.1.61}$$

where the trapped electron concentrations $n_{t_i}(n, E)$ have been eliminated from the static charge density ρ as outlined after Eq. (2.1.56). From (2.1.60, 61) the electron concentration can be eliminated, resulting in a first-order differential equation for E:

$$\frac{dE}{dz} = \frac{4\pi}{\epsilon_s} \rho\left(\frac{j_0}{e\mu_n E}, E\right) . \tag{2.1.62}$$

Integrating (2.1.62) with the boundary condition $E(0) = 0$ gives the spatial profile $E(z)$ for fixed j_0 in an implicit form:

$$z = \frac{\epsilon_s}{4\pi} \int_0^E \rho\left(\frac{j_0}{e\mu_n E}, E\right)^{-1} dE . \tag{2.1.63}$$

This is a generalization of the standard treatment of space charge limited (SCL) currents [Ref. 2.37, p. 47] to a situation with M trap levels. In the trap-free case $\rho(n, E) = e(N_D^* - n)$ holds, and (2.1.63) simplifies to the familiar single-carrier SCL result:

$$z = \frac{\epsilon_s}{4\pi e N_D^*} \int_0^E \left(1 - \frac{E_0}{E}\right)^{-1} dE = \frac{\epsilon_s}{4\pi e N_D^*}\left[E + E_0 \ln\left(1 - \frac{E}{E_0}\right)\right]$$

$$E_0 := j_0/(e\mu_n N_D^*) . \tag{2.1.64}$$

Qualitatively, E rises monotonically from its value $E = 0$ at the contact to its bulk value E_0, while n, by (2.1.60), drops monotonically from $+\infty$ to its bulk value. In the bulk n and E assume the homogeneous steady state values given by $\rho(n, E) = 0$ and $j_0 = e\mu_n nE$. If the sample is sufficiently long, its properties are dominated by the homogeneous bulk values, and the analysis given in the present chapter is a reasonable approximation.

The injecting contact is modeled in the simplified theory by an infinite reservoir of carriers with $n \to \infty$, and $E \to 0$. In a more rigorous analysis [Ref. 2.37, §9] which correctly describes the boundary conditions at *both* contacts, diffusion has to be taken into account. The governing equations for the steady state are then

$$\frac{dn}{dz} = (j_0 - e\mu_n nE)/(eD_n)$$

$$\frac{dE}{dz} = \frac{4\pi}{\epsilon_s} \rho(n, E) . \tag{2.1.65}$$

This is a system of two nonlinear first-order differential equations, which can be conveniently analysed in the (E, n) phase plane, but this will not be done here.

2.1.4 Critical Behavior

In Sect. 2.1.2 we have encountered a variety of critical points and lines for the two-level model. We shall now classify and elaborate the critical behavior and draw the analogy with equilibrium phase transitions, as outlined in Sect. 1.2.1.

In the (X_1, X_1^*) control parameter planes the following topologically different cases occur:

(A) *Isolated critical point* (C_1, C_2 in Fig. 2.5). In the (n, n_{t_1}) phase plane this corresponds to three coalescing singular points: two stable nodes and a saddle point.

(B) *Line of critical points* (l_1, l_2 in Fig. 2.10, l in Fig. 2.12). In the corresponding phase plane two singular points merge: a stable node and a saddle point. One of the singular points lies on a coordinate axis ($n = 0$), the other one crosses over from the nonphysical region ($n < 0$) to the physical quadrant ($n > 0$).

(C) *Tricritical point* (C_{t_1}, C_{t_2} in Fig. 2.10, C_t in Fig. 2.12). In the corresponding phase plane three singular points merge: two stable nodes and a saddle point, but one of the stable nodes comes from the nonphysical region ($n < 0$). The saddle point lies on the coordinate axis ($n = 0$).

The terminology alludes to the analogous situation in thermodynamic equilibrium phase transitions: at a critical point (cf., Fig. 1.6 for the Van der Waals' gas and Fig. 1.7 for the Weiss ferromagnet) two stable phases become indistinguishable. An entire line of critical points (*λ-line*) occurs, e.g., in the $P-T$ diagram of liquid ^{4}He; it separates the two regimes of the normal fluid and of the superfluid. When the λ-line is crossed, a second-order phase transition occurs in which the order parameter (i.e., the wave-function ψ_0 of the superfluid condensate, where $|\psi_0|^2$ is the macroscopic condensate density) changes from zero to nonzero values. A tricritical point occurs, e.g., in ^{3}He $-$ ^{4}He mixtures [2.38]; it is a point in the control parameter space where a first-order phase transition line changes into a single λ-line or line of critical points [2.39, 40]. (If the control parameter space of the temperature T and the difference in chemical potential between the ^{3}He and ^{4}He components Δ is extended by the experimentally inaccessible thermodynamic adjoint of the order parameter ψ_0, then at the *tricritical* point *three* critical lines meet, which explains the terminology.)

The critical behavior in our semiconductor model shows up in the steady state electron concentration as a function of the external parameters, and hence in the current density – field characteristics (Fig. 2.8, 11, 13). It depends upon the way in which the trajectory $t(E_0)$ defined by the field dependence of the impact-ionization coefficients intersects or approaches the (tri)critical points or lines in the (X_1, X_1^*) plane. This point has been discussed for general equilibrium phase transitions from a unified geometrical point of view by *Griffiths* and *Wheeler* [2.41]. The position of the bistability domain and the critical hypersurfaces in the (X_1, X_1^*) plane change themselves when additional g-r constants are varied. Thereby the function $n(E_0)$ is changed even for fixed trajectories $t(E_0)$, as shown in Figs. 2.8, 11, 13. Since the most important physical control parameters are the applied electric field E_0 governing X_1 and X_1^*, and the optical (or, alternatively, thermal) generation intensity governing X^* and X_1^s, it is convenient to introduce a physical (g, E_0) control parameter plane for fixed $t(E_0)$, see Fig. 2.17. Here two generic situations arise: the general model (Fig. 2.8) leads to a cusp-type bistability regime with a critical point C at a critical field E_c and a critical optical intensity g_c (Fig. 2.17a). If certain generation coefficients

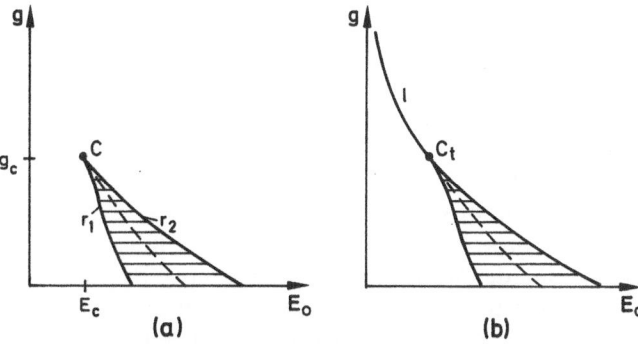

Fig. 2.17a, b. Physical control parameter plane of optical (or thermal) generation intensity g and applied electric field E_0 for (a) the general two-level model and (b) the simplified model with $X_1^S = 0$ or $X^* = 0$ (schematic). X_1 and X_1^* increase with increasing E_0, and X^* or X_1^S (depending upon the optical frequency) increase with increasing g. C is a critical point, l is a critical line, r_1 and r_2 are the boundaries of the bistability domain corresponding to the holding and the threshold fields, respectively, and C_t is the tricritical point. (---) represent the coexistence curves

(X_1^S or X^*) are zero (Figs. 2.11, 13), a critical line l appears and the critical point becomes a tricritical point C_t (Fig. 2.17b). We shall show in Chap. 4 by additionally invoking diffusion that there is a line of spatial coexistence within the bistability domain (dashed), which extends the critical line l in Fig. 2.17b beyond the tricritical point. The (tri)critical point C (C_t), the critical line l, and the bistability boundary r_1, r_2 are connected with zero eigenvalue bifurcations of type (A3) (supercritical pitchfork bifurcation), (A2) (transcritical bifurcation), and (A1) (saddle-node bifurcation), respectively, as discussed in Sect. 1.2.3, cf., the respective phase portraits of coalescing singular points in Fig. 1.12.

The current-field characteristics in Figs. 2.8a, b, 11, 13 correspond to sections in the (g, E_0) plane parallel to the E_0 axis, shifted to larger g with increasing X_1^S or X^*. In Fig. 2.17a, at $g = g_c$ the $j_0 - E_0$ characteristic (Fig. 2.8a, b) has a point of inflection with vertical tangent, analogous to a critical isotherm of the Van der Waals' system. In case of Fig. 2.17b, the $j_0 - E_0$ characteristics (Figs. 2.11, 13) have a negative slope $\lim_{j_0 \to 0} dj_0/dE_0 < 0$ below C_t, an infinite slope at C_t, and a positive slope above C_t, corresponding to a change from first-order to second-order transitions.

The generic behavior of $n(E_0, g)$ which produces these current-field characteristics can be classified according to the universality classes (A–C) defined above by expanding the steady state equation for n (2.1.36) in the neighborhood of the (tri)critical point (line) into one of the following universal forms:

(A) $m^3 + \tau m + c\tau + h = 0$, $\qquad\qquad$ (2.1.66a)

(B) $m^2 - (c\tau + h)m = 0$, $\qquad\qquad$ (2.1.66b)

(C) $m^3 + \tau m^2 - (c\tau + h)m = 0$, $\qquad\qquad$ (2.1.66c)

where

$$m \sim n - n_{\rm c}$$

$$h \sim E_0 - E_{\rm c}$$

$$\tau \sim g - g_{\rm c} \tag{2.1.67}$$

denote dimensionless deviations from the (tri)critical values, and $c > 0$ is a constant. Equation (2.1.66a) is the universal form of standard mean-field models near the critical point [2.42], where m is the order parameter (magnetization in case of a ferromagnet, volume in case of a Van der Waals' gas), τ is the control parameter corresponding to temperature, and h is the ordering field (magnetic field in case of a ferromagnet, pressure in case of a Van der Waals' gas). We conclude that in the semiconductor instability the applied electric field and the optical (or thermal) generation take the role of ordering field and temperature, respectively.

The critical exponents β, γ, δ can now be defined in analogy with (1.2.5):

$$\left.\begin{array}{l} m \sim |\tau|^\beta \quad \left(\begin{array}{ll} \tau < 0, h = -c\tau & \text{for (A), (C)} \\ \quad\quad h = 0 & \text{for (B)} \end{array}\right) \\[4mm] \left(\dfrac{\partial m}{\partial h}\right)_\tau \sim |\tau|^{-\gamma} \quad (\tau < 0, h = -c\tau) \\[4mm] h \sim |m|^\delta \quad (\tau = 0) \end{array}\right\} m \neq 0 . \tag{2.1.68}$$

Note that the exponents depend upon the path of approach to the critical point in the (τ, h) plane. The results following from (2.1.66) are listed in Table 2.3. The exponents β, γ, δ are physically observable in the current-light (or temperature) characteristic, the static differential conductivity versus light, and the current-field characteristic, respectively. In all three cases (A–C) the scaling law [1.97]

$$\gamma = \beta(\delta - 1) \tag{2.1.69}$$

is satisfied.

Next, we shall investigate [2.43] the critical exponent α following from the generalization of the specific heat suggested by *Schlögl* [2.42]:

$$\Gamma_2 := \langle b^2 \rangle_{\rm c} \sim |\tau|^{-\alpha} , \tag{2.1.70}$$

Table 2.3. Static critical exponents of the single-carrier two-level model. [For definitions see (2.1.68)]

Exponent	Class (A)	(B)	(C)	Physical observation
β	$\frac{1}{2}$	1	1	$\Delta j_0 \sim |\Delta g|^\beta$
γ	1	0	1	$\sigma_{\rm diff} \sim |\Delta g|^{-\gamma}$
δ	3	1	2	$\Delta j_0 \sim |\Delta E_0|^{1/\delta}$

$\Delta j_0, \Delta E_0, \Delta g$ denote the difference of the current density, electric field, and optical (thermal) generation from their critical values; $\sigma_{\rm diff}$ is the static differential conductivity.

where $b = -\ln \rho$ is the bit number of the probability distribution ρ, and Γ_2 is its variance (second cumulant), see (1.2.10). For systems with a distinct time-scale separation of microscopic short-time and macroscopic long-time behavior an adequate nonequilibrium distribution can be obtained by applying the full reversible Liouville dynamics in Gibbs phase space to a local equilibrium distribution over a time which is short on the long time-scale and long on the short time-scale (the *Mori* construction [2.42, 1.25]). Since Γ_2 is invariant with respect to reversible motion, we may evaluate Γ_2 with the local equilibrium distribution.

As an appropriate generalized canonical local equilibrium distribution for the semiconductor we choose

$$\rho = Z^{-1} \exp(-\beta_e E_e - \beta_L E_t) , \qquad (2.1.71)$$

where $\beta_e := 1/(kT_e)$, $\beta_L := 1/(kT_L)$, T_e and T_L are the electron and the lattice temperature, respectively, E_e and E_t are the total energy of the conduction-band electrons and the trapped electrons, respectively, and Z is the partition function.

For general probability distributions $\rho = Z^{-1} \exp(-\sum_k \lambda_k M_k)$ the following generalized fluctuation-dissipation theorem holds, see (1.2.6)

$$\langle \Delta M_i \Delta M_j \rangle = -\frac{\partial}{\partial \lambda_i} \langle M_j \rangle . \qquad (2.1.72)$$

From this we find

$$\langle b^2 \rangle_c = \langle (\ln \rho - \langle \ln \rho \rangle)^2 \rangle$$

$$= \sum_{ij} \lambda_i \lambda_j \langle (M_i - \langle M_i \rangle)(M_j - \langle M_j \rangle) \rangle$$

$$= -\sum_{ij} \lambda_i \lambda_j \frac{\partial}{\partial \lambda_i} \langle M_j \rangle . \qquad (2.1.73)$$

In particular, for the distribution (2.1.71) this becomes

$$\Gamma_2 = -\beta_e^2 \frac{\partial}{\partial \beta_e} \langle E_e \rangle - 2\beta_e \beta_L \frac{\partial}{\partial \beta_L} \langle E_e \rangle - \beta_L^2 \frac{\partial}{\partial \beta_L} \langle E_t \rangle$$

$$= \frac{1}{k} \frac{\partial}{\partial T_e} \langle E_e \rangle + \frac{2}{k} \frac{T_L}{T_e} \frac{\partial}{\partial T_L} \langle E_e \rangle + \frac{1}{k} \frac{\partial}{\partial T_L} \langle E_t \rangle . \qquad (2.1.74)$$

The mean energy of the conduction electrons is

$$\langle E_e \rangle = \int_{E_c}^{\infty} E D_c(E) [\exp(E - E_{Fn})/(kT_e) + 1]^{-1} dE$$

$$\approx \tfrac{3}{2} kT_e n , \qquad (2.1.75)$$

where a nondegenerate parabolic conduction band has been assumed, and $n \approx N_c \exp[(E_{Fn} - E_c)/(kT_e)]$ by (1.2.18a) has been used. Similarly, the mean energy of the trapped electrons is

$$\langle E_t \rangle = \sum_{j=1}^{M} E_j n_{t_j} \,, \tag{2.1.76}$$

where E_j is the energy of the jth trap level. The mean carrier concentrations n, n_{t_j} are given as a function of the control parameters E_0 and g by the g-r rate equation in the steady state. The field E_0 and the thermal-generation intensity can be related to the electron temperature T_e and the lattice temperature T_L respectively. Therefore

$$\Gamma_2 = \frac{3}{2}n + \frac{3}{2}T_e \frac{dE_0}{dT_e}\left(\frac{\partial n}{\partial E_0}\right)_g + 3T_L\frac{dg}{dT_L}\left(\frac{\partial n}{\partial g}\right)_{E_0}$$

$$+ \frac{1}{k}\sum_{j=1}^{M} E_j \frac{dg}{dT_L}\left[\left(\frac{\partial n_{t_j}}{\partial g}\right)_n + \left(\frac{\partial n_{t_j}}{\partial n}\right)_g\left(\frac{\partial n}{\partial g}\right)_{E_0}\right]. \tag{2.1.77}$$

The first term $\frac{3}{2}n$, which describes the classical equilibrium specific heat of the nondegenerate electron gas, always remains finite; the other contributions, however, which describe the change of the carrier densities by generalized susceptibilities, diverge at the critical point. We can use the universal equations of state (2.1.66) with (2.1.67) to calculate the singularities of the bit number variance Γ_2:

$$\left(\frac{\partial n}{\partial E_0}\right)_g \sim \left(\frac{\partial m}{\partial h}\right)_\tau \sim |\tau|^{-\alpha'}$$

$$\left(\frac{\partial n}{\partial g}\right)_{E_0} \sim \left(\frac{\partial m}{\partial \tau}\right)_h \sim |\tau|^{-\alpha''} \tag{2.1.78}$$

where the approach to the critical point is chosen along $\tau < 0$, $h = -c\tau$. From (2.1.68) we find $\alpha' = \gamma$ which gives $\alpha' = 1$ for class (A) and (C) and $\alpha' = 0$ for class (B). Direct calculation of α'' from (2.1.66) yields

$$\left(\frac{\partial m}{\partial \tau}\right)_h \begin{cases} \sim |\tau|^{-1} & \text{(A)} \\ = \text{const} & \text{(B)} \\ \sim |\tau|^{-1} \,. & \text{(C)} \end{cases} \tag{2.1.79}$$

Thus the critical exponent of Γ_2 is $\alpha = 1$ for classes (A) and (C) (Curie-law singularity) and $\alpha = 0$ for class (B) (finite discontinuity). The same exponents α are obtained for other generalized local equilibrium distributions, e.g., if the radiation field (of energy E_s) is explicitly included by an additional term $-\beta_s E_s$ or if the quasi-Fermi levels are included by adding $+\beta_e E_{Fn}n + \sum_i \beta_L E_{F_i} n_{t_i}$ (grand canonical distribution).

A dynamical critical phenomenon is the critical slowing-down of the relaxation time of carrier density fluctuations $\delta n \sim \exp(\lambda t)$ where λ is an eigenvalue of the Jacobian matrix A (2.1.40). Both at the boundaries of the bistability domain r_1, r_2 and at the (tri)critical points and lines C, C_t, l (Fig. 2.17) $\det\{A\} = 0$ holds because of the coalescence of two or more singular points, and therefore λ tends to zero. In the generic case ($\det\{A\} \sim E_0 - E_c$) the relaxation time diverges as

$$\tau_{\text{relax}} = |\lambda|^{-1} \approx |\text{tr}\{A\}/\det\{A\}| \sim |E_0 - E_c|^{-1} \,,$$

where E_c here corresponds either to the critical field or the threshold or holding field. Thus critical slowing-down could be observed as a sharp increase in the switching times at threshold.

In conclusion, transitions between different stable homogeneous steady states show a close analogy to equilibrium phase transitions of first and of second order, where the electric field E_0 corresponds to the thermodynamic adjoint of the order parameter (e.g., pressure or magnetic field), and the optical or thermal generation intensity g corresponds to the temperature. If one of the generation coefficients X^* or X_1^S is zero, a line of critical points and a tricritical point appear in the (E_0, g) plane. The transition changes from first to second order when g is increased beyond the tricritical point. For nonzero X^* and X_1^S an isolated critical point exists. Critical exponents and critical slowing-down are readily found.

2.1.5 Cyclotron-Resonance Induced Phase Transitions

The application of a magnetic field perpendicular to the electric field ("Faraday configuration") introduces an additional control parameter. The magnetic field (magnetic induction B) splits the conduction-band edge into Landau levels $N = 0$, $1, 2, 3, \ldots$ spaced by an energy $\hbar\omega_c = \hbar eB/m^*$, and splits and shifts the donor levels (e.g., the $2p$ state is split into $2p_{-1}$, $2p_0$, $2p_{+1}$). Cyclotron resonance occurs if the spacing between two Landau levels matches the frequency of an incident far-infrared (FIR) laser irradiation, which induces transitions from $N = 0$ to $N = 1$. The depleted $N = 0$ Landau level is then repopulated thermally or by impact ionization of neutral donors. Experimentally, this can be detected by a resonant spike in the photoconductivity or the optical absorption as a function of the magnetic field. Magneto-impurity resonance between $1s$ and $2p_+$ donor levels is also possible, and has been detected in high-purity n-GaAs at 4.2 K by cw (continuous wave) nonlinear FIR magneto-absorption [2.44] and by FIR magneto-photoconductivity [2.45]; hereby an unusually long effective lifetime of the $2p \to 1s$ impurity transitions (0.5 μs "off-resonance", 50 ns "on-resonance") has been found, and explained in terms of the g-r kinetics of the two impurity levels and the $N = 0$ Landau band.

In case of cyclotron-resonance induced FIR photoconductivity, the photoconductive signal is strongly enhanced close to the threshold of impact ionization. In an investigation on n-GaAs at low temperatures [2.46] it was shown that photoconductivity due to low-power FIR excitation of cyclotron resonance probes the critical behavior of a generalized susceptibility of the second-order nonequilibrium phase transition induced by impact ionization of shallow donors. In a subsequent paper high-power FIR laser irradiation under cyclotron-resonance conditions was investigated [2.47], and it was established that, besides the electric field E, the optical-excitation probability of cyclotron resonance σF is an additional control parameter of the nonequilibrium phase transition, where F is the photon flux density, and σ is the cyclotron-resonance absorption cross section depending upon the magnetic field and the FIR frequency. By analyzing the $\sigma F-E$ plane of control parameters it was found that the FIR irradiation shifts the threshold of the impact-ionization instability E_c (i.e., the critical point) to lower values, thus generating a whole line of critical points in the $\sigma F - E$ plane, similar to the λ-line in superfluid Helium (Sect. 2.1.4). On crossing the critical line from the low-conducting state at constant E the photoconductivity shows a threshold-like behavior due to the combined action of cyclotron-resonance excitation and impact ionization. This novel highly nonlinear photoconductive mechanism gives further insight in the kinetics of electrons bound to shallow donors and may be useful as a threshold detector and optical correlator opening up a new field of nonlinear FIR optoelectronics. Furthermore, the impact-ionization coefficient X as a function of the electric-field strength and the lifetime of electrons in the $N = 1$ Landau level could be determined experimentally.

The measurements were carried out on a high-purity n-GaAs epitaxial layer with alloyed Au-Sn ohmic strip contacts on opposite edges of the sample in order to get a homogeneous electric field. The sample was mounted in a metallic light pipe and immersed in liquid helium at the center of a superconducting solenoid. Cyclotron resonance was excited by the $\lambda = 570\,\mu$m line of a CH_3OH laser pumped by a pulsed CO_2 laser. The duration of the laser pulses was 300 µs being much larger than any expected relaxation time, thus steady state conditions during optical excitation could be presumed. Photoconductivity was measured in a Faraday configuration by application of a standard load resistor circuit. The load resistance was chosen in all cases to be much smaller than the sample resistance. Therefore, and because the mobility of n-GaAs at low temperatures is not appreciably affected by cyclotron resonance absorption [2.48], it follows that $\Delta V/V \sim \Delta n$ where ΔV and Δn are the changes due to irradiation of the voltage across the sample and the free-electron concentration, respectively.

The photoconductivity signal at the center of the resonance is shown in Fig. 2.18 for various fixed electric-field strengths below the critical field $E_c^{(0)}$ as a function of the intensity $I = \hbar\omega F$. The photo-signal obviously sets in at a threshold intensity $\hbar\omega F_c$ and saturates at high intensities. Both the intensity threshold and the saturation intensity decrease with rising electric field. Figure 2.19 shows the cyclotron-resonance line for three different electric-field strengths and in each case for various laser intensities. For the lowest electric field (Fig. 2.19a) and for low intensities the cyclotron resonance shows up as a Lorentzian shaped line of halfwidth $\Delta B = 15$ mT as it is usually observed in high-purity n-GaAs applying low-power lasers. With increasing intensity the linewidth broadens and the lineshape deviates from a Lorentzian. This effect is more drastically shown at higher electric-field strengths (Fig. 2.19b). In particular, close to $E_c^{(0)}$ (Fig. 2.19c) the lineshape does not resemble a Lorentzian at all. This strange behavior of the lineshape is an immediate consequence of the fact that cyclotron-resonance excitation critically controls the sample conductance and leads to second-order phase transitions along the critical line $[E_c, (\sigma F)_c]$. At constant intensity and field the optical-transition probability σF varies with the magnetic induction B like a Lorentzian centered at the resonance field B_{CR}. Approaching B_{CR} the signal vanishes or is very small at low but nonzero temperatures as long as $\sigma F < (\sigma F)_c$, where $(\sigma F)_c$ is independent of the magnetic induction but depends upon E_c. When, upon further increase of B, σF crosses the threshold $(\sigma F)_c$, the sample is converted into the high-conducting phase and the conductivity rapidly increases on further proceeding to B_{CR}.

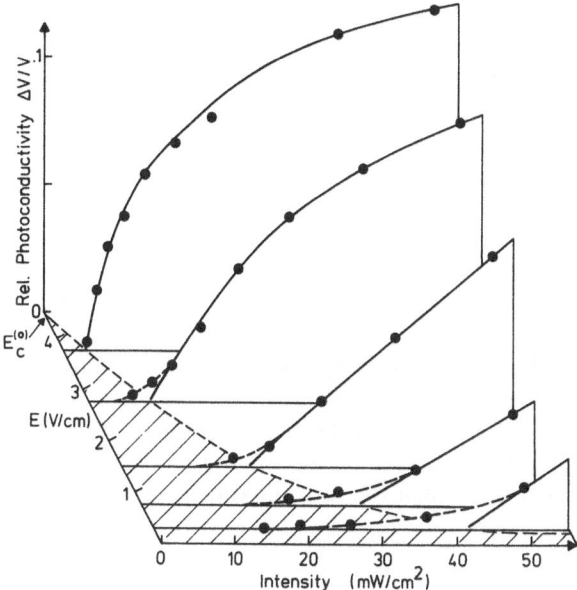

Fig. 2.18. Measured relative photoconductive signal $\Delta V/V \sim \Delta n$ at 4.2 K for n-GaAs in the center of cyclotron resonance for various electric-field strengths E as a function of the effective optical intensity in the sample. Effective donor concentration $N_D - N_A = 8.3 \times 10^{13}$ cm^{-3}, compensation $N_A/N_D = 0.7$. [2.47]

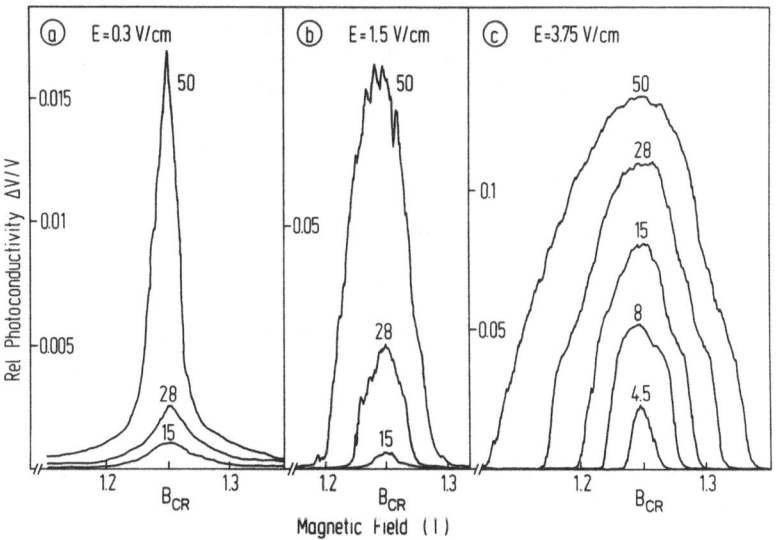

Fig. 2.19. Measured cyclotron-resonance induced photoconductivity versus magnetic induction for three different electric bias fields (a)–(c) and various irradiation intensities in n-GaAs. Numbers identifying the curves give the effective optical intensity in the sample in units of mW/cm^2. [2.47]

Thus, when the control parameter σF is modulated by the magnetic induction B according to a Lorentzian $\sigma(B)$, the photoconductivity (the order parameter) shows a distinctly non-Lorentzian structure with a basewidth ΔB, which is given by the condition $\sigma(B_{CR} - \Delta B/2)F = (\sigma F)_c$. Hence $\Delta B \sim [\sigma(B_{CR})F/(\sigma F)_c - 1]^{1/2}$ increases with increasing F and with decreasing $(\sigma F)_c$, corresponding to rising electric field strength. Cyclotron resonance around the critical line of a nonequilibrium phase transition exhibits a strongly nonlinear modulation effect. The situation is quite similar to tuning a laser through resonance, and the observed photoconductivity lines closely resemble the tuning curves for different pumping rates of a single-mode laser, which is a more familiar example of a nonequilibrium phase transition [1.6].

In order to describe the E- and F-dependence of the cyclotron-resonance induced photoconductivity we apply a three-level model including the donor ground state and the two lowest Landau levels. The concentrations of electrons bound to donors and in the $N = 0$ and $N = 1$ Landau levels and the density of ionized donors are denoted by n_D, n_0, n_1, and p_D respectively. Then the rate equations are given by [2.46, 47]:

$$\dot{n}_1 = X_2^S n_0 - T_2^S n_1$$

$$\dot{n}_0 = X_1^S n_D + X n n_D + T_2^S n_1 - T_1^S n_0 p_D - X_2^S n_0 \ , \tag{2.1.80}$$

where the g-r coefficients are defined in Fig. 2.20. We assume that the $N = 1$ Landau level is populated by optical transitions solely, ignoring thermal excitations, and take into account stimulated emission to allow for saturation of the cyclotron-resonance absorption. Hence $X_2^S = \sigma F$ and $T_2^S = \tau_1^{-1} + \sigma F$ where τ_1 is the lifetime of electrons in the $N = 1$ Landau level. In the steady state and under the local neutrality condition $N_D^* = n_D + n_0 + n_1$ the free-electron concentration $n(F) = n_0 + n_1$ by (2.1.80) follows from

$$[X_1^S + (X_1^S + nX)\sigma F/T_2^S + nX](N_D^* - n) - T_1^S n(N_A + n) = 0 \ . \tag{2.1.81}$$

At low temperature the probability of impact ionization nX exceeds that of thermal ionization of the donors X_1^S for electric-field strengths even well below E_c, therefore X_1^S will be neglected. In this case for $F = 0$ the stable solution is $n = 0$ as long as $X < X_c^{(0)} := T_1^S N_A/N_D^*$ and thus photoconductivity is due to $\Delta n = n(F)$ probing the

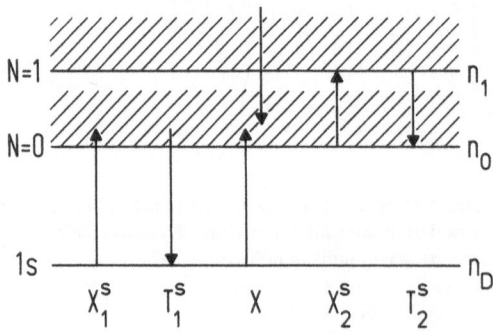

Fig. 2.20. Schematic energy diagram of 1s donor ground state and $N = 0$ and $N = 1$ Landau levels. The electron transitions considered are labeled by the respective g-r coefficients. [2.46]

free-electron concentration. Under this condition (2.1.81) has two stable solutions

$$n(F) = 0 \quad \text{for } \sigma F < (\sigma F)_c$$

$$n(F) = \Delta n = \eta N_D^* \tau_{\text{eff}} (\sigma F - (\sigma F)_c)(1 + F/F_s)^{-1} \quad \text{for } \sigma F > (\sigma F)_c \ , \qquad (2.1.82)$$

which represent the order parameter in the two phases and exhibit the observed threshold-like behavior of the photo-signal. The critical optical-excitation probability depends on X via

$$(\sigma F)_c = \tau_1^{-1}(X_c^{(0)} - X)/(2X - X_c^{(0)}) \qquad (2.1.83)$$

and determines the critical line in the σF–E plane when X is given as a function of E. Cyclotron-resonance induced transitions are possible in the range $X_c^{(0)}/2 < X < X_c^{(0)}$, where $(\sigma F)_c$ diverges at $X_c^{(0)}/2$ because of the saturation of cyclotron resonance and vanishes at $X_c^{(0)}$. In (2.1.82) $F_s = (\sigma \tau_{\text{eff}})^{-1}$ is the saturation photon flux density, $\tau_{\text{eff}} = \tau_1[(2X + T_1^S)/(X + T_1^S)]$ is an effective lifetime of the electrons in the conduction band and $\eta = (2X - X_c^{(0)})/(2X + T_1^S)$ is a dimensionless quantum efficiency.

The experimentally observed intensity dependence of the photo-signal has been fitted by the relation (2.1.82) (see solid lines in Fig. 2.18), which yields $(\sigma F)_c$ and F_s for various electric-field strengths. At lower fields and low laser intensities the condition $X_1^S \ll nX$ obviously is not fulfilled (dashed lines). Here the photo-signal is nonzero already at low intensity $\sigma F < (\sigma F)_c$ due to electrons thermally excited into the $N = 0$ Landau level via X_1^S. At higher intensities the signal proceeds superlinearly into the transition region (solid line). Also in this case $(\sigma F)_c$ can be estimated from the measurements by extrapolation of the high-intensity data. The softening of the sharp second-order phase transition by thermal generation is analogous to the smoothing out of the laser threshold by spontaneous emission, or the suppression of the ferromagnetic second-order phase transition by a magnetic field. The second-order phase transition at $(\sigma F)_c$ becomes sharper, i.e., $(dn/dF)_{(\sigma F)_c}$ becomes larger as E increases towards $E_c^{(0)}$ and $n^{-1}(dn/dF)_{(\sigma F)_c}$ diverges as

$$\frac{1}{n}\left(\frac{dn}{dF}\right)_{(\sigma F)_c} \sim |E - E_c^{(0)}|^{-1} \quad \text{for } E \to E_c^{(0)} \ . \qquad (2.1.84)$$

This identifies a critical exponent $\tilde{\gamma} = 1$ of the generalized susceptibility $\chi \sim n^{-1} dn/dF$, in agreement with findings in the low FIR power regime [2.46]. Note, however, that $(dn/dF)_{(\sigma F)_c}$ remains finite for all E.

In Fig. 2.21 the critical optical transition probability $(\sigma F)_c$ obtained by these procedures is plotted as a function of the electric-field strength E, and it separates both phases in the σF–E plane. It can be seen that $(\sigma F)_c$ vanishes at $E = E_c^{(0)} = 4.25$ V/cm, corresponding to $X = X_c^{(0)}$. The inverse of $(\sigma F)_c$ has been proved to extrapolate to zero at $E_1 = 0.25$ V/cm, corresponding to $X = X_c^{(0)}/2$. By (2.1.83) the impact-ionization coefficient X in units of $X_c^{(0)}$ as a function of E has been evaluated. The result, shown also in Fig. 2.21, may be fitted, with excellent agreement, by the simple relation $X = X_0 \exp(-E_i/E)$ derived from Shockley's "Lucky Electron"

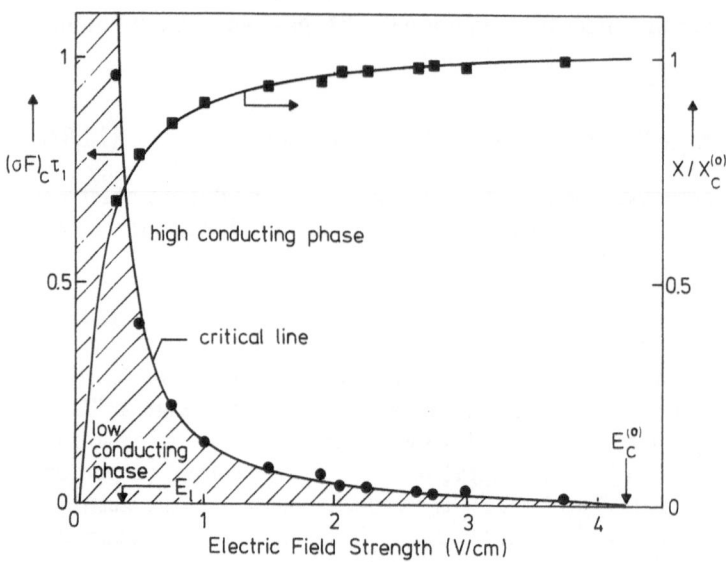

Electric Field Strength (V/cm)

Fig. 2.21. The critical line of the optical transition probability $(\sigma F)_c$ in units of τ_1^{-1} and the effective impact-ionization probability per electron X in units of $X_c^{(0)}$ as functions of the electric field strength. Circles and squares are derived from the measurements. The solid lines are calculated with the Shockley formula. [2.47]

Model [1.131] with $E_i = 0.18$ V/cm and $X_0/X_c^{(0)} = X_0 N_D^*/(T_1^S N_A) = 2.0$. The inverse of $T_1^S N_A$ is just the lifetime τ_0 of the electrons in the $N = 0$ Landau level at zero electric field. This time constant has been estimated from the decay of the current through the sample after application of short electric pulses. For the sample in [2.47], $\tau_0 \approx 5$ ns was found, which yields $X_0 N_D^* \approx 10^8$ s^{-1}. The lifetime of the electrons in the $N = 1$ Landau level τ_1 can easily be derived from the saturation intensity for X close to $X_c^{(0)}$, in this case $\tau_{eff} = \tau_1(1 + N_A/N_D)$ being free from the inherent uncertainties of X and T_1^S. For the saturation intensity $I_s = \hbar \omega F_s = 9$ mW/cm^2 was obtained, which yields an electron lifetime $\tau_1 = 1.9$ ns.

The three-level model worked out here is a reasonable approximation as long as the population of the $N = 2$ Landau level may be neglected compared to that of $N = 1$. The good agreement between the calculated photoconductivity and the experimental results displayed in Fig. 2.18 indicates that this condition is satisfied, most probably because of an even shorter electron lifetime in the $N = 2$ Landau level. In general, optical excitation of higher Landau levels increases the saturation intensity and lowers $X_c(F \to \infty)$ below $X_c^{(0)}/2$. In a more refined model, the effective impact-ionization rate $X n n_D$, which represents an average over different Landau bands, should be replaced by more detailed expressions, and more sophisticated expressions for the field dependence of the impact-ionization coefficients [1.135, 138] could be used. However, even such refined rate-equation models will not alter the basic features of the threshold behavior of the photoconductivity, since the behavior of equilibrium as well as nonequilibrium systems near critical points falls into only a few universality classes, as discussed in Sect. 2.1.4.

In conclusion, a critical behavior of the FIR photoconductivity due to the cyclotron resonance, and a drastic deviation of the cylotron-resonance lineshape from a Lorentzian has been observed in n-GaAs at low temperatures by applying a high-power cw FIR laser. Both effects may consistently be explained in terms of second-order nonequilibrium phase transitions which correspond to the crossing of a line of critical points in the control parameter plane of the optical-excitation probability σF and the electric field E.

2.2 Two-Carrier Models

In this section we will deal with g-r mechanisms involving both electrons and holes, and discuss their relevance to threshold switching in amorphous and crystalline materials. In our survey on g-r processes in Sect. 1.2.2 we have singled out four "autocatalytic reactions" as key processes for g-r instabilities: band-to-band impact ionization Y_1, Y_2 and band-to-trap impact ionization X_1, X_4.

2.2.1 Models with Band-Band Impact Ionization

The simplest mechanism based on band-to-band impact ionization [2.49] is shown schematically in Fig. 2.22. It represents an intrinsic semiconductor subject to band-to-band recombination (B^S), generation (Y^S), impact ionization (Y_1), linear decay (l; e.g., trapping which does not affect the equality $n \approx p$), and Auger recombination (B_1). The rate equation for the carrier density is given by

$$\dot{n} = Y^S + (Y_1 - l)n - B^S n^2 - B_1 n^3 , \tag{2.2.1}$$

where the intrinsic charge neutrality condition $n = p$ has been used. For $Y^S > 0$, (2.2.1) always has a unique steady state solution $n > 0$, but for $Y^S = 0$ (low temperature, no photogeneration) the possibility of a second-order phase transition appears. The stable steady state solutions are:

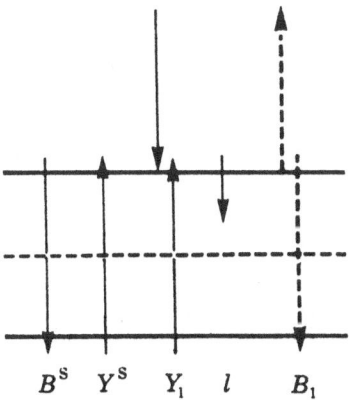

$$B^S \quad Y^S \quad Y_1 \quad l \quad B_1$$

Fig. 2.22. Intrinsic two-carrier model for g-r induced second-order phase transitions, involving band-band recombination (B^S), generation (Y^S), impact ionization (Y_1), linear decay (l), and Auger recombination (B_1)

$$n = \begin{cases} 0 & \text{for } Y_1 \leq l \\ \dfrac{B^{\mathrm{S}}}{2B_1}\{[1 + 4B_1(Y_1 - l)/(B^{\mathrm{S}})^2]^{1/2} - 1\} & Y_1 \geq l \; . \end{cases} \qquad (2.2.2)$$

The second-order transition from a nonconducting to a conducting state occurs at $Y_1 = l$; it is connected with the threshold of avalanche breakdown. Eqs. (2.2.1, 2) retain their form if the symmetric processes Y_2 and B_2 (Fig. 1.10) are added. If Auger recombination (dashed arrows in Fig. 2.22) is neglected, the threshold of the phase transition is unchanged:

$$n = \begin{cases} 0 & Y_1 \leq l \\ (Y_1 - l)/B^{\mathrm{S}} & Y_1 \geq l \; . \end{cases} \qquad (2.2.3)$$

The critical behavior of the correlation length, and stochastic fluctuations have also been treated for this model [2.50].

An unstable NDC state, in addition to a stable steady state, arises in an extrinsic model (effective doping N_{D}^*) if recombination via deep impurity centers of density $N_{\mathrm{t}} < N_{\mathrm{D}}^*$ is explicitly taken into account [2.51]. The processes considered are shown in Fig. 2.23a. The steady state solutions resulting from the rate equations

$$\dot{n} = Y_1 n + Y_2 p - T_1^{\mathrm{S}} n(N_{\mathrm{t}} - n_{\mathrm{t}})$$

$$\dot{p} = Y_1 n + Y_2 p - T_2^{\mathrm{S}} p n_{\mathrm{t}} \qquad (2.2.4)$$

with the charge neutrality condition

$$N_{\mathrm{D}}^* - n - n_{\mathrm{t}} + p = 0 \qquad (2.2.5)$$

are shown schematically in Fig. 2.23b for the symmetric case $Y_1 = Y_2$. There exists a stable low-conductivity state (solid line) with $n \approx N_{\mathrm{D}}^* - N_{\mathrm{t}}$, $p \approx 0$, $n_{\mathrm{t}} \approx N_{\mathrm{t}}$ for $Y_1 = Y_2 \leq Y_{\mathrm{th}} := N_{\mathrm{t}}[(T_1^{\mathrm{S}})^{-1/2} + (T_2^{\mathrm{S}})^{-1/2}]^{-2}$, and an unstable NDC state (dashed

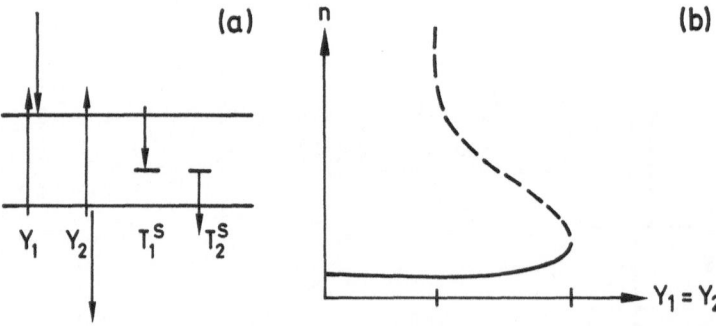

Fig. 2.23a, b. Extrinsic two-carrier model, involving band-band impact ionization (Y_1, Y_2) and electron (T_1^{S}) and hole capture (T_2^{S}). (a) Schematic band diagram. (b) Steady state electron concentration n versus control parameter $Y_1 = Y_2$ (schematic). The solid and the dashed lines represent stable and unstable states, respectively

line) for $Y_h := N_t[2(1/T_1^S + 1/T_2^S)]^{-1} < Y_1 = Y_2 < Y_{th}$ where $T_1^S < T_2^S$ has been assumed. If the impact-ionization coefficients $Y_1 = Y_2$ exceed the threshold value Y_{th}, a steady state no longer exists. This is a result of the neglect of other recombination processes.

This model has been advanced as a mechanism for threshold switching in amorphous chalcogenides [2.51]. However, the model does not give a second stable homogeneous steady state, i.e., no spatially homogeneous bistability and SNDC characteristics are produced. Only if spatially inhomogeneous states are considered, one can obtain additional branches of the current-voltage characteristics (Sect. 4.2.2).

Models for dielectric breakdown in insulators based upon band-to-band impact ionization in combination with recombination or drift have also been studied [2.52–54].

2.2.2 Models with Band-Trap Impact Ionization

Next, we shall discuss two-carrier g-r mechanisms based upon band-to-trap impact ionization [2.7, 8] and show that these can be applied to explain threshold switching [2.55, 56]. We assume an extrinsic semiconductor with a completely ionized effective donor density N_D^*, and $N_t > N_D^*$ deep traps. The g-r processes considered are shown schematically in Fig. 2.24. The simplest model includes B^S, X_1, and X_4 only. The rate equations are

$$\dot{n} = [X_1 N_D^* - X_1 n - (B^S - X_1)p]n$$

$$\dot{p} = [X_4 P_D - X_4 p - (B^S - X_4)n]p , \qquad (2.2.6)$$

where we have used the charge neutrality conditions (2.2.5) to eliminate n_t, and set

$$P_D := N_t - N_D^* . \qquad (2.2.7)$$

The physical region of the (n, p) phase plane is given by

$$p - P_D \leq n \leq p + N_D^* , \qquad (2.2.8)$$

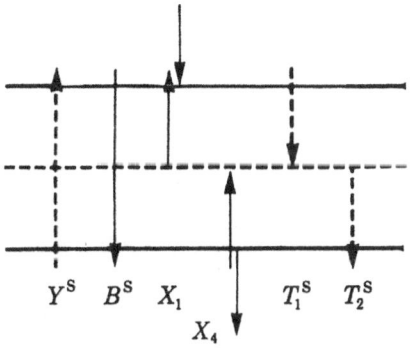

Fig. 2.24. Two-carrier model for bistability and first-order phase transitions, involving band-trap impact ionization (X_1, X_4), electron (T_1^S) and hole capture (T_2^S), band-band generation (Y^S) and recombination (B^S). The dashed processes are not necessary for phase transitions

which follows from $n_t = N_D^* - n + p \geq 0$, $p_t = P_D + n - p \geq 0$. The steady states and their stability range are as follows:

(i) $n = p = 0$, $n_t = N_D^*$ (always unstable) ,

(ii) $n = 0$, $p = P_D$, $n_t = N_t$ $(X_1 < B^S P_D/N_t)$,

(iii) $p = 0$, $n = N_D^*$, $n_t = 0$ $(X_4 < B^S N_D^*/N_t)$,

(iv) $n = (X_1 N_t - B^S P_D)X_4/D_0$,

$\qquad\qquad p = (X_4 N_t - B^S N_D^*)X_1/D_0$ $(X_1 + X_4 > B^S)$,

$\qquad\qquad D_0 := (X_1 + X_4 - B^S)B^S$.   (2.2.9)

The stability readily follows from linearization of (2.2.6) around the steady state

$$\begin{pmatrix} \delta\dot{n} \\ \delta\dot{p} \end{pmatrix} = A \begin{pmatrix} \delta n \\ \delta p \end{pmatrix} , \qquad\qquad (2.2.10)$$

where

$$A := \begin{pmatrix} X_1 N_D^* - 2X_1 n - (B^S - X_1)p & -(B^S - X_1)n \\ -(B^S - X_4)p & X_4 P_D - 2X_4 p - (B^S - X_4)n \end{pmatrix} .$$

Substitution of the steady state values (2.2.9) yields four matrices $A^{(i)}, \ldots, A^{(iv)}$, from which the following eigenvalues λ_1, λ_2 are obtained:

(i) $\lambda_1 = X_1 N_D^*$, $\lambda_2 = X_4 P_D$

(ii) $\lambda_1 = X_1 N_t - B^S P_D$, $\lambda_2 = -X_4 P_D$

(iii) $\lambda_1 = -X_1 N_D^*$, $\lambda_2 = X_4 N_t - B^S N_D^*$

(iv) $\lambda_{1,2} = \dfrac{1}{2} T[1 \pm (1 - 4\tilde{D}/T^2)^{1/2}]$

$\qquad$ where $\tilde{D} := (X_1 + X_4 - B^S)B^S np \gtrless 0$, $T := -X_1 n - X_4 p < 0$

$\qquad$ from $A^{(iv)} = \begin{pmatrix} -X_1 n & -(B^S - X_1)p \\ -(B^S - X_4)n & -X_4 p \end{pmatrix}$.   (2.2.11)

Hence, in the phase plane, solution (i) always represents an unstable node, solutions (ii) and (iii) are stable nodes or saddle points, depending upon the sign of λ_1 (in case ii) and λ_2 (in case iii), and solution (iv) lies in the physical phase regime if $X_1 > B^S P_D/N_t$ and $X_4 > B^S N_D^*/N_t$, and is then a stable node. The steady states (ii) (p-conducting phase) and (iii) (n-conducting phase) have a rectangular overlap of their stability ranges in the (X_1, X_4) control parameter plane given by

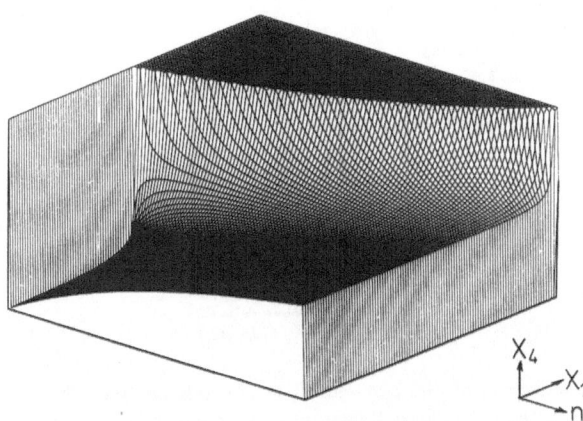

Fig. 2.25. Steady-state electron concentration $n(X_1, X_4)$ for the g-r mechanism of Fig. 2.24 with $B^S = 10^{-10}$ cm^3 s^{-1}, $N_t = 1.5 N_D^* = 3 \times 10^{15}$ cm^{-3}, $Y^S = T_1^S = T_2^S = 0$, and 10^{-13} cm^3 s$^{-1} \leq X_1 \leq 10^{-10}$ cm^3 s^{-1}, $0 \leq X_4 \leq 10^{-10}$ cm^3 s^{-1}, $0 \leq n \leq N_D^*$

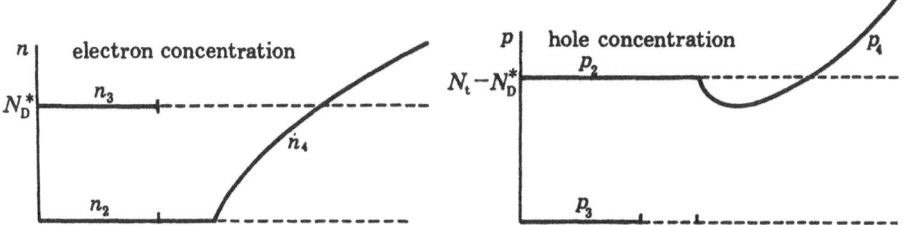

Fig. 2.26. Steady state concentrations n and p for the g-r mechanism of Fig. 2.24 with $Y^S = T_1^S = T_2^S = 0$ versus electric field (schematic). Unstable states are dashed

$$0 < X_1 < (1 - N_D^*/N_t)B^S , \qquad 0 < X_4 < (N_D^*/N_t)B^S . \qquad (2.2.12)$$

It is instructive to use state diagrams of the steady state electron density as a function of the control parameters X_1 and X_4. The multivalued surface $n(X_1, X_4)$ is represented in the (n, X_1, X_4) state space in Fig. 2.25. The hatched planes correspond to the stable states (ii) $(n = 0)$ and (iii) $(n = N_D^*)$; they are joined by the curved surface (iv). First-order phase transitions may occur if the boundary of the rectangle of bistability between (ii) and (iii) is reached due to an increase in the applied electric field which defines a path in the (X_1, X_4) control parameter plane. Solution (iv) is reached by a second-order transition upon further increase of the electric field. The electron and hole concentrations along such a path which crosses first the line $X_4 = B^S N_D^*/N_t$ and then the line $X_1 = B^S P_D/N_t$ is shown schematically in Fig. 2.26. The bistability between (iii) and (ii) and the second-order transition from (ii) to (iv) can be seen.

If band-band generation Y^S is added to the rate equations (2.2.6), the nonconducting state $(n = p = 0)$ is no longer a steady state solution. In some ranges of the rate constants there are three steady states, two of which are stable (Fig. 2.27). The $n(X_1, X_4)$ surface analogous to Fig. 2.25 is shown in Fig. 2.28. The rectangular bistability range corresponding to $Y^S = 0$ is rounded off, and the second-order phase transitions of Fig. 2.26 are smoothed over. Analytical solutions in special cases have been given elsewhere [2.8].

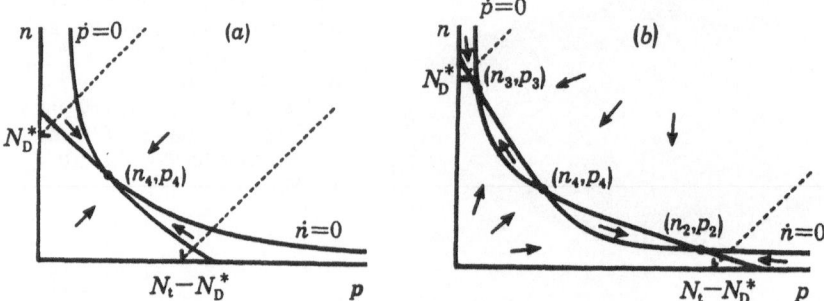

Fig. 2.27a, b. Phase portrait for the g-r mechanism of Fig. 2.24 with $T_1^S = T_2^S = 0$ (schematic). The physical region ($n, p, n_t, p_t \geq 0$) is confined by the dotted lines. Either one (**a**) or two (**b**) stable steady states are possible

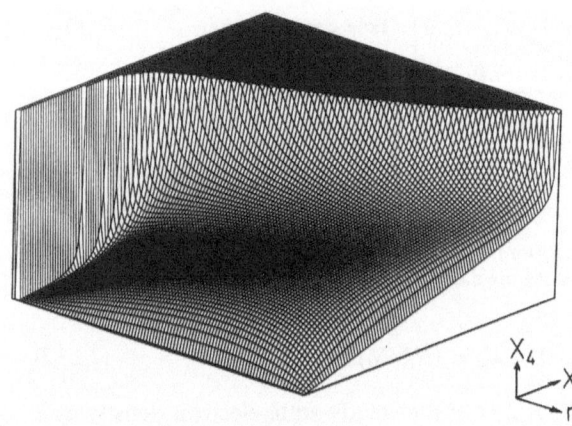

Fig. 2.28. Steady state electron concentration $n(X_1, X_4)$, as in Fig. 2.25, but with $Y^S = 4 \times 10^{+18}$ cm^{-3} s^{-1}

Model for Threshold Switching

If band-trap recombination ($T_1^S, T_2^S \neq 0$) is added, the rate equations (2.2.6) take the form:

$$\dot{n} = [X_1 N_D^* - T_1^S P_D - (X_1 + T_1^S)n - (B^S - X_1 - T_1^S)p]n$$

$$\dot{p} = [X_4 P_D - T_2^S N_D^* - (X_4 + T_2^S)p - (B^S - X_4 - T_2^S)n]p \ . \tag{2.2.13}$$

The steady state solutions and their stability ranges are

(i) $n_1 = 0, \quad p_1 = 0$ $(F_a, F_b < 0)$

(ii) $n_2 = 0, \quad p_2 = F_b/(X_4 + T_2^S)$ $(F_b, F_c > 0)$

(iii) $n_3 = F_a/(X_1 + T_1^S), \quad p_3 = 0$ $(F_a, F_d > 0)$

(iv) $n_4 = -F_c/D, \quad p_4 = -F_d/D$ $(D > 0) \ , \tag{2.2.14}$

where

$$F_a := N_D^* X_1 - T_1^S P_D \; ; \qquad F_b := P_D X_4 - T_2^S N_D^*$$

$$F_c := B^S F_b - C \; , \qquad F_d := B^S F_a - C$$

$$C := (X_1 X_4 - T_1^S T_2^S) N_t \; , \qquad D := B^S (X_1 + X_4 + T_1^S + T_2^S - B^S) \; .$$

We shall also use $A := P_D/N_D^* = N_t/N_D^* - 1$, which is a measure of the degree of compensation. Inspection of the stability criteria shows that solutions (ii) and (iii) can be stable for the same values of the control parameters. A necessary and sufficient condition for such an overlap in the (X_1, X_4) plane is

$$\frac{B^S}{N_t} > \frac{T_1^S}{N_D^*} + \frac{T_2^S}{P_D} \; . \tag{2.2.15}$$

The (X_1, X_4) control parameter plane is represented schematically in Fig. 2.29. The boundary curves of the stability regions of the four solutions are given by $F_a = 0$, $F_b = 0$, $F_c = 0$, $F_d = 0$, and $D = 0$. It is readily seen from (2.2.14) that the first four of these conditions are simultaneously fulfilled at a point Q where $X_1 = T_1^S A$ and $X_4 = T_2^S/A$. If (2.2.15) holds, the curves $F_c = 0$ and $F_d = 0$ cross again at a second point R where $X_1 = AB^S/(A + 1) - T_2^S$ and $X_4 = B^S/(A + 1) - T_1^S$ and the solutions (ii) and (iii) overlap as shown in Fig. 2.29. From (2.2.14) it can be seen that the line $D = 0$ also passes through the point R cutting the X_1- and X_4-axes at $B^S - T_1^S - T_2^S$. However, since the condition that n_4 and p_4 are positive provides a stronger

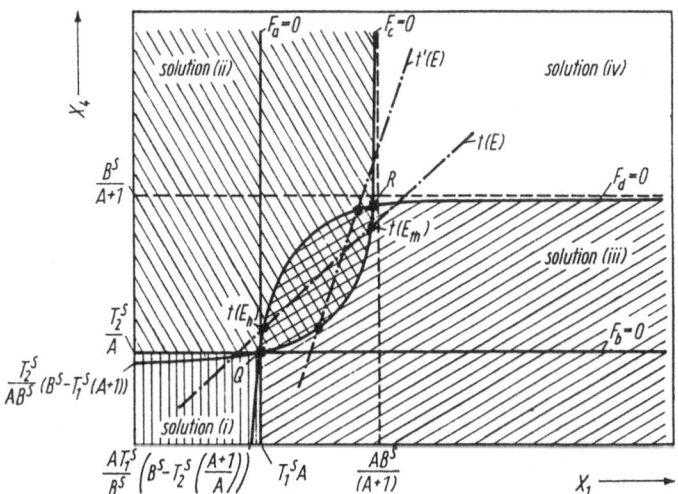

Fig. 2.29. Schematic plot of the stability regions of the steady state solutions (i)–(iv) in the (X_1, X_4) control parameter plane for the g-r mechanism of Fig. 2.24 with $Y^S = 0$. The trajectories $t(E)$ and $t'(E)$ correspond to different field dependencies of the impact-ionization coefficients X_1, X_4. The special points Q and R are discussed in the text. [2.56]

constraint, namely, $F_c < 0$, $F_d < 0$, the line $D = 0$ is omitted from Fig. 2.29. An overlap of other solutions is not possible in this model.

When the lines $F_a = 0$ ($F_b < 0$), or $F_b = 0$ ($F_a < 0$), or $F_c = 0$ ($F_d < 0$, $F_b > 0$), or $F_d = 0$ ($F_c < 0$, $F_a > 0$) are crossed, second-order phase transitions (i) $\rightarrow$ (iii), (i) $\rightarrow$ (ii), (ii) $\rightarrow$ (iv), or (iii) $\rightarrow$ (iv), respectively, occur.

For control parameters corresponding to the point Q the two stable solutions (ii) and (iii) merge into one stationary point at the origin, while at R (ii) and (iii) merge into a degenerate stationary line; Q and (in a degenerate sense) R are tricritical points.

As the electric field E is increased from low values the impact-ionization coefficients X_1 and X_4 increase and the system describes a trajectory across Fig. 2.29. Two examples, $t(E)$ and $t'(E)$, are shown as straight lines, though a trajectory will normally be a curve. If this trajectory crosses the overlap region, the necessary conditions for S-type switching may be produced. If the trajectory enters it via $F_d = 0$ and leaves via $F_c = 0$, or if it enters via $F_c = 0$ and leaves via $F_d = 0$ the usual type of switching characteristic is found. The first case corresponds to trajectory $t(E)$ in Fig. 2.29 with solutions (i), (ii), (iii), (iv) following each other as the field is raised. A jump occurs at $t(E_{\rm th})$ as one passes from (ii) to (iii). The appropriate field is interpreted as the threshold field $E_{\rm th}$. On reducing the field the sequence is simply reversed, but the low-field part of solution (iii) is utilized so that the jump from (iii) to (ii) occurs at a lower field, at $t(E_{\rm h})$. This field is interpreted as the holding field $E_{\rm h}$. The second case corresponds to trajectory $t'(E)$. If T_1^S, T_2^S, or B^S were field dependent then the structure of Fig. 2.29 would change with field. However, it is to be expected that the field dependences of the autocatalytic processes X_1 and X_4 which are taken into account here, will dominate at least for the overlap region. Thus one may consider the structure of Fig. 2.29 as representative for a range of electric fields extending at least from $E_{\rm h}$ to $E_{\rm th}$.

The associated steady state surface $n(X_1, X_4)$ is shown in Fig. 2.30. For a typical field dependence of X_1 and X_4, corresponding to trajectory $t(E_0)$ in Fig. 2.29, the electron and hole concentrations are plotted versus the electric field E in Fig. 2.31.

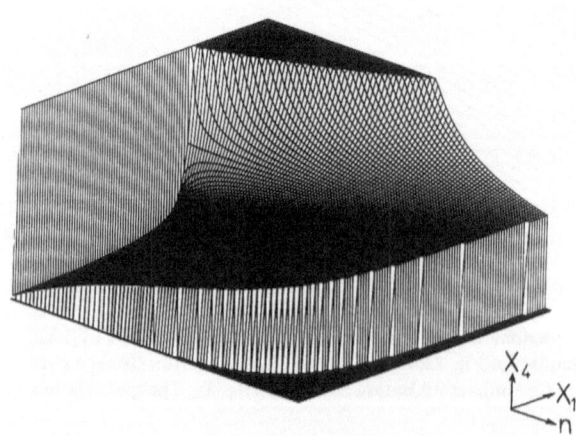

Fig. 2.30. Steady state electron concentration $n(X_1, X_4)$, as in Fig. 2.25, but with $T_1^S = T_2^S = 10^{-12}$ cm^3 s^{-1}, and 3×10^{-12} cm^3 s$^{-1} \leq X_1 \leq 0.7 \times 10^{-10}$ cm^3 s^{-1}

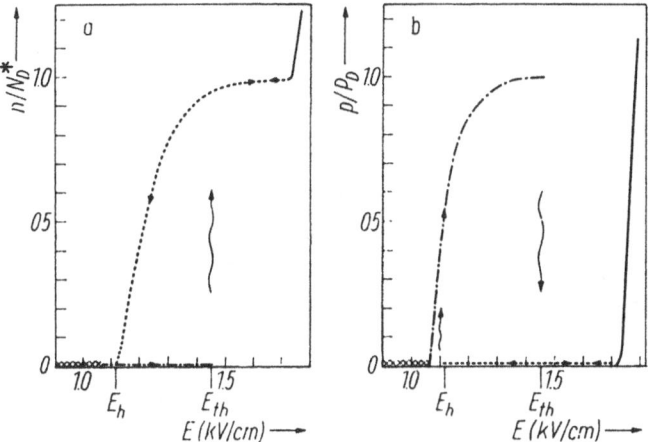

Fig. 2.31a, b. Carrier concentrations versus field in the four regimes of stability for the numerical values of Table 2.4, a1–a3. xxxx solution (i), $-\cdot-\cdot-\cdot$ solution (ii), $\cdots\cdots$ solution (iii), —— solution (iv). Holding (E_h) and threshold (E_{th}) fields are also shown. There is a first-order phase transition at both E_h and E_{th}; however, that at E_h in (a) does not show up for the numbers chosen

Current-Field Characteristics

The current $I(E)$ is plotted in Fig. 2.32 as the dimensionless quantity

$$\frac{I(E)}{e\mu_n N_D^* EF} = \frac{n(E)}{N_D^*} + \frac{p(E)}{N_D^*} b \, , \qquad (2.2.16)$$

where $b := \mu_p/\mu_n$ is the ratio of mobilities, and is assumed field independent, and F is the cross-sectional area of the current flow. The current-field characteristics displayed show that the switching model allows a good deal of flexibility despite the relatively small variations in the control parameters employed (see Table 2.4), and despite the assumption that B^S, T_1^S, T_2^S and b are field independent.

It is desirable, in view of the various alternatives, to introduce a precise nomenclature. On increasing an electric field, its value just before a jump is called a "threshold field", while the value just before a jump on decreasing the field is called a "holding field". We shall call a switch an (up, down)-switch if the jump at threshold is to a higher current, while it is to a lower current at the holding field. The normal switch is of the (up, down)-variety (Fig. 2.32c) but our model also leads to a twisted form of S-type switching i.e., the (up, up)-variety (Fig. 2.32a2, a3, b2, d). It also leads to (down, up)-switching, which one might call reverse switching (Fig. 2.32a1, b1).

Figure 2.32a is derived from Fig. 2.31 for three different values of the mobility ratio b, and is described qualitatively by the trajectory $t(E)$ in Fig. 2.29. It shows how variations in the mobility ratio b can lead from (down, up)-switching (Fig. 2.32a1) to (up, up)-switching (Fig. 2.32a2, a3). For sufficiently small b normal (up, down)-switching would be produced. Fig. 2.32b shows curves corresponding to Fig. 2.32a when the significances of the order parameters n and p are reversed. Fig. 2.32c assumes a slightly modified field dependence for the ionization coefficients

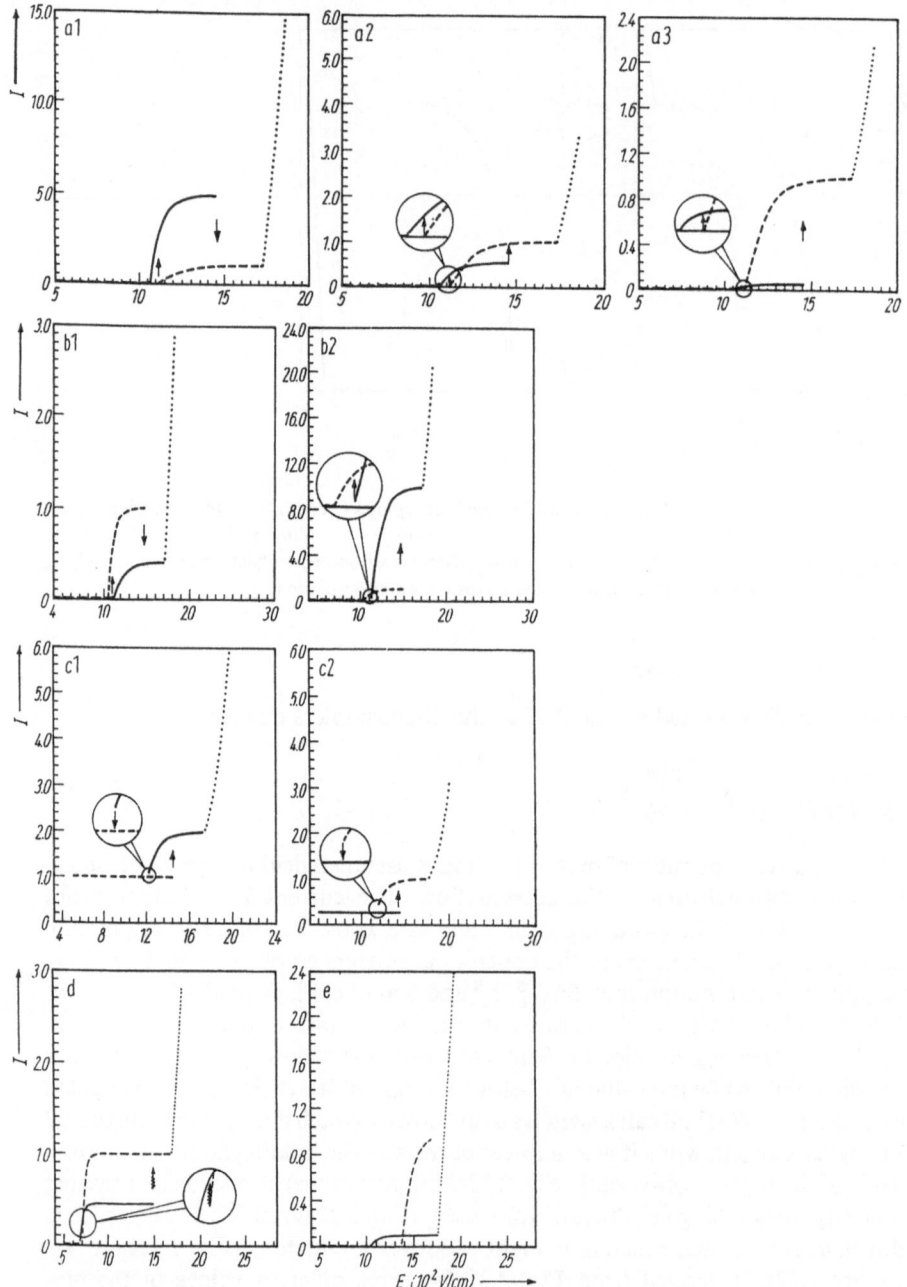

Fig. 2.32. Current-field characteristics for the g-r coefficients given in Table 2.4. The four stability regions are marked as follows: —— solution (i), —— solution (ii), ---- solution (iii), ····· solution (iv). The current is plotted in units of $e\mu_n N_D^* EF$, where F is the cross section of the current flow

Table 2.4. Data used in Fig. 2.32 [2.2, 16, 57]. $N_D^* = 2 \times 10^{15}$ cm⁻³ is always assumed

Fig.	X_1 [cm³ s⁻¹]	X_4 [cm³ s⁻¹]	T_1^S [cm³ s⁻¹]	T_2^S [cm³ s⁻¹]	B^S [cm³ s⁻¹]	N_t [cm⁻³]	$b = \mu_p/\mu_n$
a1 a2 a3	$3.0 \times 10^{-5} \exp\left(-\dfrac{2 \times 10^4}{E}\right)$	$3 \times 10^{-5} \exp\left(-\dfrac{2.25 \times 10^4}{E}\right)$	1.0×10^{-12}	1.0×10^{-14}	1.0×10^{-10}	3.0×10^{15}	{10.0 1.0 0.1
b1 b2	$3 \times 10^{-5} \exp\left(-\dfrac{2.23 \times 10^4}{E}\right)$	$3 \times 10^{-5} \exp\left(-\dfrac{2.0 \times 10^4}{E}\right)$	1.0×10^{-14}	1.0×10^{-12}	1.0×10^{-10}	6.0×10^{15}	{0.2 5.0
c1	$3 \times 10^{-5} \exp\left(-\dfrac{2.5 \times 10^4}{E}\right) + 5 \times 10^{-11}$	$3 \times 10^{-5} \exp\left(-\dfrac{2.0 \times 10^4}{E}\right)$	1.0×10^{-14}	1.0×10^{-12}	1.0×10^{-10}	6.0×10^{15}	1.0
c2	$3 \times 10^{-5} \exp\left(-\dfrac{2 \times 10^4}{E}\right)$	$3 \times 10^{-5} \exp\left(-\dfrac{2.5 \times 10^4}{E}\right) + 6 \times 10^{-11}$	1.0×10^{-12}	1.0×10^{-14}	1.0×10^{-10}	2.5×10^{15}	1.0
d	$3 \times 10^{-5} \exp\left(-\dfrac{2.0 \times 10^4}{E}\right)$	$3 \times 10^{-5} \exp\left(-\dfrac{2.25 \times 10^4}{E}\right)$	1.0×10^{-17}	1.0×10^{-19}	1.0×10^{-10}	3.8×10^{15}	0.5
e	$3 \times 10^{-5} \exp\left(-\dfrac{2.35 \times 10^4}{E}\right)$	$3 \times 10^{-5} \exp\left(-\dfrac{2.25 \times 10^4}{E}\right)$	1.0×10^{-12}	1.0×10^{-14}	1.0×10^{-10}	4.0×10^{15}	0.1

(in the region of the field E_h and E_{th}). Both Fig. 2.32c1, which corresponds to trajectory $t'(E)$ in Fig. 2.29 and Fig. 2.32c2, in which the significances of the order parameters n and p are reversed relative to Fig. 2.32c1, show situations closely resembling the experimentally found (up, down) S-type switching. Fig. 2.32d shows the effect of very small values of the recombination coefficients T_1^S and T_2^S in reducing the holding field. In the limit $T_1^S = T_2^S = 0$ (see above) $E_h = 0$, and solution (i) is never stable. Both solutions (ii) and (iii) can be stable from $X_1 = X_4 = 0$, and the switching behavior is lost.

The curve shown in Fig. 2.32e is the result of a trajectory which crosses $F_d = 0$ on both entering and leaving the overlap region (rather than entering via $F_d = 0$, and leaving via $F_c = 0$ as for Fig. 2.32a). One then has the curious phenomenon that solution (iii), although overlapping solution (ii), is actually completely inaccessible by electric-field manipulations alone, which provide merely the main curve shown in Fig. 2.32e running continuously through solutions (i), (ii), and (iv). However, it is still possible to use this system as a switch by injecting carriers and thus inducing transitions to solution (iii).

Threshold switching is a widely occurring phenomenon in amorphous and crystalline semiconductors, insulators, organic films, and liquids [2.58–84]. There are wide variations in the switching characteristics of these materials, and a number of thermal (e.g., [1.1, 2.67, 76, 82], see also the critical review [2.61]) and electronic [e.g., 2.51, 55, 56, 77] mechanisms have been proposed. The controversy in case of the amorphous chalcogenides has recently been resolved by the experimental demonstration that the threshold switching process is initiated electronically rather than thermally [2.79]. Within the context of the model discussed here, we adopt the view that a sudden rise in carrier concentration is responsible for switching, and that this is induced by passing between the stability regions of different solutions of the rate equations. Switching thus represents a g-r induced nonequilibrium phase transition [2.55, 56].

Our model is rather flexible in the sense that for a different choice of g-r parameter quite different switching characteristics can be reproduced, see Fig. 2.32. The numerical parameters of Table 2.4 typically correspond to n-type semiconductors near room temperatures, cf., the order of magnitude of B^S [2.2], T_1^S and T_2^S [2.57], X_1 and X_4 [2.16].

We note three points:

(a) Equation (2.2.13) holds for an n-type semiconductor with acceptor-like deep recombination centers (i.e., the neutral centers may capture an electron, upon which they become negatively charged). Without changing the rate equations (2.2.13), we can also apply the model to a p-type semiconductor of effective acceptor density $N_A - N_D > 0$ with donor-like recombination centers of concentration N_t (i.e., the neutral centers may lose an electron, upon which they become positively charged), simply by redefining $N_D^* := N_D + N_t - N_A, P_D := N_A - N_D$, provided that $N_D < N_A < N_D + N_t$ holds. Of course, both the n- and the p-type model include the special case that there are no extra compensating acceptors or donors besides the N_t recombination centers, i.e., $N_A = 0$ (n-type model) or $N_D = 0$ (p-type model).

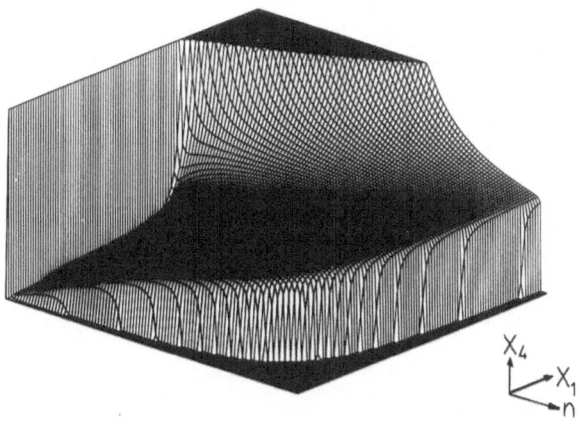

Fig. 2.33. Steady state electron concentration $n(X_1, X_4)$ as in Fig. 2.25, but with $T_1^S = T_2^S = 10^{-12}$ cm^3 s^{-1}, $X_1^S = X_2^S = 20$ s^{-1}, $Y^S = 4 \times 10^{16}$ cm^{-3} s^{-1}, 10^{-13} cm^3 s$^{-1} \leq X_1 \leq 0.7 \times 10^{-10}$ cm^3 s^{-1}

(b) By varying the rate of progress along the trajectory in the (X_1, X_4) plane with increasing field, the holding field E_h and the threshold field E_{th} may be scaled to higher values ($\approx 10^5$ V/cm in the chalcogenide glasses [2.74]) for less strongly field-dependent impact-ionization coefficients, or lower fields ($\approx 5 \times 10^2$ V/cm in strongly doped compensated Ge [2.5]) for more strongly field-dependent coefficients, without significant change in the other switching characteristics (switching and delay times, see next section).

(c) The model may also be elaborated by the addition of further g-r processes, such as the thermalization of carriers from traps. The presence of extra generation processes smoothes out the second-order phase transition from solution (i) to (ii) and from (iii) to (iv) in Figs. 2.31, 32, but the model retains its essential switching features for suitable numerical values. The electron concentration in state (ii) and the hole concentration in state (iii) are then no longer zero. The surface $n(X_1, X_4)$ is shown for this case in Fig. 2.33.

In our model, switching occurs between a p-conducting state (ii) and an n-conducting state (iii). In the standard case where the hole mobility is much smaller than the electron mobility (due to the larger hole effective mass), this results in a p-type OFF state (ii) and an n-type ON state (iii), as shown in Fig. 2.32a3, c2. Switching in chalcogenide glasses has indeed been found to occur between a p-type OFF state and an n-type ON state [2.51, 62]. Physically, in state (ii) most of the electrons are captured by traps, hence the majority of traps are occupied, and the steady state g-r traffic in state (ii) is predominantly between the trap level and the valence band. In state (iii) most of the holes are captured by traps, hence the majority of traps is not occupied, and the g-r traffic is predominantly between the trap level and the conduction band. In state (i) the g-r traffic between the traps and both the conduction band and the valence band is practically zero, while in the ambipolar state (iv) it is nonzero. The bistable states (ii) and (iii) are quite similar to the bistable states in double-injecting pin diodes [2.37] (Sect. 1.1.2), but the carriers in our model are not provided by injection of electrons and holes from the two contacts, but rather by electron and hole impact ionization in the bulk.

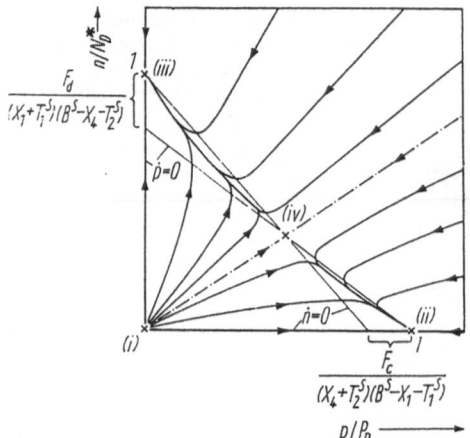

Fig. 2.34. Phase portrait of electron concentration versus hole concentration, corresponding to a typical point in the overlap region of Fig. 2.29. This point (X_1, X_4) corresponds to the electric field E_1 in the system before and after the switching pulse. Numerical parameters (in cm³ s⁻¹): $B^S = 10^{-10}$, $T_1^S = 10^{-12}$, $T_2^S = 10^{-14}$, $X_1 = 3 \times 10^{-11}$, $X_4 = 6 \times 10^{-11}$. Further $N_t = 1.5 N_D^*$

2.2.3 Dynamics of Threshold Switching Transitions

The qualitative description given in Sect. 2.2.2 in terms of the control parameter plane (Fig. 2.29) enables discussion of the steady state properties of switching, but it is not suitable for discussions of the dynamics of a switching event as characterized by the switching time τ_s, the delay time τ_d, and the decay time τ_{dec}. These can be described qualitatively in terms of phase portraits, which are obtained by plotting the integral curves of (2.2.13) in (n, p) phase space for fixed values of the control parameters [2.56]. That given in Fig. 2.34 corresponds to a typical point in the overlap region of Fig. 2.29. Here the steady state solutions (ii) and (iii) provide stable nodes, the solution (i) is an unstable node, and solution (iv) is a saddle point (i.e., unstable) as indicated by the flow-lines in Fig. 2.34. The region of phase space attracted towards solution (ii) is divided from the region attracted towards solution (iii) by a separatrix (dash-dotted) running through the unstable node and the saddle.

The dynamics of switching in our model will be discussed by supposing that the initial field $E_1 < E_{th}$ is raised abruptly to $E_2 > E_{th}$ for some time τ, and then restored to E_1, and that during this process the stable solutions are in turn (ii) and (iii), then (iii), then (ii) and (iii). This case corresponds to the trajectory $t(E)$ in Fig. 2.29. When (ii) and (iii) are stable, a regime such as indicated in Fig. 2.34 holds while another phase portrait (Fig. 2.35) would be applicable when only solution (iii) is stable. The numerical choices incorporated in Fig. 2.32a, c2, d would be appropriate for such switching. The dynamics which decide the final steady state solution of the system are as follows. The notation "prime" for evaluation at field E_1 and "double prime" for evaluation at E_2 is adopted.

Assume that the system is initially in the stable steady state (ii) given by $n_2 = n_2' = 0$; $p_2 = p_2'$ at the field E_1 (Fig. 2.34). When the field is raised to E_2, this is no longer a stable state of the reaction equations (2.2.13). Therefore, the system moves along a flow-line of Fig. 2.35 in a manner described by the time-dependent equations (2.2.13). If, when the field is restored to E_1, the separatrix in Fig. 2.34 has been crossed, the new steady state solution (iii) will be reached along a flow-line of

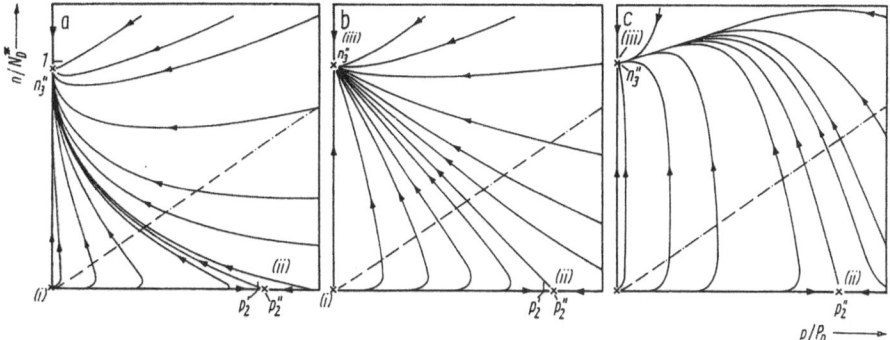

Fig. 2.35a–c. Phase portraits corresponding to typical points outside the overlap region of Fig. 2.29 but in the stability region of solution (iii). These points correspond to the electric field E_2 during the switching pulse. The chain line represents the separatrix and p_2' the steady state (ii) of Fig. 2.34, at field E_1. The values of T_1^s, T_2^s, B^s, and N_t are those of Fig. 2.32a. The values of X_1 and X_4 are (in cm^3 s^{-1}): (**a**) $X_1 = 4.86 \times 10^{-11}$, $X_4 = 9.18 \times 10^{-12}$ (corresponding to $E = 1500$ V/cm); (**b**) $X_1 = 8.31 \times 10^{-11}$, $X_4 = 1.67 \times 10^{-11}$ (corresponding to $E = 1563$ V/cm); (**c**) $X_1 = 2 \times 10^{-10}$, $X_4 = 4.54 \times 10^{-11}$ (corresponding to $E = 1679$ V/cm $= 15\%$ overvoltage)

Fig. 2.34. If the separatrix in Fig. 2.34 has not been crossed the system will revert to solution (ii). Herein lies the origin of the delay time τ_d which, in our model, is the minimum length of time τ required for the system to cross the separatrix.

Another feature of this model is that, if solution (ii) at field E_1 lies close to solution (ii) at field E_2, as it will be for small overvoltages, the initial movement of the system is very sensitive to fluctuations in the system. If there were no fluctuations, the system would follow the flow-line from p_2' to the saddle p_2'' in Fig. 2.35, and remain there, since this is a stationary, though unstable, state. Of course, an infinitesimally small fluctuation in concentration or in the recombination-generation coefficients will force the system away from the saddle (ii) towards the stable node (iii). Since the flow-lines are traversed more slowly the closer they are to a steady state (stable or unstable), the delay time depends upon the flow-line taken and thus upon fluctuations. A further consequence is that the system will spend most of its time "near" solutions (ii) and (iii) in Fig. 2.35 where n and p are small. Thus the delay time may be described as the amount of time spent by the system near state (ii), just as it arises experimentally [2.51, 60]. The origin of the statistical regime for τ_d found in [2.51] for small overvoltages is also explained by this mechanism.

We will now mention three additional points which may be helpful.

(a) Strictly, each phase portrait applies locally to a small region of the material, but for a uniform material in which the field is reasonably constant one can use the concept of an overvoltage as has been done above. This is $(E_2 - E_{th})L$ where L is the length of the sample. However, the semiconductor may also develop a transverse spatial inhomogeneity ("current filament"). Switching induced by transverse spatially localized fluctuations and by the nucleation of current filaments will be discussed in Chap. 5.

(b) The switching time τ_s, in our model, is the time taken for the system to move from solution (ii) to solution (iii) at a given overvoltage minus τ_d.

(c) The decay time τ_{dec} describes the return step from solution (iii) to solution (ii) at E_{h}, when the field is lowered for a time τ from $E_1 > E_{\text{h}}$ to $E_2 < E_{\text{h}}$, in a completely analogous manner to τ_{s}.

Approximate Expressions for τ_{s} and τ_{d}

The general problem of calculating switching times from (2.2.13) according to the interpretation given above can only be solved numerically. However, it is possible to generate some simple expressions for τ_{s} and τ_{d} on the basis of a linearized system of equations for which time-dependent solutions for $n(t)$ and $p(t)$ may be obtained explicitly. Linearizing (2.2.13) around the steady state one obtains for the deviations δn and δp from the steady state

$$\begin{pmatrix} \delta \dot{n} \\ \delta \dot{p} \end{pmatrix} = A \begin{pmatrix} \delta n \\ \delta p \end{pmatrix} , \tag{2.2.17}$$

where, near solution (ii),

$$A = A_2 := \begin{pmatrix} -F_{\text{c}}/(X_4 + T_2^{\text{S}}) & 0 \\ -p_2(B^{\text{S}} - X_4 - T_2^{\text{S}}) & -F_{\text{b}} \end{pmatrix} , \tag{2.2.18}$$

and, near solution (iii),

$$A = A_3 := \begin{pmatrix} -F_{\text{a}} & -n_3(B^{\text{S}} - X_1 - T_1^{\text{S}}) \\ 0 & -F_{\text{d}}/(X_1 + T_1^{\text{S}}) \end{pmatrix} . \tag{2.2.19}$$

This holds for all values of the electric field.

Consider a switch from solution (ii) to solution (iii) at E_{th} as in Fig. 2.32a. Assume that at $t = 0$ the system is in a field E_1 in solution (ii). The eigenvalues of A_2 in (2.2.18) are

$$\lambda_1 = -F_{\text{c}}/(X_4 + T_2^{\text{S}}) \qquad \text{and} \qquad \lambda_2 = -F_{\text{b}}$$

with corresponding eigenvectors $(1, -\xi)$ and $(0, 1)$, respectively, where

$$\xi := \frac{(B^{\text{S}} - X_4 - T_2^{\text{S}})F_{\text{b}}}{(X_4 + T_2^{\text{S}})F_{\text{b}} - F_{\text{c}}} .$$

A general solution of (2.2.17) near solution (ii) is therefore

$$\delta n(t) = a e^{\lambda_1 t} , \qquad \delta p(t) = -a\xi e^{\lambda_1 t} + b e^{\lambda_2 t} \tag{2.2.20}$$

where a and b are constants determined by the initial conditions. Since both F_{c}' and $F_{\text{b}}' > 0$, λ_1' and $\lambda_2' < 0$, and (2.2.20) confirms that solution (ii) is stable at field E_1. For $t > 0$ the field is increased to E_2. Since $F_{\text{c}}'' < 0$ and $F_{\text{b}}'' > 0$, $\lambda_1'' > 0$ and $\lambda_2'' < 0$, so that solution (ii) becomes a saddle.

As a first approximation it is assumed that the switching is influenced by fluctuations in the carrier concentrations of magnitude δn_0 and δp_0 arising in a negligibly small time compared to τ_{d}. With initial condition $\delta n(0) = \delta n_0$, and

$\delta p(0) = \Delta p_2 := p_2' - p_2'' + \delta p_0$ in (2.2.20), this yields

$$n(t) = \delta n_0 e^{\lambda_1'' t} \, , \qquad p(t) = p_2'' - \delta n_0 \zeta'' e^{\lambda_1'' t} + (\Delta p_2 + \delta n_0 \zeta'') e^{\lambda_2'' t} \, . \qquad (2.2.21)$$

Equation (2.2.21) represents a solution which moves away from $n = \delta n_0$, $p = p_2' + \delta p_0$ at $t = 0$. For small overvoltage F_c'' is small and $|\lambda_1''| \ll |\lambda_2''|$; therefore, the term dependent upon λ_2'' in (2.2.21) can be neglected in the calculation of τ_d since it represents fast decay to p_2'' at constant n. Approximately, therefore, the system moves away from the saddle (ii) along the line $n = -(p - p_2'')/\zeta''$. These results are assumed to hold for a large enough region extending from solution (ii) in a phase portrait appropriate to field E_2 (Fig. 2.35) to include the separatrix of Fig. 2.34 (which is appropriate to the field E_1).

The separatrix may be roughly approximated by a line $n = (F_c'/F_d')p$ through the unstable node and the saddle in Fig. 2.34. Since the system will move very rapidly through regions of phase space with large $\dot{n}$ and $\dot{p}$ as it crosses the separatrix it is to be anticipated that τ_d will be insensitive to errors introduced by such a linear approximation. The closer E_1 is to E_{th} the better this approximation is. As E_1 approaches E_{th} the stable node (ii) and the saddle (iv) merge. In this case the separatrix coincides with the p-axis.

The time taken for the system to reach the separatrix is therefore

$$\tau_d = \frac{1}{\lambda_1''} \ln \left| \frac{n_s}{\delta n_0} \right| \, , \qquad \text{where}$$

$$n_s := p_2'' F_c' / (F_c' \zeta'' + F_d') \, . \qquad (2.2.22)$$

For a constant degree of compensation N_t/N_D^*, $\lambda_1'' \sim N_D^*$; and so approximately $\tau_d \sim N_D^{*-1}$. Further, the smaller the overvoltage the closer F_c'' and λ_1'' lie to zero, thus leading to larger values of τ_d. An estimate of the explicit dependence of τ_d upon the overvoltage $L\Delta E$ with $\Delta E := E - E_{th}$ can be obtained for T_1^S, $T_2^S \ll X_1$, X_4 and $\Delta E \ll E_{th}$. From (2.2.14) it then follows with the Shockley formula (2.1.45) $X_1(E) = X_1^0 \exp(-E_i/E) \approx X_1(E_{th})\exp(\Delta E \cdot E_i/E_{th}^2)$:

$$\lambda_1'' \approx X_1 N_t - B^S P_D \approx B^S P_D[\exp(\Delta E \cdot E_i/E_{th}^2) - 1] \, ,$$

where $\lambda_1''(E_{th}) = 0$ has been used. Hence for $E_{th}/E_i \ll \Delta E/E_{th} \ll 1$:

$$\tau_d \sim \exp(-\Delta E \cdot E_i/E_{th}^2)$$

holds, which explains the exponential dependence of τ_d upon the overvoltage measured, e.g., in amorphous chalcogenides [2.51].

The switching time at some field E_2 may be estimated in a similar way. It is assumed that the pulse length $\tau > \tau_s + \tau_d$ in order that the switch may be completed at field E_2 (if $\tau_d + \tau_s > \tau > \tau_d$ the switch would be completed at field E_1). The eigenvalues of A_3 in (2.2.19) at field E_2 are

$$\lambda_m'' = -F_d''/(X_1'' + T_1^S) < 0 \quad \text{and} \quad \lambda_M'' = -F_a'' < 0 \, . \qquad (2.2.23)$$

The corresponding eigenvectors are $(-\eta, 1)$ and $(1, 0)$, respectively, where

$$\eta := \frac{(B^S - X_1'' - T_1^S)F_a''}{(X_1'' + T_1^S)F_a'' - F_d''} \ . \tag{2.2.24}$$

These expressions lead to a solution of the linearized rate equations (2.2.17) near solution (iii), analogous to (2.2.20) which is assumed to be adequate to describe τ_s. Since solution (iii) is a stable node, all trajectories ending at solution (iii), save two, are tangent to the eigenvector corresponding to the eigenvalue with the *smaller* magnitude. (Fig. 1.11c) This observation allows three cases to be distinguished corresponding to Fig. 2.35a, b, and c respectively.

(a) $|\lambda_M''| < |\lambda_m''|$

If $F_d'' > F_a''(X_1'' + T_1^S)$ in (2.2.23), $|\lambda_M''| < |\lambda_m''|$. For $X_1'' \gg T_1^S$, as in Table 2.4, this corresponds to $X_4'' < (B^S - X_1'')N_D^*/N_t$ which is satisfied for small X_4'', and small overvoltage. A typical phase portrait is shown in Fig. 2.35a. The switching time τ_s may be approximated by the time the system takes to pass from n_s at the separatrix to within δn_∞ and δp_∞ of the final steady state (iii). The trajectory from near solution (ii) to solution (iii) becomes tangent to the n-axis near (iii), see Fig. 2.35a, and the system moves slowest there, so that the largest contribution to τ_s results from motion along the n-axis. That part of the solution depending upon λ_m'' may therefore be neglected, then

$$\tau_s \approx \frac{1}{|\lambda_M''|} \ln \left| \frac{n_3'' - n_s}{\delta n_\infty} \right| \ . \tag{2.2.25}$$

where n_s is defined in (2.2.22). With increasing overvoltage X_1'', and thus $|\lambda_M''|$, increase.

(b) $|\lambda_m''| < |\lambda_M''|$; $\eta > 0$

With increasing overvoltage, F_d'' decreases and F_a'' increases so that for moderate overvoltages (Fig. 2.35b) $X_1'' < B^S - T_1^S$ and $F_d'' < F_a''(X_1'' + T_1^S)$, so that $|\lambda_m''| < |\lambda_M''|$ and $\eta > 0$. If the field dependence of X_1 and X_4 were such that at E_{th}, $X_4 > (B^S - X_1)N_D^*/N_t$ (assuming $X_1 \gg T_1^S$), then case (b) would apply to small overvoltages also.

The main contribution to τ_s is now via the slowly decaying part of the solution along the eigenvector $(-\eta, 1)$ corresponding to λ_m. In analogy with case (a) the λ_M solution can now be neglected leading to

$$\tau_s \approx \frac{1}{|\lambda_m''|} \ln \left| \frac{n_3'' - n_s}{\delta n_\infty} \right| \ . \tag{2.2.26}$$

This approximation is best for $D = 0$ when $\eta = \xi^{-1} = n_3''/p_2''$. In this case the trajectory, from (ii) to (iii), and the eigenvectors $(-\eta, 1)$ and $(1, \xi)$ are collinear.

(c) $|\lambda_m''| < |\lambda_M''|$; $\eta < 0$

This holds for $X_1'' > B^S - T_1^S$, i.e. for large overvoltage (Fig. 2.35c). The trajectory

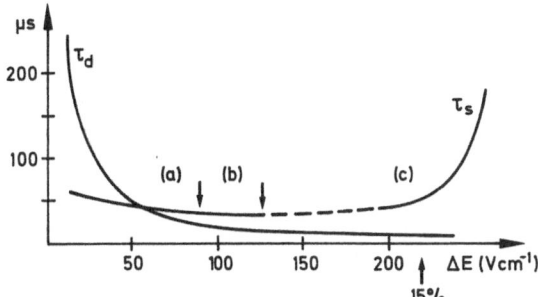

Fig. 2.36. Delay time τ_d and switching time τ_s versus overvoltage $\Delta E = E - E_{th}$, calculated with the approximate analytical formulas (2.2.22) and (2.2.25–27). The regimes of validity of (2.2.25), (2.2.26), (2.2.27) are symbolically indicated by (a), (b), and (c), respectively; the dashed line is an interpolation. The point corresponding to 15% overvoltage is also indicated. The numerical parameters are as in Table 2.4a, with $E_1 = 1400$ Vcm^{-1}, $E_{th} = 1458$ Vcm^{-1}, $\delta n_0 = 10^{-3} N_D^*$, $\delta n_\infty = 10^{-2} N_D^*$

leaving (ii) approaches (iii) along the eigenvector $(|\eta|, 1)$. This corresponds to an overshoot in n as can be seen in Fig. 2.35c. The linear approximation is not very good in this case, but a crude estimate for τ_s may be obtained by replacing $(n_3'' - n_s)$ in (2.2.26) by the quantity

$$n_0 := (p_2'' - n_3'' \xi)|\eta|/(1 + \xi|\eta|) ,$$

which is an approximation to the overshoot in n. It corresponds to the intersection of the lines defined by the eigenvectors $(|\eta|, 1)$ and $(1, -\xi)$. Now

$$\tau_s \approx \frac{1}{|\lambda_m''|} \ln \left| \frac{n_0}{\delta n_\infty} \right| . \tag{2.2.27}$$

With increasing overvoltage the system moves closer to the line $F_d'' = 0$ in Fig. 2.29, hence $|\lambda_m''|$ becomes small, resulting in an increasing "settling time" for τ_s.

The approximate analytical expressions (2.2.22, 25–27) for τ_d and τ_s are plotted versus the overvoltage in Fig. 2.36 for the numerical parameters of Fig. 2.32 a (see Table 2.4) with $\delta n_0 = 10^{-3} N_D^*$, $\delta n_\infty = 10^{-2} N_D^*$, $N_D^* = 2 \times 10^{15}$ cm^{-3}. They are in good agreement with the numerical computations [2.55]. A comparison with typical experimental values is given in Table 2.5. Note that the switching and delay times τ_s, τ_d and the threshold and holding fields E_{th}, E_h can be scaled independently in the theoretical calculations by varying N_D^* or E_i, respectively. For instance, $N_D^* = 10^{19}$ cm^{-3} would yield $\tau_s \approx 10^{-9}$ s, a value typical of chalcogenide glasses.

The treatment of fluctuations given here is clearly inadequate; however, since τ_d is fairly insensitive to δn_0 in a reasonable range of statistical fluctuations, the above expression should still provide useful estimates. The quantity δn_∞ has a direct interpretation in experiment where, as a result of fluctuations, the theoretical steady state may never be exactly attained, and it is necessary to fix some more or less arbitrary limit such as the 10% and 90% points used in [2.85]. A more sophisticated treatment should include the delay-time statistics. Such a treatment could be based upon a Langevin approach by adding stochastic forces to the deterministic equa-

Table 2.5. Comparison of experimental and theoretical switching parameters

Reference		τ_s [s]	τ_d [s]	E_{th} [Vcm^{-1}]	E_h [Vcm^{-1}]	Remarks
Experimental	[2.5]		$10^{-3} - 10^{-4}$ (at 15% overvoltage)	300		Strongly-doped compensated Ge (4 K)
	[2.85]	$10^{-5} - 10^{-4}$	10^{-5}	$10^4 - 10^5$	5×10^3 $-5 \times 10^{+4}$	α-Si (77 and 300 K)
	[2.66]		$10^{-7} - 10^{-6}$	4×10^5		α-Se (probably room temperature)
	[2.51]	$\leqslant 10^{-10}$	$\leqslant 10^{-5}$	$\sim 10^5$		Chalcogenides (room temp.)
Numerically calculated	[2.55]	4×10^{-5} (at 15% overvoltage)	1.4×10^{-5}	1458	1120	Parameters as in Table 2.4a

tions (2.2.13), as it has, for instance, been done for an analogous problem in superfluorescence [2.86, 87]. The probability distribution of delay times has been measured as a function of the overvoltage for amorphous Bor(α-B) at 77 K [2.85], and its variance has been found to increase substantially for small overvoltages (in the so-called statistical regime [2.51]).

2.2.4 Auger Recombination Induced Tristability

We shall now show that additional overlaps of the stability regimes of solutions (ii) and (iv), and of (iii) and (iv), and an overlap of all three stable solutions (ii), (iii), and (iv) can be produced if Auger recombination (T_1, T_4) is added to the processes of Fig. 2.24 [2.88], as sketched in Fig. 2.37. Auger recombination becomes relevant for high carrier densities since its rate is cubic in the concentrations. The rate equations are

$$\dot{n} = [X_1 N_D^* - T_1^S P_D - (X_1 + T_1^S + T_1 P_D)n - (B^S - X_1 - T_1^S)p$$
$$+ T_1 np - T_1 n^2]n$$
$$\dot{p} = [X_4 P_D - T_2^S N_D^* - (X_4 + T_2^S + T_4 N_D^*)p - (B^S - X_4 - T_2^S)n$$
$$+ T_4 np - T_4 p^2]p \; . \tag{2.2.28}$$

The steady state solutions are qualitatively similar to those given in (2.2.14) for the model of Fig. 2.24. There is a nonconducting state (i) with $n = p = 0$, a p-conducting state (ii) with $n = 0$, $p > 0$, and an n-conducting state (iii) with $n > 0$, $p = 0$. In contrast to (2.2.14) there may, however, be up to three ambipolar states with $n > 0$, $p > 0$. If so, one of these three is stable (iv), and the other two are unstable. The stability regions of these four solutions are sketched schematically in the (X_1, X_4) control parameter space in Fig. 2.38. A comparison with Fig. 2.29 reveals that the critical lines of second-order phase transitions l_1, l_2, l_3, l_4 and the overlap region

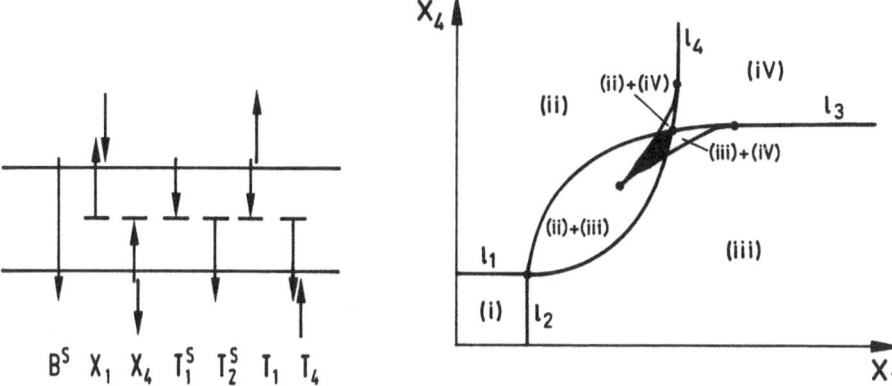

Fig. 2.37. Two-carrier model for tristability, involving Auger recombination (T_1, T_4) in addition to the g-r processes of Fig. 2.24

Fig. 2.38. Schematic plot of the stability regions of the steady state solutions (i)–(iv) in the (X_1, X_4) control parameter plane for the g-r mechanism of Fig. 2.37. The tristability region of solutions (ii), (iii), and (iv) is shaded. l_1, l_2, l_3, l_4 are critical lines of second-order phase transitions

of solutions (ii) and (iii) are still present. Additionally, two new bifurcation lines occur, forming the boundaries of the bistability domains of solutions (ii) and (iv), and of (iii) and (iv), and of the tristability domain (shaded) of (ii), (iii), and (iv). The steady state surface $n(X_1, X_4)$ is shown in Fig. 2.39. Note the ripple in the curved surface of solution (iv), which is necessary for tristability.

If the generation processes Y^S or X_1^S, X_2^S are added to the processes of Fig. 2.37, the second-order phase transitions are again smoothed out, and the critical lines l_1, l_2, l_3, l_4 disappear from the control parameter space (Fig. 2.40), but the domains of bistability (hatched) and tristability (shaded) are still present. There can be up to five physical solutions, three of which can be stable. The bifurcation set, i.e., the bifurcation lines separating the different regimes of bi- and tristability is strongly reminiscent of the "butterfly catastrophe", which is one of the universal unfoldings of gradient systems well known in catastrophe theory [1.19–22]. At the critical points C_0, C_2, C_3 one unstable and two stable steady states coalesce: (ii) and (iii), (iii) and (iv), (ii) and (iv), respectively. C_1 might be called an "anticritical point", since it is associated with the coalescence of one stable and two unstable steady states. The bifurcation which occurs at C_1 is a subcritical pitchfork bifurcation, as shown in Fig. 1.12 (A4).

The steady state surface $n(X_1, X_4)$ corresponding to Fig. 2.40 is shown in Fig. 2.41. Again the sharp edges of the second-order transitions are rounded off. The rippled surface of tristability can be clearly seen. Depending upon the path in the control parameter plane which a particular field dependence of X_1 and X_4 picks, a variety of bifurcations can occur. An example of such an $n(E)$ relation is plotted in Fig. 2.42. It consists of two S-shaped sections with an overlapping range of tristability; it may lead to double SNDC or more complicated current-field characteristics. The spatially inhomogeneous, filamentary steady states which can result from such a configuration will be discussed in Sect. 4.5.

Fig. 2.39

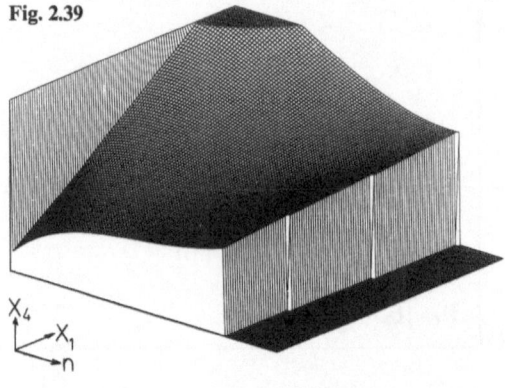

Fig. 2.40

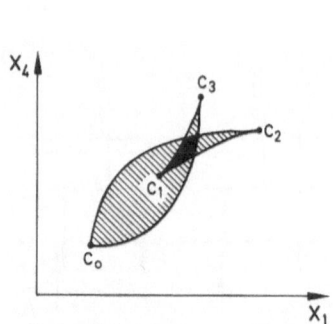

Fig. 2.41

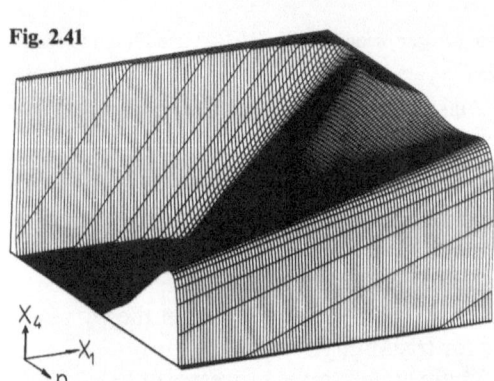

Fig. 2.39. Steady state electron concentration $n(X_1, X_4)$, as in Fig. 2.25, but with $T_1 = T_4 = 1.5 \times 10^{-26}$ cm^6 s^{-1}, and 2.8×10^{-11} cm^3 s^{-1} $\leq X_1 \leq 3 \times 10^{-11}$ cm^3 s^{-1}, 2.6×10^{-11} cm^3 s^{-1} $\leq X_4 \leq 3.1 \times 10^{-11}$ cm^3 s^{-1}, $0 \leq n \leq 0.5 N_D^*$

Fig. 2.40. Schematic plot of the control parameter plane for the g-r mechanism of Fig. 2.37 with additional generation processes Y^s, X_1^s, X_2^s. The bistability regions are hatched, the tristability region is shaded. C_0, C_2, C_3 are critical points, C_1 is an anticritical point

Fig. 2.41. Steady state electron concentration $n(X_1, X_4)$, as in Fig. 2.25, but with $T_1 = T_4 = 1.5 \times 10^{-26}$ cm^6 s^{-1}, $Y^s = 4 \times 10^{16}$ cm^{-3} s^{-1}, and 2.8×10^{-11} cm^3 s^{-1} $\leq X_1, X_4 \leq 2.85 \times 10^{-11}$ cm^3 s^{-1}, $0.001 N_D^* \leq n \leq 0.4 N_D^*$

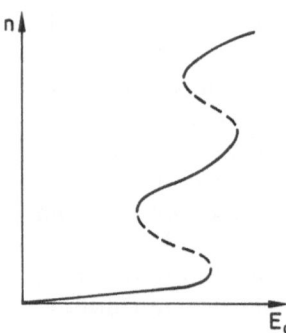

Fig. 2.42. Typical plot of the steady state electron concentration n versus the electric field E_0 for a trajectory crossing the tristability region in Fig. 2.40 (schematic)

Experimental evidence of tristability between n-type, p-type, and ambipolar steady states as predicted by the present model is so far lacking. However, intricate current-field characteristics with several sections and combinations of SNDC and NNDC have been reported in the literature [1.40–43], in particular in the context of many-valley semiconductors, see e.g., [Ref. 1.33, p. 147].

To summarize Sect. 2.2, we note that coupled band-to-trap impact ionization of electrons and holes is the basis of a large, rather flexible class of models giving bistability, tristability, and SNDC, which can be applied to explain the statics and dynamics of threshold switching in a variety of materials. The novel viewpoint of threshold switching as a g-r induced first-order nonequilibrium phase transition is elaborated by discussing control parameter planes, phase portraits, and state diagrams. Some, but not all, of the material presented in Sects 2.1.1, 2.2.1, and 2.2.2 has been reviewed in [Ref. 2.19, part 2 §8.6] and in [1.114].

2.3 Excitonic Models

In this section we shall discuss the possibility of bistability induced by g-r processes involving excitons. It is well known that electrons and holes can become bound to each other by virtue of Coulomb interaction and form excitons [2.88–95]. The excitons may move freely through the crystal or become bound at impurity centers. In Table 2.6 a variety of g-r processes involving excitons are listed [2.96–106]; there are other processes involving bi-excitons and more complicated excitonic molecules [2.89].

Table 2.6. Excitonic generation-recombination processes

Process		Rate	Reference
$\gamma \rightarrow X$	Optical generation of excitons	E_1	
$e + h \rightarrow X$	Electron-hole pairing	$E_2 np$	[2.96]
$\gamma + X \rightarrow 2X$	Stimulated exciton generation	$E_3 x$	[2.108]
$e + h + X \rightarrow 2X$	Stimulated exciton generation	$E_4 npx$	[2.49]
$X \rightarrow \gamma$	Radiative recombination	$F_1 x$	[2.90, 96]
$X \rightarrow e + h$	Exciton ionization	$F_2 x$	[2.96, 100]
$2X \rightarrow X + \gamma$		$F_3 x^2$	
$2X \rightarrow X + e + h$		$F_4 x^2$	[2.89]
$2X \rightarrow e + h$	Auger recombination of excitons	$F_5 x^2$	
$X + e \rightarrow e + \gamma$		$F_6 nx$	[2.97]
$X + h \rightarrow h + \gamma$		$F_7 px$	[2.90]
$X + e \rightarrow 2e + h$	Impact ionization of excitons	$F_8 nx$	[2.99, 104–106]
$X + h \rightarrow e + 2h$	Impact ionization of excitons	$F_9 px$	[2.99, 104–106]
$X + N_0 \rightarrow N_x$	Exciton capture at neutral trap		
$N_x \rightarrow N_0 + \gamma$	Recombination of bound excitons		[2.101, 103]
$h + N_0 \rightarrow N_+$	Formation of bound exciton		
$e + N_+ \rightarrow N_x$	Formation of bound exciton		
$N_x \rightarrow N_+ + e$	Auger recombination of bound exciton		
$N_x \rightarrow N_- + h$	Auger recombination of bound exciton		[2.98, 102]
$N_x \rightarrow N_0 + X$	Detrapping of bound exciton		
$N_x \rightarrow N_0 + e + h$	Ionisation of bound exciton		

γ, X, e, h, N_0, N_+, N_-, N_x denote photons, excitons, electrons, holes, neutral, positively, and negatively charged impurity centers, and excitons bound at impurities, respectively.

2.3.1 Stimulated Exciton Creation

As we have seen earlier on, autocatalytic processes are a necessary ingredient for g-r induced bistability. Since excitons are of boson character at not too high densities, stimulated exciton creation processes should exist in analogy with the stimulated emission of photons, which is a consequence of the boson character of photons. The stimulated exciton generation processes $\gamma + X \to 2X$ and $e + h + X \to 2X$ listed in Table 2.6 represent autocatalytic reactions, where the exciton (X) stimulates its own generation from a photon (γ) or a free electron-hole pair ($e + h$). Although experimental evidence of these processes is so far lacking, the reverse processes – Auger recombination of excitons – have been observed [2.89].

Generation-recombination models based on the process $e + h + X \to 2X$ have been shown to exhibit second-order phase transitions [2.49], and limit-cycle oscillations [2.107] (Sect. 6.2.3), but no bistability was found. Here we shall use the autocatalytic process $\gamma + X \to 2X$, in addition to $e + h + X \to 2X$ in a g-r model which yields bistability and first-order nonequilibrium phase transitions [2.108]. We consider an intrinsic semiconductor ($n = p$), subject to the following g-r processes (see Table 2.6):

(i) $X \to e + h \to \gamma$

(ii) $X + \gamma' \rightleftharpoons 2X$

(iii) $X + e + h \to 2X$.

Other g-r processes are neglected for simplicity. The rate equations for the density of free carriers (n) and excitons (x) are:

$$\dot{n} = F_2 x - B^S n^2 - E_4 n^2 x \tag{2.3.1a}$$

$$\dot{x} = (E_3 - F_2)x + E_4 n^2 x - F_3 x^2 . \tag{2.3.1b}$$

In the steady state it follows from (2.3.1a):

$$n^2 = \frac{F_2 x}{B^S + E_4 x} . \tag{2.3.2}$$

Substitution into (2.3.1b) yields the following steady states:

$$x_1 = n_1 = 0$$

$$x_{2,3} = \frac{E_3 E_4 - F_3 B^S}{2F_3 E_4}\left[1 \pm \left(1 - \frac{4F_3 E_4(F_2 - E_3)B^S}{(E_3 E_4 - F_3 B^S)^2}\right)^{1/2}\right] . \tag{2.3.3}$$

The g-r coefficients E_3 and F_2 are natural control parameters, since the rate constant of stimulated exciton generation E_3 is proportional to the optical irradiation intensity at the exciton resonance frequency, and the thermal exciton ionization rate constant F_2 can be controlled by varying the temperature. The (E_3, F_2) control

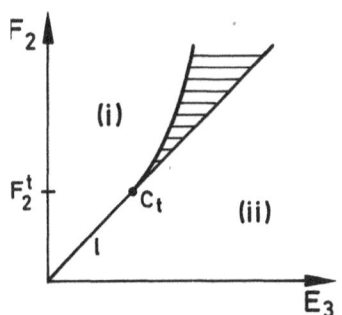

Fig. 2.43. Control parameter plane of the stimulated exciton generation coefficient E_3 and the exciton ionization coefficient F_2 (schematic). The hatched area denotes the bistability region; l and C_t are a critical line and a tricritical point, respectively

Fig. 2.44a–c. Qualitative behavior of the steady state concentrations n and x versus the control parameter E_3 for three values of the control parameter F_2 in Fig. 2.43 (schematic): (a) $F_2 > F_2^t$, (b) $F_2 = F_2^t$, (c) $F_2 < F_2^t$

parameter plane is shown schematically in Fig. 2.43. The steady state $x_1 = n_1 = 0$ (i) is stable for $F_2 > E_3$, and unstable otherwise. For $F_2 < E_3$ one of the solutions $x_{2,3}$ is positive and stable (ii), while the other is negative, i.e., nonphysical. For

$$E_3 < F_2 < (E_3 E_4 + F_3 B^S)^2/(4F_3 B^S E_4) \tag{2.3.4}$$

and

$$E_3 > F_3 B^S/E_4$$

(hatched area in Fig. 2.43) both x_2 and x_3 are positive, x_3 [+ sign in (2.3.3)] being stable, x_2 unstable, such that bistability between x_1 and x_3 occurs. The critical line l is given by $F_2 = E_3 < F_3 B^S/E_4$ and is associated with second-order phase transitions from (i) to (ii); C_t is a tricritical point. The qualitative form of the exciton and electron densities versus E_3 ($\sim$ optical intensity) is shown in Fig. 2.44 for three values of F_2 ($\sim$ temperature), corresponding to sections of the (F_2, E_3) plane across the bistability region (a), the tricritical point (b), and the critical line (c).

The nonequilibrium phase transition achieved from a dilute phase ($n \approx 0$) to a dense phase ($n > 0$) bears some similarity to the electron-hole (e-h) condensation of nonequilibrium carriers that has been observed in indirect and, more recently, direct semiconductors [2.91–93], although this phenomenon is based upon the coherence and correlations of the carriers in a dense system, and therefore requires a more sophisticated microscopic description in terms of correlation and exchange energies. Hydrodynamic theories and Becker-Döring type nucleation theories of this nonequilibrium phase transition from a dilute exciton or e-h gas to a dense metallic e-h liquid have also been developed [2.109]. At low densities and low temperatures the excitons form an ideal Bose gas while at higher densities a degenerate e-h plasma is created which performs a first-order phase transition to a degenerate e-h Fermi liquid via the nucleation of e-h droplets. The phase diagram in the temperature-density plane is similar to that of gas-liquid equilibrium phase transitions, containing a coexistence region and a critical point, but additional phase transition lines ("Mott transition") are also present [2.93]. Still, one might speculate that it may be useful to study simple rate equation models of the type (2.3.1) for the onset of such cooperative phenomena in excitonic systems, just as in laser physics the simple classical rate-equation approach gives an indication of the intensities and the threshold of the laser phase transition though it cannot describe the coherence properties of the laser radiation.

2.3.2 Bound-Exciton Recombination and Optical Bistability

In Sect. 2.2.1 we discussed mechanisms based on band-to-band impact ionization which give rise to second-order phase transitions or NDC, but not to bistability of homogeneous steady states. We will now show that bistability can be induced by band-to-band impact ionization if an additional recombination channel of the free carriers via bound excitons is taken into account [2.108]. The following processes are considered, see Table 2.6 (the rate constants are given in brackets):

(i) $\gamma \rightleftharpoons e + h$ (optical generation and recombination of electron-hole pairs; Y^S, B^S)

(ii) $e \rightarrow 2e + h$ (impact ionization of electrons; Y_1)

(iii) $e + h \rightarrow X \rightarrow \gamma'$ (recombination via excitons; E_2, F_1)

(iv) $X + N_0 \rightarrow N_x \rightarrow N_0 + \gamma''$ (bound-exciton recombination; k'; k'')

(v) $2X \rightarrow e + h$ (Auger recombination of excitons; F_5) .

For an intrinsic semiconductor ($n = p$) with N_t deep traps the rate equations for the densities of free carriers n, free excitons x, and bound excitons n_x are:

$$\dot{n} = Y^S + Y_1 n - (B^S + E_2)n^2 + F_5 x^2 \tag{2.3.5a}$$

$$\dot{x} = E_2 n^2 - 2F_5 x^2 - F_1 x - k'x(N_t - n_x) \tag{2.3.5b}$$

$$\dot{n}_x = k'x(N_t - n_x) - k''n_x . \tag{2.3.5c}$$

In the steady state, (2.3.5c) gives a Michaelis-Menten law of saturable excitonic recombination:

$$n_x = \frac{qx}{1 + qx} N_t , \qquad k'x(N_t - n_x) = \frac{kx}{1 + qx} \tag{2.3.6}$$

with $k := k'N_t$, $q := k'/k''$. The steady state concentrations n, x are given by the intersection of the two null-isoclines $\dot{n} = 0$ (2.3.5a) and $\dot{x} = 0$ (2.3.5b):

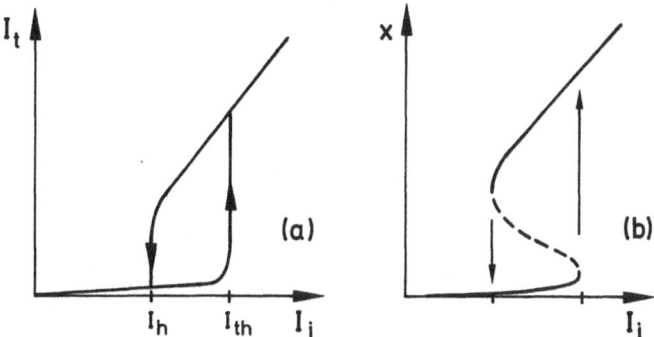

Fig. 2.45. (a) Transmitted light intensity I_t versus incident light intensity I_i in case of optical bistability (schematic). I_{th} and I_h denote the threshold and the holding intensity, respectively. (b) Steady state exciton density ($\sim I_{out}$) versus input intensity I_i for the bistable excitonic g-r mechanism (2.3.5) (schematic)

$$n = [2(B^S + E_2)]^{-1}\{Y_1 + [Y_1^2 + 4(B^S + E_2)(Y^S + F_5 x^2)]^{1/2}\} \qquad (2.3.7a)$$

$$n = E_2^{-1/2}\left(F_1 x + 2F_5 x^2 + \frac{kx}{1 + qx}\right)^{1/2} \qquad (2.3.7b)$$

which yields an algebraic equation for x. The possibility of bistability arises when the curve (2.3.7b) has a point of inflection,

$$8F_5 k > F_1^2 q , \qquad (2.3.8)$$

and when the other g-r coefficients are chosen appropriately.

Let us now speculate about a potential application of the model (2.3.5) to the problem of optical bistability. A nonlinear absorptive or dispersive medium embedded in an optical cavity may exhibit bistability and hysteresis of the transmitted light intensity I_t as a function of the incident light intensity I_i, as sketched in Fig. 2.45a [2.110–113]. Optical bistability is an example of a first-order nonequilibrium phase transition, and is of great interest for possible device applications, such as optical switches, optical transistors, or limiters in integrated optics. The nonlinearity of the medium and the optical feedback of the cavity provide the physical basis of optical bistability. The nonlinearity may be due to a nonlinear refractive index (*dispersive* optical bistability) or to nonlinear absorption processes (*absorptive* optical bistability). In semiconductors, dispersive optical bistability was first observed in GaAs [2.114] due to a free-exciton resonance, and in InSb [2.118] due to bandgap resonant saturation. In GaAs [2.114] an input laser beam at 770–870 nm was used in the temperature range 5–120 K; the holding intensity was ≈ 1 mW/μm^2, the switch-on time < 1 ns, and the switch-off time ≈ 40 ns. Excitons are involved in optical bistability in a variety of systems [2.114–117]. In particular, a bound-exciton resonance was noted in [2.117]. Spatial structures like longitudinal propagating excitation discontinuities (kinks) and spatial phase coexistence similar to the phenomena which arise in g-r induced nonequilibrium phase transitions (Chaps 4, 5) have also been found [2.120–122], as well as chaotic oscillations [2.123–125].

The bistable excitonic g-r model underlying (2.3.5) may be viewed as a simple novel mechanism for optical bistability [2.108]. In order to demonstrate this we consider the generation coefficients Y^S and Y_1 as functions of the incident light intensity I_i: The optical generation rate is $Y^S = \kappa I_i$, where κ is the band-to-band absorption coefficient. The impact-ionization coefficient Y_1, in case of an optically induced avalanche, depends in a complex, saturable way upon the incident-light intensity [2.126, 127]. Thus I_i can be used as a control parameter. The steady state exciton and electron densities depend upon the control parameter as shown schematically in Fig. 2.45b provided that the other g-r coefficients are in appropriate ranges. Since the excitons and the electron-hole pairs recombine radiatively, the light output intensity at the appropriate frequencies (of γ or γ') may be assumed proportional to the respective recombination rates:

$$I_{\text{out}} \sim F_1 x \qquad \text{or} \qquad \sim B^S n^2 \ .$$

Thus Fig. 2.45b may be visualized as the input-output characteristic of the light intensity, exhibiting optical bistability and hysteresis.

In conclusion, Sect 2.3 contains some novel, exciton-based g-r mechanisms for bistability. We have speculated that this might be of interest in the context of electron-hole droplet formation and optical bistability, although no direct experimental evidence is available so far. The area of excitonic g-r models has only been touched briefly, and a variety of other g-r mechanisms may exist which could give rise to nonequilibrium phase transitions, bistability, or even chaos.

3. Small Fluctuations from the Homogeneous Steady State

In this chapter we analyze the response to small spatial and temporal fluctuations of the homogeneous steady state. We shall show that under certain conditions, in particular when negative differential conductivity occurs, these fluctuations can grow and lead to the bifurcation of filamentary or domain-like spatial structures or self-sustained oscillations. We show that a variety of electromagnetic modes can exist, depending upon the orientation of the field fluctuation and the wave vector relative to the uniform field. Only some of these modes couple to the g-r instability and are associated with the bifurcation of spatial or temporal dissipative structures. Whether a current filament or an electric-field domain bifurcates, depends upon the details of the uniform current density-field characteristic. In particular, we point out a novel possibility of moving domains associated with an anomalous tilted S-shaped characteristic. Analytical conditions for the Hopf bifurcation of limit-cycle oscillations are also derived. The analysis in this chapter is confined to single-carrier g-r mechanisms, as discussed in Sect. 2.1.

3.1 Linear Modes of One-Carrier Models

The linear modes of small current and electric-field fluctuations from a homogeneous steady state have been investigated for general NNDC and SNDC elements [3.1–5], using the concept of a macroscopic dynamic differential conductivity or conductivity tensor. It has been shown that in the regime of negative differential conductivity some of these modes become undamped and lead to the formation of moving field domains in case of NNDC, and stationary current filaments in case of SNDC. In this section we shall give a more explicit treatment [3.6] appropriate to single-carrier g-r SNDC mechanisms of the type discussed in Sect. 2.1.3, in order to obtain sufficiently detailed results. Later on, in Sects 3.4, 3.5, we will consider a more general situation which also allows for drift nonlinearities, such as the intervalley electron transfer (Gunn effect), which is associated with NNDC. This approach will emphasize the duality between NNDC and SNDC based upon the nonmonotonic behavior of the constitutive equations for the current density and the charge density, respectively. In the whole chapter the charge carriers are assumed to be electrons; analogous results apply for holes. Our linear-mode analysis covers general g-r mechanisms involving M trap levels (ground state and excited states of bound electrons) [3.6]. The particular expressions which result for the specific two-level mechanism of Sect. 2.1.2 [3.7] are listed in Table 3.1.

Table 3.1. Linear mode expressions for the two-level g-r mechanism of Fig. 2.4 (Sect. 2.1.2) p. 48

Defining equation	Expression
(3.1.12)	$B(v) = \tau_M \begin{pmatrix} -X^* - X_1 N_D^* v & T^* \\ X^* - T_1^S N_D^* v & -T^* - X_1^S - T_1^S N_D^* v - X_1^* N_D^* v \end{pmatrix}$ $c = \tau_M \begin{pmatrix} 0 \\ T_1^S N_D \end{pmatrix}$
(3.1.14)	$\rho = \dfrac{\tau_M^2}{\Delta(v)}(av^3 + bv^2 + cv + d)$ $a = -(T_1^S + X_1^*) X_1 N_D^*$ $b = [X_1^* X_1 N_D^* - (X_1^S + T_1^S N_A + T^*)X_1 - X^* X_1^* - T_1^S(T^* + X^*)] N_D^*$ $c = [(X_1^S + T^*)X_1 N_D^* + X^* X_1^* N_D^* - X_1^S X^* - T_1^S N_A(T^* + X^*)]$ $d = X_1^S X^*$ $\Delta = \det\{B\} = \tau_M^2 [(X_1^S + T_1^S N_D^* v + X_1^* N_D^* v)(X^* + X_1 N_D^* v) + T^*(T_1^S + X_1)N_D^* v]$
(3.1.24)	$G(\lambda) = \lambda^2 + \lambda\theta + \Delta$ $\theta = -\text{tr}\{B\} = \tau_M[X^* + T^* + X_1^S + (T_1^S + X_1 + X_1^*)N_D^* v]$
(3.1.28)	$H(\lambda) = \lambda^2 - \lambda\,\text{tr}\{\bar{A}\} + \det\{\bar{A}\}$ $\bar{A} = \tau_M \begin{pmatrix} -X^* + X_1 N_D^*(v_{i_1} - v) & T^* + X_1 N_D^* v_{i_1} \\ X^* + X_1^* N_D^* v + T_1^S[N_D^*(v_{i_1} + v_{i_2} - v) - N_D] & -T^* - X_1^S + X_1^* N_D^*(v_{i_2} - v) + T_1^S[N_D^*(v_{i_1} + v_{i_2} - v) - N_D] \end{pmatrix}$

Table 3.1 (continued)

$$F(\lambda) = -\lambda\alpha_1 - \alpha_2$$

(3.1.27)

$$\alpha_1 = -(f_1 + f_2)$$

$$\alpha_2 = -\tau_M[f_1(T^* + X^* + X_1^s + X_1^* N_D^* v) + f_2(T^* + X^* + X_1 N_D^* v)] = (\bar{A}_{22} - \bar{A}_{21})f_1 + (\bar{A}_{11} - \bar{A}_{12})f_2$$

(3.1.20)

$$f_1 = -\tau_M(\partial X_1/\partial\varepsilon_0)N_D^* v v_{t_1}$$

$$f_2 = -\tau_M(\partial X_1^*/\partial\varepsilon_0)N_D^* v v_{t_2}$$

(3.1.13)

$$v_{t_1}(v,\varepsilon) = \tau_M^2 \frac{T^*}{\Delta(v)} T_1^S N_D v$$

$$v_{t_2}(v,\varepsilon) = \tau_M^2 \frac{X^* + X_1 N_D^* v}{\Delta(v)} T_1^S N_D v$$

The above formulas correspond to a model where the donors act as traps. If the traps are independent of the donors, then N_D and N_A should be replaced by N_t and $N_t - N_D^*$, respectively, in the above formulas (cf., Table 2.1, p. 42), where N_t is the trap concentration

3.1.1 Linearized Transport Equations

We shall first discuss the steady state solutions of the nonlinear transport equations, which are not necessarily homogeneous, and then linearize these transport equations around the steady state. Throughout this and the next chapters we shall use dimensionless variables. To this purpose we normalize all concentrations by the effective donor concentration

$$N_D^* := N_D - N_A :$$

$$v := n/N_D^* \tag{3.1.1}$$

$$\boldsymbol{v}_t := \boldsymbol{n}_t/N_D^* \equiv (v_{t_1}, \dots, v_{t_M}) \tag{3.1.2}$$

and introduce dimensionless time and space variables

$$\tau := t/\tau_M \tag{3.1.3}$$

$$\xi := x/L , \tag{3.1.4}$$

where

$$\tau_M := \epsilon_s/(4\pi e \mu_0 N_D^*) \tag{3.1.5}$$

is an effective dielectric-relaxation time, and

$$L_D := (D_0 \tau_M)^{1/2} \tag{3.1.6}$$

is an effective Debye length. The terms D_0, μ_0, and ϵ_s are, respectively, the low-field electron diffusion constant, low-field mobility, and static dielectric constant. Further, we introduce the dimensionless electric field ε, using the Einstein relation (1.2.15):

$$\varepsilon := (\mu_0 L_D/D_0)E = (eL_D/kT)E . \tag{3.1.7}$$

The continuity equations for the free and trapped electron concentrations (1.1.5, 7) are in dimensionless form given by

$$\frac{\partial}{\partial \tau} v - \boldsymbol{\nabla} \cdot (\varepsilon v + \boldsymbol{\nabla} v) = \varphi_0(v, \boldsymbol{v}_t, \varepsilon) \tag{3.1.8}$$

$$\frac{\partial}{\partial \tau} \boldsymbol{v}_t = \boldsymbol{\varphi}_t(v, \boldsymbol{v}_t, \varepsilon) \tag{3.1.9}$$

where we assume throughout Sects. 3.1–3.3 that $\mu_n = \mu_0$, $D_n = D_0$, independently of the field. The concentrations and fields are coupled via Maxwell's equation for the charge density (1.1.8):

$$\boldsymbol{\nabla} \cdot \varepsilon = 1 - v - \sum_{i=1}^{M} v_{t_i} , \tag{3.1.10}$$

φ_0, φ_t are the g-r rates which are – for nondegenerate statistics – polynomial functions of the concentrations. They depend implicity upon the magnitude of the electric field ε through the rate constants, e.g., impact-ionization coefficients. Since the total number of electrons is conserved by g-r processes,

$$\varphi_0 + \sum_{i=1}^{M} \varphi_{t_i} = 0 \tag{3.1.11}$$

holds. The g-r rates usually considered – see, e.g., (1.2.25) – are linear in the concentration of impurity electrons, so that we can write

$$\varphi_t(v, v_t) = B(v)v_t + cv \tag{3.1.12}$$

with a v-dependent matrix B and a constant "vector" c. The dependence of $B(v)$ and c upon ε has been suppressed for brevity.

Hence, in the steady state, v_t can be eliminated from the transport equations by (3.1.12):

$$v_t(v, \varepsilon) = -B(v)^{-1}cv = -\frac{\text{adj}\{B\}}{\varDelta(v)}cv \tag{3.1.13}$$

where $\varDelta(v) := \det\{B\}$ and $(\text{adj}\{B\})_{ij}$ is $(-1)^{i+j}$ times the determinant of the matrix obtained by deleting the jth row and the ith column of B (adjunct of B). The static charge density is then

$$\rho(v, \varepsilon) := 1 - v - \sum_{i=1}^{M} v_{t_i}(v, \varepsilon) = \frac{P(v)}{\varDelta(v)} \tag{3.1.14}$$

where

$$P(v) := (1 - v)\varDelta(v) + v \sum_{i,j=1}^{M} [\text{adj}\{B(v)\}]_{ij}c_j$$

is a polynomial in v.

The spatially homogeneous steady states are determined by the local neutrality condition

$$\rho(v, \varepsilon) = 0 , \tag{3.1.15}$$

which is an implicit equation for v as a function of ε.

The electric field ε is then just given by the externally applied homogeneous field ε_0. According to our basic assumption we consider g-r mechanisms which give rise to an S-shaped current density-field characteristic, i.e., bistability of the homogeneous steady states in some field range $\varepsilon_h < \varepsilon_0 < \varepsilon_{th}$. Thus ρ as a function of v has three non-negative zeros for ε_0 fixed in the range $\varepsilon_h < \varepsilon_0 < \varepsilon_{th}$ (Fig. 3.1). Examples of such mechanisms have been discussed in Sect. 2.1.

Next, we shall consider small fluctuations from the steady state. For convenience we introduce a formal vector notation for the homogeneous steady state:

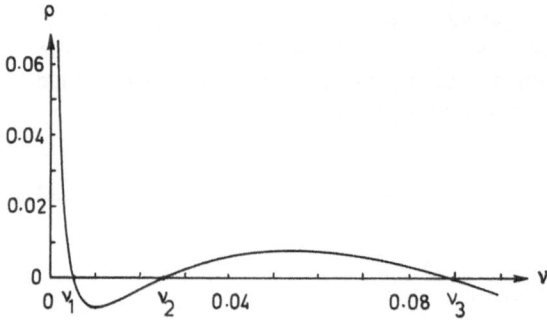

Fig. 3.1. Static local charge density ρ versus electron concentration v for a value of the field ε_0 within the bistability range $\varepsilon_h < \varepsilon_0 < \varepsilon_{th}$. The numerical parameters correspond to the two-level model shown in Fig. 2.4 and are listed in Table 3.2, p. 118, with $\varepsilon_0 = 8.4$ (ρ is in units eN_D^* and v in units N_D^*). The spatially homogeneous steady states are v_1, v_2, v_3

$$\phi_0 := (\varepsilon_0, v, v_t) \ .$$

In order to investigate the stability of ϕ against small space- and time-dependent pertubations

$$\delta\phi(\xi, \tau) := \phi(\xi, \tau) - \phi_0 \ ,$$

we linearize the transport equations (3.1.8–10) around ϕ_0 and set

$$\delta\phi(\xi, \tau) = (\delta\varepsilon(\xi), \delta v(\xi), \delta v_t(\xi))\exp \lambda\tau \ , \tag{3.1.16}$$

where λ is an eigenvalue of the linearized transport equations. We obtain from the linearized rate equations (3.1.9):

$$\lambda\delta v_t = B(v)\delta v_t + d\delta v + f\frac{\varepsilon_0 \cdot \delta\varepsilon}{|\varepsilon_0|} \ , \tag{3.1.17}$$

and from Maxwell's equation (3.1.10)

$$\nabla \cdot \delta\varepsilon = \delta\rho(\xi, \tau) = -\left(\delta v + \sum_{i=1}^{M} \delta v_{t_i}\right) \ , \qquad \text{where} \tag{3.1.18}$$

$$d := \left[\frac{\partial}{\partial v} B(v)\right] v_t \bigg|_{\text{st.st.}} + c \qquad \text{and} \tag{3.1.19}$$

$$f := \frac{\partial}{\partial\varepsilon} \varphi_t(v, v_t, \varepsilon) \bigg|_{\text{st.st.}} \ . \tag{3.1.20}$$

Instead of linearizing the continuity equation (3.1.8) itself, we linearize the total current density which is equivalent to a first integral. This can be seen by adding all equations (3.1.8, 9) and using (3.1.10, 11):

$$\nabla \cdot \delta j^{tot} = 0 \ , \qquad \text{where} \qquad\qquad\qquad (3.1.21)$$

$$\delta j^{tot} := \frac{\partial}{\partial \tau} \delta \varepsilon + v \delta \varepsilon + \varepsilon \delta v + \nabla \delta v$$

$$= (\lambda + v) \delta \varepsilon + (\varepsilon + \nabla) \delta v \qquad\qquad\qquad (3.1.22)$$

is the total current density fluctuation composed of the displacement current density $\lambda \varepsilon$ and the conduction current density. The latter consists of the drift current density $v \delta \varepsilon + \varepsilon \delta v$ and the diffusion current density $\nabla \delta v$.

Decomposing the field fluctuation $\delta \varepsilon$ into a component $\delta \varepsilon_\parallel$ parallel to, and a component $\delta \varepsilon_\perp$ perpendicular to the applied field ε_0, we observe that the f-term in (3.1.17) vanishes for transverse perturbations ($\delta \varepsilon_\parallel = 0$). Next, we eliminate δv_t from the linearized system (3.1.17, 18, 22) by inverting (3.1.17):

$$\delta v_t = -[B(v) - \lambda]^{-1}(d\delta v + f\delta\varepsilon_\parallel)$$

$$= -\frac{\text{adj}\{B - \lambda\}}{G(\lambda)}(d\delta v + f\delta\varepsilon_\parallel) \ , \qquad\qquad (3.1.23)$$

where

$$G(\lambda) := \det\{B - \lambda\} \ . \qquad\qquad\qquad (3.1.24)$$

The dynamical charge density fluctuation $\delta\rho(\xi, \tau)$ in (3.1.18) can then be expressed in terms of δv and $\delta\varepsilon_\parallel$ only:

$$\nabla \cdot \delta\varepsilon = -\frac{H(\lambda)}{G(\lambda)}\delta v - \frac{F(\lambda)}{G(\lambda)}\delta\varepsilon_\parallel \qquad \text{where} \qquad\qquad (3.1.25)$$

$$H(\lambda) := G(\lambda) - \sum_{i,j}^{M} \{\text{adj}(B - \lambda)\}_{ij} d_j \ , \qquad \text{and} \qquad (3.1.26)$$

$$F(\lambda) := -\sum_{i,j=1}^{M} \{\text{adj}(B - \lambda)\}_{ij} f_j \ . \qquad\qquad (3.1.27)$$

Although the charge density fluctuation in (3.1.25) does not contain the trapped electron concentrations δv_t explicitly, it includes *all M linear modes of the trapped electrons* through the λ-dependence. Thus the independent temporal evolution (in the linear regime near steady state) of the occupancy of all trap levels is fully taken into account. This is more general than the often invoked assumption of a pseudo-steady state of the traps (generalized Shockley-Read-Hall kinetics [3.8]) and is in fact essential for the description of the g-r instability.

We will now give a physical interpretation of (3.1.26). It is straightforward to show that the right-hand side of (3.1.26) can be expressed as the characteristic polynomial

$$H(\lambda) = \det\{\tilde{A} - \lambda\} \qquad\qquad\qquad (3.1.28)$$

of the g-r matrix

$$\tilde{A}_{ij} := B_{ij} - d_i \; . \tag{3.1.29}$$

From (3.1.17) it follows that the matrix $\tilde{A}$ describes spatially homogeneous, charge-neutral g-r fluctuations:

$$\frac{\partial}{\partial \tau} \delta v_t = \tilde{A} \delta v_t \quad \text{with} \quad \delta v + \sum_{i=1}^{M} \delta v_{t_i} = 0 \quad \text{and} \quad \delta \varepsilon_{\parallel} = 0 \; . \tag{3.1.30}$$

Hence the zeros $\lambda_1, \ldots, \lambda_M$ of $H(\lambda)$ are the eigenmodes of such neutral fluctuations. They are completely determined by the g-r mechanism. In most cases they are real and can be ordered $\lambda_1 > \cdots > \lambda_M$. For fluctuations of stable homogeneous steady states the largest eigenvalue satisfies $\lambda_1 < 0$, whereas for unstable homogeneous steady states $\lambda_1 > 0$ holds.

The zeros $\lambda_1^\infty > \cdots > \lambda_M^\infty$ of $G(\lambda)$ describe fluctuations with $\delta v = 0$, $\delta \varepsilon_{\parallel} = 0$ by (3.1.17, 24). In Chap. 2 we have seen that the g-r instability is due to the autocatalytic production of conduction electrons by impact ionization. Hence the instability is associated with a change in electron concentration $\delta v \neq 0$. This is the physical motivation for assuming in the following that all λ_i^∞ are negative, and $\lambda_1^\infty < \lambda_1$. A typical eigenvalue spectrum corresponding to the two-level mechanism discussed in Sect. 2.1.2 is shown in Fig. 3.2.

From (3.1.22, 25) we can eliminate either $\delta \varepsilon$ or δv in order to obtain a single eigenvalue equation. We shall first eliminate $\delta \varepsilon$, which provides us with an equation for the carrier-concentration fluctuations δv. It is the governing equation for the onset of the g-r instability. Later on in Sect. 3.4, we shall eliminate δv and couple the resulting constitutive current-field relation to Maxwell's equations in order to study the possibility of electromagnetic modes.

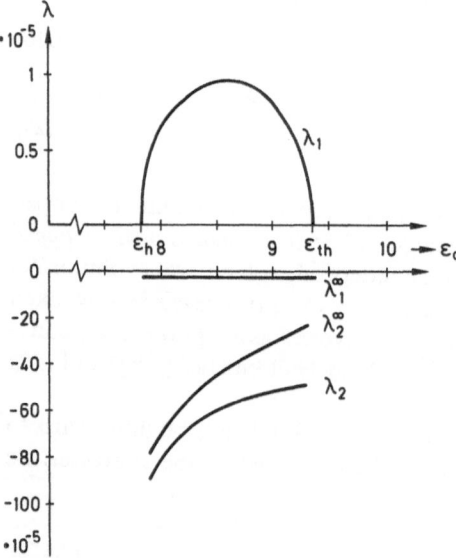

Fig. 3.2. Eigenvalue spectrum λ_1, λ_2 of $\tilde{A}$ (homogeneous, charge-neutral fluctuations) in units of τ_M^{-1} versus the control parameter ε_0 (electric field), for the numerical parameters of the two-level model of Fig. 2.4 listed in Table 3.2. In the bistability range ($\varepsilon_h < \varepsilon_0 < \varepsilon_{th}$) λ_1 is positive, while λ_2 is negative. λ_1^∞, λ_2^∞ are the eigenvalues of B (fluctuations with $\delta v = 0$, $\delta \varepsilon_{\parallel} = 0$), and are always negative

In the homogeneous steady state, (3.1.22) can be resolved for $\delta\varepsilon$ and substituted into (3.1.25), using (3.1.21). We obtain the governing equation for small concentration fluctuations:

$$-\frac{1}{\lambda + v}(\varepsilon_0\nabla_\| + \Delta)\delta v = -\frac{H(\lambda)}{G(\lambda)}\delta v - \frac{F(\lambda)}{G(\lambda)}\cdot\frac{1}{\lambda + v}[\delta j_\|^{\text{tot}} - (\varepsilon_0 + \nabla_\|)\delta v] \quad (3.1.31)$$

or

$$\left[\tilde{V}(\lambda) - \Delta - \varepsilon_0\nabla_\| - \frac{F(\lambda)}{G(\lambda)}(\varepsilon_0 + \nabla_\|)\right]\delta v = -\frac{F(\lambda)}{G(\lambda)}\delta j_\|^{\text{tot}} , \quad (3.1.32)$$

where

$$\tilde{V}(\lambda) := (\lambda + v)\frac{H(\lambda)}{G(\lambda)} , \quad (3.1.33)$$

and the subscript $\|$ denotes components parallel to ε_0.

3.1.2 Stability and Differential Conductivity

Let us now point out a remarkable relation between the g-r eigenvalues $\lambda_1, \ldots, \lambda_M$ and the static charge density $\rho(v, \varepsilon_0)$. It results in a connection of the temporal evolution and the stability of homogeneous steady states with the purely static properties of the current-field characteristic.

First we note from (3.1.24, 26) that

$$\prod_{i=1}^{M} \lambda_i = \det\{\tilde{A}\} = H(0) \quad (3.1.34)$$

and

$$\prod_{i=1}^{M} \lambda_i^\infty = \det\{B\} = \Delta = G(0) . \quad (3.1.35)$$

The static ($\lambda = 0$) charge density fluctuations by (3.1.25) become

$$\delta\rho = -\frac{\det\{\tilde{A}\}}{\Delta}\delta v - \frac{F(0)}{\Delta}\delta\varepsilon_\| . \quad (3.1.36)$$

On the other hand, the static charge density $\rho(v, \varepsilon_0)$ can be expanded for small changes in v and ε_0:

$$\delta\rho(v, \varepsilon_0) = \left(\frac{\delta\rho}{\delta v}\right)_{\varepsilon_0}\delta v + \left(\frac{\partial\rho}{\partial\varepsilon_0}\right)_v\delta\varepsilon_\| . \quad (3.1.37)$$

A comparison of (3.1.36, 37) using (3.1.34) shows that

$$\prod_{i=1}^{M} \lambda_i = H(0) = -\Delta \left(\frac{\partial \rho}{\partial v}\right)_{\varepsilon_0} \qquad \text{and} \qquad (3.1.38)$$

$$F(0) = -\Delta \left(\frac{\partial \rho}{\partial \varepsilon_0}\right)_v . \qquad (3.1.39)$$

From (3.1.35) it follows that sign $\Delta = (-1)^M$ since all λ_i^∞ are negative. Therefore, by (3.1.38)

$$\text{sign} \prod_{i=1}^{M} \lambda_i = (-1)^{M+1} \text{sign} \left(\frac{\partial \rho}{\partial v}\right)_{\varepsilon_0} . \qquad (3.1.40)$$

If $(\partial\rho/\partial v)_{\varepsilon_0} > 0$, then $\prod_{i=1}^{M} \lambda_i$ is negative for even M and positive for odd M. In both cases there is at least one *unstable* mode $\lambda_1 > 0$. Conversely, if all λ_i are negative except for one, $\lambda_1 > 0$, then $(\partial\rho/\partial v)_{\varepsilon_0}$ must be positive. If, however, all λ_i are negative, then $(\partial\rho/\partial v)_{\varepsilon_0} < 0$. Since $\rho(v)$ is a continuous function, the sign of the slope $(\partial\rho/\partial v)_{\varepsilon_0}$ alternates for consecutive zeros, cf., Fig. 3.1. Therefore, at fixed ε_0, between two stable homogeneous steady states there always lies an unstable steady state (Fig. 3.3).

Thus, the instability of the homogeneous state can be read off directly from the $j_0 - \varepsilon_0$ characteristic. Note that we have not considered the influence of an external circuit; as we shall see later, the coupling to resistive or reactive circuits can drastically influence the above results on stability.

We now proceed to derive a relation between $(\partial\rho/\partial v)_{\varepsilon_0}$ and the static differential conductivity σ_{diff} from (3.1.38, 39). For homogeneous steady states the (dimensionless) static differential conductivity is defined by

$$\sigma_{\text{diff}} := \frac{dj_0}{d\varepsilon_0} = \frac{d}{d\varepsilon_0} [v(\varepsilon_0)\varepsilon_0] = v(\varepsilon_0) + \frac{dv(\varepsilon_0)}{d\varepsilon_0} \varepsilon_0 . \qquad (3.1.41)$$

The homogeneous steady state $v(\varepsilon_0)$ is given by

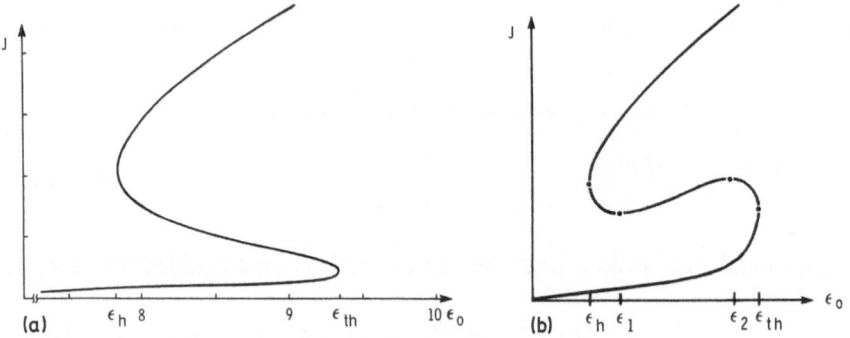

Fig. 3.3a, b. Static current density-field characteristic for a one-carrier SNDC mechanism based on a $\rho(v)$ relation of the form shown in Fig. 3.1. (a) Normal SNDC (numerical parameters as in Table 3.2). (b) Anomalous tilted SNDC (schematic). Note that the middle branch of the characteristic is always unstable, even if it has positive differential conductivity ($\varepsilon_1 < \varepsilon_0 < \varepsilon_2$ in b)

$0 = \rho(v(\varepsilon_0), \varepsilon_0)$.   Hence

$$0 = \frac{d}{d\varepsilon_0} \rho(v(\varepsilon_0), \varepsilon_0) = \left(\frac{\partial \rho}{\partial v}\right)_{\varepsilon_0} \frac{dv(\varepsilon_0)}{d\varepsilon_0} + \left(\frac{\partial \rho}{\partial \varepsilon_0}\right)_v$$

$$= -\frac{H(0)}{\Delta} \frac{dv(\varepsilon_0)}{d\varepsilon} - \frac{F(0)}{\Delta} , \tag{3.1.42}$$

where we have used (3.1.38, 39). We obtain with (3.1.41):

$$\sigma_{\text{diff}} = v(\varepsilon_0) - \frac{F(0)}{H(0)} \varepsilon_0 . \tag{3.1.43}$$

Using (3.1.39) and the static charge density

$$\rho(v, \varepsilon_0) = 1 - v - \sum_{i=1}^{M} v_{t_i}(v, \varepsilon_0), \text{ we can write}$$

$$F(0) = \Delta \sum_{i=1}^{M} \left(\frac{\partial v_{t_i}}{\partial \varepsilon_0}\right)_v = -\Delta \left(\frac{\partial p_t}{\partial \varepsilon_0}\right)_v , \tag{3.1.44}$$

where $p_t := N_t - \sum_{i=1}^{M} v_{t_i}$ is the normalized concentration of empty traps. From (3.1.13) it is evident that $v_t(v, \varepsilon_0)$ depends explicitly upon ε_0 through the g-r coefficients contained in the g-r rates $\varphi_t(v, v_t, \varepsilon_0)$. Since an electric field normally acts to empty the traps, we can in general assume

$$\left(\frac{\partial p_t}{\partial \varepsilon_0}\right)_v > 0 . \tag{3.1.45}$$

In Table 3.1 all the relevant expressions are listed for the two-level model of Sect. 2.1.2, and it is obvious that $\Delta > 0$, $F(0) < 0$. Hence from (3.1.44) it follows explicitly that (3.1.45) holds. The differential conductivity (3.1.43) can be restated, using (3.1.38, 44) as

$$\sigma_{\text{diff}} = v(\varepsilon_0) - \left(\frac{\partial \rho}{\partial v}\right)_{\varepsilon_0}^{-1} \left(\frac{\partial p_t}{\partial \varepsilon_0}\right)_v \varepsilon_0 \tag{3.1.46}$$

which follows also directly from (3.1.42, 41). We conclude from (3.1.46, 45) that the differential conductivity can become negative only if $(\partial \rho/\partial v)_\varepsilon > 0$, which is connected with an unstable, positive g-r eigenvalue λ_1 by (3.1.40). Stable homogeneous steady states, by (3.1.40), cannot satisfy $\sigma_{\text{diff}} < 0$. By (3.1.46) the turning points of S-shaped current density-field characteristics, where $\sigma_{\text{diff}} \to \infty$, are characterized by $(\partial \rho/\partial v)_\varepsilon \to 0$, i.e., a vanishing g-r eigenvalue λ_1, and hence by a saddle-node bifurcation of the solution branches.

We note that the middle branch of an S-shaped current density-field characteristic, which is unstable against g-r fluctuations according to the preceding discussion, need not have NDC through its complete range $\varepsilon_h < \varepsilon_0 < \varepsilon_{\text{th}}$. Figure 3.3b shows an example where the middle branch has NDC only for $\varepsilon_h < \varepsilon_0 < \varepsilon_1$ and

$\varepsilon_1 < \varepsilon_0 < \varepsilon_{th}$. For $\varepsilon_1 < \varepsilon_0 < \varepsilon_2$ the differential conductivity is *positive* although $(\partial \rho / \partial v)_{\varepsilon_0} > 0$. This is possible if the concentration of empty traps p_t depends only weakly upon ε_0 at fixed v, compare (3.1.46). This case occurs, for instance, if the impact-ionization coefficients are weakly dependent on ε_0. Similar anomalous S-type current-field characteristics have also been found in carrier-density instabilities based on the Poole-Frenkel effect [3.9], in anisotropy instabilities in many-valley semiconductors [3.10], and in a nerve-axon model in neurobiology [3.11]. The consequences of this anomalous behaviour will be discussed in Sect. 3.3. The main results of the present section are the relations (3.1.38, 46) between the eigen-values λ_i, the static charge density $\rho(v, \varepsilon_0)$, and the static differential conductivity $dj_0/d\varepsilon_0$.

3.2 Filamentary Instability

We shall now compute the linear modes of the eigenvalue equation (3.1.32) for *transverse* fluctuations and show that they lead to an instability of the homogeneous NDC state and to the bifurcation of current layers or filaments [3.7].

3.2.1 The Spectrum

In this section we specialize the linearized equations derived in Sect. 3.1.1 for transverse fluctuations such that δv varies in space only in the direction perpendicu-lar to ε_0:

$$\nabla_{\|} \delta v = 0 \ . \tag{3.2.1}$$

Equation (3.1.32) is reduced to

$$[\tilde{V}(\lambda) - \nabla_{\perp}^2] \delta v + \frac{F(\lambda)}{G(\lambda)} (\delta j_{\|}^{tot} - \varepsilon_0 \delta v) = 0 \ . \tag{3.2.2}$$

A fluctuation of the form (3.2.1) induces a ripple $\delta v(\xi)$ in the carrier concentration, and hence a transverse modulation of the drift current density. We shall see in Sect. 3.4 that for slow fluctuations, as they occur near the onset of instabilities ("critical slowing-down"), the change $\delta \varepsilon_{\|}$ in ε_0 can be neglected. Therefore $\delta j_{\|}^{tot} = \varepsilon_0 \delta v$, by (3.1.22). The eigenvalue equation (3.2.2) becomes

$$[\tilde{V}(\lambda) - \nabla_{\perp}^2] \delta v = 0 \ . \tag{3.2.3}$$

The auxiliary function $\tilde{V}(\lambda)$ is plotted for the two-level model of Sect. 2.1.2 in Fig. 3.4. The solution of (3.2.3) depends upon the boundary conditions. If the device is a rectangular parallelepiped of dimensions L_ξ, L_η, L_ζ we can attempt to separate (3.2.3) in Cartesian coordinates ξ, η. For simplicity let us assume that the fluctuations are constant in the η-direction. We are then dealing with an effectively one-dimensional geometry, which gives rise to plane-layered struc-tures, as shown in Fig. 3.5a.

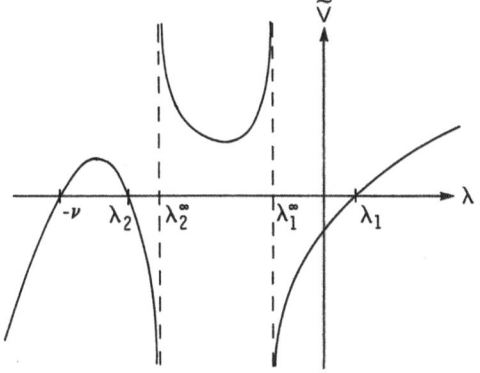

Fig. 3.4. Auxiliary function $\tilde{V}(\lambda) := (\lambda + v)H(\lambda)/G(\lambda)$ describing transverse fluctuations (schematic). The form shown corresponds to an unstable point of the SNDC characteristic: $\lambda_1 > 0$. The spectrum is given by $k_\perp^2 = -\tilde{V}(\lambda)$

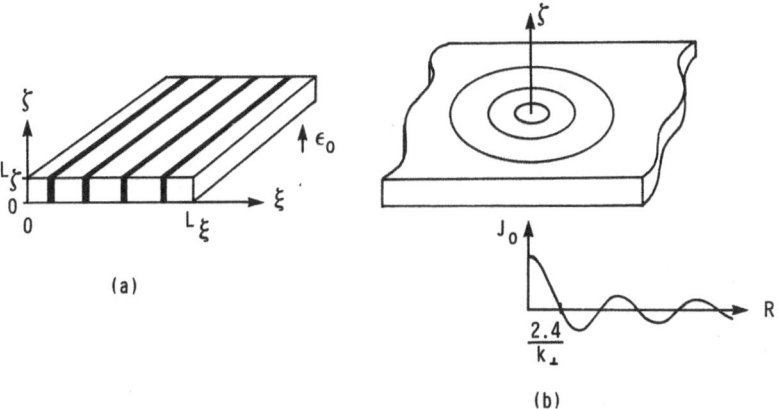

Fig. 3.5. Sketch of (a) plane-layered, (b) cylindrical spatial structures of the carrier density in the sample as a result of transverse fluctuations. In (a) the sample is a rectangular parallelepiped of dimensions L_ξ, L_η, L_ζ. In (b) it is a thin film; the inset shows the Bessel function $J_0(k_\perp R)$ describing the radial normal modes

For periodic boundary conditions $\delta v(0) = \delta v(L_\xi)$ we can set

$$\delta v(\xi) = e^{ik_\perp \xi} , \qquad k_\perp = \frac{2\pi}{L_\xi}n , \qquad n \in \mathbb{Z} . \tag{3.2.4a}$$

For homogeneous Dirichlet boundary conditions $\delta v(0) = \delta v(L_\xi) = 0$ we use

$$\delta v(\xi) = \sin k_\perp \xi , \qquad k_\perp = \frac{\pi}{L_\xi}n , \qquad n \in \mathbb{N} . \tag{3.2.4b}$$

For homogeneous Neumann boundary conditions

$$\frac{d}{d\xi}\delta v(0) = \frac{d}{d\xi}\delta v(L_\xi) = 0 , \qquad \text{we use}$$

$$\delta v(\xi) = \cos k_\perp \xi , \qquad k_\perp = \frac{\pi}{L_x}n , \qquad n \in \mathbb{N} . \tag{3.2.4c}$$

All three sets of eigenfunctions (3.2.4a, b, c) lead to the dispersion relation

$$\tilde{V}(\lambda) + k_\perp^2 = 0 \tag{3.2.5}$$

when substituted into (3.2.3).

If the two transverse dimensions of the device are both large, as for example in thin semiconductor films, it is often useful to solve (3.2.3) in polar coordinates R, φ. Equation (3.2.3) becomes

$$\tilde{V}(\lambda)\delta v - \frac{1}{R}\frac{\partial}{\partial R}\left(R\frac{\partial}{\partial R}\delta v\right) - \frac{1}{R^2}\frac{\partial^2}{\partial\varphi^2}\delta v = 0 \tag{3.2.6}$$

which, with the variable separation

$$\delta v(R, \varphi) = \delta\tilde{v}(R)e^{im\varphi}, \qquad m = 0, \pm 1, \pm 2, \ldots \tag{3.2.7}$$

yields the Bessel differential equation

$$\left[\frac{d^2}{dR^2} + \frac{1}{R}\frac{d}{dR} - \frac{m^2}{R^2} - \tilde{V}(\lambda)\right]\delta\tilde{v}(R) = 0 . \tag{3.2.8}$$

For $\tilde{V}(\lambda) > 0$ there is no solution which is regular both at $R = 0$ and $R = \infty$. For $\tilde{V}(\lambda) < 0$ the regular solutions are the mth order Bessel functions of the first kind [3.12]:

$$\delta\tilde{v}(R) = J_m(k_\perp R) . \tag{3.2.9}$$

Only the solutions with $m = 0$ correspond to cylindrically symmetric modes and will be considered. For $\delta v(R, \varphi) = J_0(k_\perp R)$, (3.2.8) again leads to the dispersion relation (3.2.5). Such concentration and current-density fluctuations ($\delta j_\parallel = \varepsilon_0 \delta v$) describe cylindrical filamentary structures, as shown in Fig. 3.5b.

The dispersion relation (3.2.5), by (3.1.33) yields

$$k_\perp = [-\tilde{V}(\lambda)]^{1/2} = \left[-(\lambda + v)\frac{H(\lambda)}{G(\lambda)}\right]^{1/2} \qquad \text{for } \tilde{V}(\lambda) < 0 . \tag{3.2.10}$$

The corresponding spectrum $\lambda(k_\perp)$ has $M + 1$ real branches, representing M coupled recombination-diffusion modes with $\lambda(0) = \lambda_1, \lambda_2, \ldots, \lambda_M$ (g-r eigenvalues) and one dielectric-relaxation mode with $\lambda(0) = -v$. In real time units, v^{-1} is the dielectric-relaxation time $\varepsilon_s/(4\pi e\mu_0 n)$. Note that usually dielectric relaxation is much faster than recombination; hence $v \gg |\lambda_i|$ ($i = 1, \ldots, M$). In Fig. 3.6 the spectrum is shown for the two-level model of Sect. 2.1.2.

For homogeneous steady states which are stable against charge-neutral g-r fluctuations, we have seen in Sect. 3.1 that all λ_i, $i = 1, \ldots, M$, are negative. The spectrum $\lambda(k_\perp)$ then contains damped modes only. When a state with infinite differential conductivity is reached (corresponding to $\varepsilon_0 = \varepsilon_h$ or $\varepsilon_0 = \varepsilon_{th}$ in the static current density-field characteristic), a soft-mode instability ($\lambda_1 = 0$) occurs. On the middle (NDC) branch of the current density-field characteristic we have $\lambda_1 > 0 >$

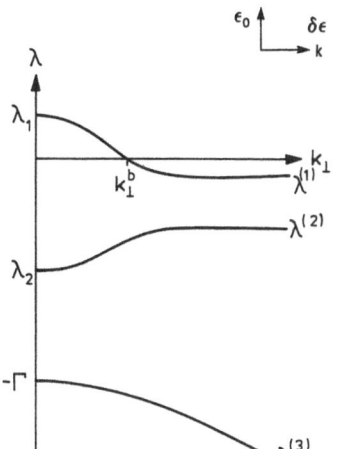

Fig. 3.6. Linear-mode spectrum of transverse fluctuations for an NDC state of a two-level SNDC mechanism (schematic). The eigenvalues λ are plotted versus the transverse wavevector $k_\perp$. $\Gamma := \nu$ is the dielectric-relaxation frequency, λ_1 and λ_2 are the g-r eigenvalues of homogeneous, charge-neutral fluctuations discussed in Sect. 2.1.2 after (2.1.39). All modes $\lambda^{(1)}(k_\perp)$ with $k_\perp < k_\perp^b$ are unstable. The inset shows the orientation of the applied static electric field ε_0, the field fluctuation $\delta\varepsilon$, and the wavevector k relative to each other

$\lambda_2 > \cdots > \lambda_M$. Hence these states are unstable against long-wavelength fluctuations with $k_\perp < k_\perp^b$ (cf., Fig. 3.6), where

$$k_\perp^b := \left(-\nu \frac{H(0)}{G(0)}\right)^{1/2} = \left(-\nu \prod_{i=1}^{M} \lambda_i \frac{1}{\Delta}\right)^{1/2} = \left[\nu \left(\frac{\partial\rho}{\partial\nu}\right)_{\varepsilon_0}\right]^{1/2} \tag{3.2.11}$$

and stable against modes with $k_\perp > k_\perp^b$.

In the long-wavelength ("hydrodynamic") limit the branches $\lambda(k_\perp)$ of the dispersion relation (3.2.10) may be expanded in terms of $k_\perp$. For NDC states we obtain a branch with undamped recombination-diffusion modes

$$\lambda^{(1)} = \lambda_1 - \tilde{D}_\perp k_\perp^2 \tag{3.2.12}$$

with the effective transverse diffusion constant

$$\tilde{D}_\perp := \left[\frac{\partial}{\partial\lambda}\tilde{V}(\lambda)\right]_{\lambda=\lambda_1}^{-1} = \frac{G(\lambda_1)}{(\lambda_1 + \nu)}\left[\frac{\partial}{\partial\lambda}H(\lambda)\right]_{\lambda=\lambda_1}^{-1}$$

$$= \prod_{i=1}^{M}(\lambda_1 - \lambda_i^\infty)\left[(\lambda_1 + \nu)\prod_{i=2}^{M}(\lambda_1 - \lambda_i)\right]^{-1}. \tag{3.2.13}$$

For typical experimental settings (cf., Table 3.2) $\tilde{D}_\perp$ is of the order of 1/1000 of the physical diffusion constant. This reflects the slowing-down of laterally diffusing carriers by multiple trapping and detrapping.

In [3.7] it was shown that for transverse fluctuations the instability at $k_\perp^b$ (3.2.11) is the only one that can occur.

3.2.2 Bifurcation of Layered or Filamentary Stationary Structures

With increasing current I, or equivalently, decreasing applied field ε_0, the eigenvalue λ_1 associated with the unstable homogeneous NDC state changes as shown in Fig.

Table 3.2. Typical material parameters corresponding to n-GaAs at 4.2 K for the two-level g-r mechanism shown in Fig. 2.4 (used in Figs. 3.1, 2, 3a, 7)

Parameter	Value	Ref.
T_1^S	10^{-6} cm^3 s^{-1}	[2.31]
T^*	10^6 s^{-1}	[2.45]
X^*	10^6 s^{-1}	
X_1^S	5×10^5 s^{-1}	[2.32] with photon flux 10^{19} cm^{-2} s^{-1}
X_1	$5 \times 10^{-8} \exp(-6/\varepsilon_0)$ cm^3 s^{-1}	[2.31]
X_1^*	$10^{-6} \exp(-1.5/\varepsilon_0)$ cm^3 s^{-1}	
$N_D^* = N_D - N_A$	10^{15} cm^{-3}	[2.31, 2.32]
N_A/N_D^*	0.5	
μ_0	500 cm^2/Vs	[2.32]
D_0	0.15 cm^2 s^{-1}	[2.32] with $eD_0 = \mu_0 kT$
ϵ_s	10	[2.32]

Physical units:
$$\tau_M = \epsilon_s/(4\pi e\mu_0 N_D^*) \approx 10^{-11} \text{ s}$$
$$L_D = (D_0\tau_M)^{1/2} \approx 10^{-6} \text{ cm}$$

Derived quantities for the unstable homogeneous steady state at $\varepsilon_0 = 8$:

ν	3.7×10^{-2}
λ_1/τ_M	0.62 μs^{-1}
λ_2/τ_M	-80 μs^{-1}
Θ/τ_M	71 μs^{-1}
Δ/τ_M^2	170 μs^{-2}
λ_1^∞/τ_M	-2.5 μs^{-1}
λ_2^∞/τ_M	-68 μs^{-1}
$\tilde{D}_\perp$	7.2×10^{-4}

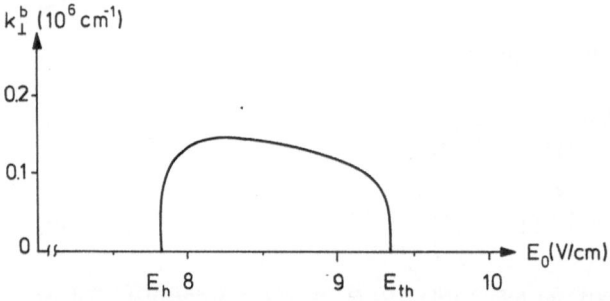

Fig. 3.7. Bifurcation line $k_\perp^b$ versus the applied field ε_0 for the two-level SNDC mechanism of Fig. 2.4 (numerical parameters as in Table 3.2). The modes $k_\perp < k_\perp^b$ are undamped, all modes $k_\perp > k_\perp^b$ are damped. The bifurcation line gives the wavevectors of bifurcating spatial structures

3.2. The marginally stable wavevector $k_\perp^b$ (3.2.11) is plotted versus ε_0 in Fig. 3.7. As ε_0 increases from ε_h, or decreases from ε_{th}, a whole family of modes $k_\perp^b(\varepsilon_0)$ becomes successively undamped. This leads to the successive bifurcation of a family of spatially inhomogeneous stationary solutions from the unstable homogeneous steady state [3.13, 14]. These are modulated perpendicular to the static field ε_0 with period $\Lambda(\varepsilon_0) = 2\pi/k_\perp^{\rm b}(\varepsilon_0)$. The family of permitted bifurcation wavevectors $k_\perp^{\rm b}$ is specified by one of the relations (3.2.4a, b, c) depending upon the boundary conditions chosen. For $L_\xi \to \infty$ the family of bifurcation solutions becomes quasi-continuous. The solution bifurcating nearest to $\varepsilon_{\rm h}$ or $\varepsilon_{\rm th}$ is the one with smallest wavevector compatible with the boundary conditions.

If the transverse dimension L_ξ of the sample is so small that $k_\perp^{\rm b}(\varepsilon_0) \geq 2\pi/L_\xi$ (for periodic boundary conditions) or $\geq \pi/L_\xi$ (for a finite plane geometry) or $\geq 2.4/R_0$ (for a finite cylindrical geometry with Dirichlet boundary conditions) cannot be satisfied for any ε_0, then the sample is stable against all inhomogeneous transverse fluctuations, and stationary filamentary structures cannot form in the ξ-direction. Which members of the bifurcating family actually can be observed, depends upon the stability of these inhomogeneous spatial structures; this will be investigated in Chap. 5. Usually, the solution corresponding to the largest period that is allowed by the boundary conditions is the most stable one; for cylindrical geometries this represents a "filament".

The bifurcation of a filamentary branch from the current-voltage characteristic of homogeneous negative differential conductivity states induced by a thermal instability has already been reported half a century ago [3.15].

3.3 Domain Instability

Let us now discuss longitudinal fluctuations in the direction of ε_0:

$$\nabla_\perp \delta v = 0 , \qquad \delta \varepsilon_\perp = 0 . \tag{3.3.1}$$

From (3.1.22) it then follows that the transverse current density fluctuation $\delta j_\perp^{\rm tot}$ vanishes everywhere. Eq. (3.1.21, 3.1) give $\nabla_\parallel \delta j_\parallel^{\rm tot} = 0$, hence $\delta j_\parallel^{\rm tot}$ is spatially constant and reflects the total current fluctuation in the external circuit. We restrict ourselves to constant current conditions: $\delta j_\parallel^{\rm tot} = 0$. Then the eigenvalue equation (3.1.32) becomes

$$\left[\tilde{V}(\lambda) - \varepsilon_0 \nabla_\parallel - \nabla_\parallel^2 - \frac{F(\lambda)}{G(\lambda)}(\varepsilon_0 + \nabla_\parallel) \right] \delta v = 0 . \tag{3.3.2}$$

Using the one-dimensional Fourier ansatz for periodic boundary conditions

$$\delta v(\zeta) = e^{ik_\parallel \zeta}$$

we obtain the complex dispersion relation

$$\tilde{V}(\lambda) - \frac{F(\lambda)}{G(\lambda)} \varepsilon_0 + k_\parallel^2 - ik_\parallel \left(\varepsilon_0 + \frac{F(\lambda)}{G(\lambda)} \right) = 0 . \tag{3.3.3}$$

For $k_\parallel \to 0$, (3.3.3) gives

$$\tilde{V}(\lambda) - \frac{F(\lambda)}{G(\lambda)} \varepsilon_0 = 0 \tag{3.3.4}$$

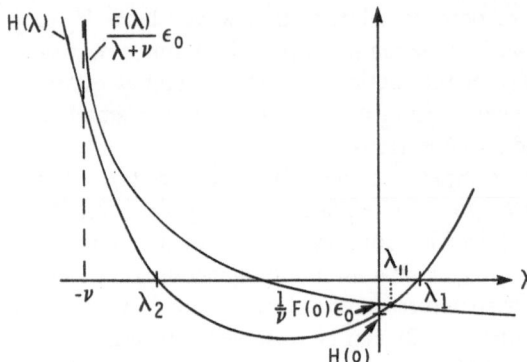

Fig. 3.8. Graphical construction of the long-wavelength limit $\lambda_\parallel$ of the longitudinal fluctuation spectrum: $\lambda_\parallel$ is the (largest) intersection of $H(\lambda)$ and $F(\lambda)\varepsilon_0/(\lambda + v)$ (schematic)

or equivalently

$$H(\lambda) = \frac{1}{\lambda + v} F(\lambda)\varepsilon_0 \ . \tag{3.3.5}$$

The solutions λ of (3.3.5) give the long-wavelength limit of the spectrum.

Obviously, the infinite-wavelength limit $\lambda(k_\parallel = 0)$ for longitudinal fluctuations is different from that of transverse fluctuations, $\lambda(k_\perp = 0)$, which is given by $\tilde{V}(\lambda) = 0$ according to (3.2.5). The difference is due to $\delta\varepsilon_\parallel \approx 0$ holding in the latter case. The term $F(\lambda)$ in (3.3.4) reflects the shift of the operating point ε_0 along the static current-field characteristic by $\delta\varepsilon_\parallel \neq 0$. In Fig. 3.8 the right-hand side and the left-hand side of (3.3.5) is plotted versus λ. The intersections of the two curves define the real solutions $\lambda(k_\parallel = 0)$. If the explicit field dependence of $\varphi_t(v, v_t, \varepsilon_0)$ is weak, then the curve $F(\lambda)\varepsilon_0/(\lambda + v)$ is close to the λ-axis, and $\lambda(k_\parallel = 0) \cong \lambda(k_\perp = 0)$. Let us denote the largest real solution $\lambda(k_\parallel = 0)$ of (3.3.5) by $\lambda_\parallel$. In the present section we exclude the possibility that (3.3.5) has complex solutions with $\text{Re}\{\lambda\} > \lambda_\parallel$. This case will be treated in Sect. 3.5, where we show that it can give rise to an oscillatory instability which is the basis of g-r induced self-sustained oscillations and chaos as discussed in Chap. 6.

We will now show that the sign of $\lambda_\parallel$ is connected with the static differential conductivity $\sigma_{\text{diff}} = v - [F(0)/H(0)]\varepsilon_0$ given by (3.1.43). For even M, according to Sect. 3.1.2, the following holds:

$$H(0) = -\Delta \left(\frac{\partial\rho}{\partial v}\right)_{\varepsilon_0} < 0 \text{ if and only if } \left(\frac{\partial\rho}{\partial v}\right)_{\varepsilon_0} > 0 \ ,$$

and $F(0) < 0$ always. Hence, by (3.1.43),

$$H(0) < F(0)\varepsilon_0/v \quad \text{holds if and only if}$$

$$\sigma_{\text{diff}} > 0 \quad \text{and} \quad \left(\frac{\partial\rho}{\partial v}\right)_{\varepsilon_0} > 0 \ . \tag{3.3.6}$$

We will see that this is essentially the condition for a moving-domain instability.

If (3.3.6) is violated then it follows that $\lambda_\parallel$ is negative (implying stability) provided that

$$\frac{dH}{d\lambda} > \varepsilon_0 \frac{d}{d\lambda}\left(\frac{F(\lambda)}{\lambda + v}\right) \qquad \text{for all } \lambda \geqq 0 . \tag{3.3.7}$$

In the two-level model the assumption (3.3.7) can readily be verified, using Table 3.1:

$$\frac{\partial H}{\partial \lambda} = 2\lambda - \lambda \operatorname{tr}\{\tilde{A}\} > 0 \tag{3.3.8}$$

since $\operatorname{tr} \tilde{A} < 0$;

$$\frac{\partial}{\partial \lambda}\left(\frac{F(\lambda)}{\lambda + v}\right) = \frac{-v\alpha_1 + \alpha_2}{(\lambda + v)^2} < 0 \tag{3.3.9}$$

since $\alpha_2/\alpha_1 < \tau_M(T^* + X^* + X_1^S + X_1^* N_D^* v) < v$.

The above result $\lambda_\parallel < 0$ means that NDC states, which are unstable against transverse fluctuations, are stable against longitudinal fluctuations in the long-wavelength limit $k_\parallel \to 0$, provided that (3.3.5) has no complex solutions.

If, on the other hand, (3.3.6) is satisfied, then (3.3.8) implies $\lambda_\parallel > 0$, i.e., instability against longitudinal fluctuations. This occurs on the positive differential conductivity portion ($\varepsilon_1 < \varepsilon_0 < \varepsilon_2$) of the middle branch of a $j_0 - \varepsilon_0$ characteristic like the one shown in Fig. 3.3b. The onset of such an instability is connected with $\lambda_\parallel = 0$, and is hence marked by $\sigma_{\text{diff}} = 0$, as follows from (3.3.5, 1.43).

In the hydrodynamic limit of small, but nonzero $k_\parallel$ we can use an expansion in terms of $k_\parallel$ to investigate the dispersion $\lambda(k_\parallel)$ following from (3.3.3). To the lowest order we set:

$$\lambda = \lambda_\parallel + \alpha k_\parallel^2 + i\beta k_\parallel . \tag{3.3.10}$$

Expanding

$$\tilde{V}(\lambda) \approx \tilde{V}(\lambda_\parallel) + (\lambda - \lambda_\parallel)V_1 + (\lambda - \lambda_\parallel)^2 V_2 \tag{3.3.11}$$

and

$$\frac{F(\lambda)}{G(\lambda)} \approx \frac{F(\lambda_\parallel)}{G(\lambda_\parallel)} + (\lambda - \lambda_\parallel)W_1 + (\lambda - \lambda_\parallel)^2 W_2 \tag{3.3.12}$$

we obtain, by equating coefficients up to order $O(k_\parallel^2)$,

$$\beta = \frac{\varepsilon_0 + F(\lambda_\parallel)/G(\lambda_\parallel)}{V_1 - W_1 \varepsilon_0} , \tag{3.3.13}$$

$$\alpha = -\frac{1 + \beta W_1 - \beta^2(V_2 - W_2 \varepsilon_0)}{V_1 - W_1 \varepsilon_0} . \tag{3.3.14}$$

The modes (3.3.10) describe propagating longitudinal charge density waves with a diffusion constant $\tilde{D}_\parallel := -\alpha$ and an effective mobility $\tilde{\mu} := \beta/\varepsilon_0$. If the field dependence of φ_t and hence $F(\lambda)$ is negligible, then $\lambda_\parallel \approx \lambda_1$, $\tilde{\mu} \approx 1/V_1 \approx \tilde{D}_\perp$ and $\tilde{D}_\parallel \approx \tilde{D}_\perp(1 - \tilde{D}_\perp^2 \varepsilon_0^2 V_2)$. For $\tilde{D}_\parallel > 0$ all modes $\lambda(k_\parallel)$ with

$$k_\parallel < k_\parallel^b := (\tilde{D}_\parallel/\lambda_\parallel)^{1/2} \tag{3.3.15}$$

are undamped in states where $\sigma_{\text{diff}} > 0$ and $(\partial\rho/\partial v)_{\varepsilon_0} > 0$ holds. The instability at $k_\parallel^b$ then leads to the successive bifurcation of a family of traveling waves with

$$\delta v, \delta\varepsilon_\parallel \sim \exp[ik_\parallel^b(\zeta + \tilde{\mu}\varepsilon_0\tau)]$$

with wavevector $k_\parallel^b(\varepsilon_0)$ and phase velocity $v(\varepsilon_0) = \tilde{\mu}(\varepsilon_0)\varepsilon_0$. They represent electric-field domains moving parallel to the applied static field ε_0. This is similar to the bifurcation of moving-field domains in the Gunn effect, where the current-field characteristic is N-shaped. In both cases the threshold for the existence of the instability is given by $\sigma_{\text{diff}} = 0$. The actual onset of the moving-domain instability is, however, determined by the boundary conditions at the contact $\zeta = 0$ [3.16, 17]. The interesting feature about the unconventional S-shaped characteristic shown in Fig. 3.3b is that it opens up the possibility of two different kinds of instabilities: at points where $\sigma_{\text{diff}} = \infty$, or equivalently $\lambda_1 = 0$, a transverse filamentary instability can occur; at points where $\sigma_{\text{diff}} = 0$, or equivalently $\lambda_\parallel = 0$, a longitudinal domain instability is possible. The latter is important, as it does not seem to have been noticed – except in [3.9] – in the existing literature, where moving domains have always been associated only with N-shaped current density-field characteristics. We note, however, that the temporal and spatial interaction of filaments and moving domains has in fact been observed in n-InSb [3.18], where two different mechanisms compete: impact ionization resulting in SNDC, and intervalley electron transfer resulting in NNDC.

3.4 Electromagnetic Modes

In this section we will present a general macroscopic theory of the dynamics of field fluctuations in media with bulk negative differential conductivity and one type of charge carriers [3.6].

3.4.1 Maxwell's Equations in Media with NDC

We assume that the medium is described by the local constitutive equations for the total current density (including displacement current)

$$j^{\text{tot}}(v, \varepsilon) = v\tilde{\mu}(\varepsilon)\varepsilon + \tilde{D}(\varepsilon)\nabla v + \frac{\partial}{\partial\tau}\varepsilon = vv(\varepsilon) + \tilde{D}(\varepsilon)\nabla v + \frac{\partial}{\partial\tau}\varepsilon \tag{3.4.1a}$$

and the charge density

$$\rho(v, \varepsilon) = 1 - v - \sum_i v_{t_i}(v, \varepsilon) \; . \tag{3.4.1b}$$

In generalizing Sect. 3.1–3 we have replaced the diffusion current ∇v by $\tilde{D}(\varepsilon)\nabla v$ with a dimensionless field-dependent diffusion coefficient $\tilde{D}(\varepsilon)$, and the drift current $v\varepsilon$ by $vv(\varepsilon) = v\tilde{\mu}(\varepsilon)\varepsilon$, where the dimensionless drift mobility $\tilde{\mu}$ may now have an arbitrary dependence upon the magnitude of ε (assuming isotropy). This allows for a nonmonotonic mobility $\mu(\varepsilon)$, as it occurs, for example, in the Gunn effect. Here the transfer of electrons from a high-mobility conduction-band valley to a low-mobility satellite valley with increasing field leads to an N-shaped $\mu(\varepsilon)$ relation, with a section of negative differential mobility (NDM), which, in turn, results in an N-shaped current density-field characteristic. Analogously, a nonmonotonic N-shaped dependence of $v + \sum_i v_{t_i}$ upon v in a certain range of ε leads to an S-shaped current density-field characteristic, as shown in Fig. 3.1. Note that in our dimensionless notation j^{tot} and ρ are measured in units of

$$J_0 := eN_D^* L_D/\tau_M$$

and $\rho_0 := eN_D^*$, respectively.

The spectrum of electromagnetic modes of fluctuations in the homogeneous steady state is given by Maxwell's equations

$$\nabla \times \delta\varepsilon = -\kappa \frac{\partial}{\partial\tau} \delta H \tag{3.4.2a}$$

$$\nabla \times \delta H = \delta j^{tot} \tag{3.4.2b}$$

$$\nabla \cdot \delta H = 0 \tag{3.4.2c}$$

$$\nabla \cdot \delta\varepsilon = \delta\rho \tag{3.4.2d}$$

coupled to the linearized constitutive equations (3.4.1a, b), cf., (3.1.22, 25),

$$\delta j^{tot} = (\lambda + \tilde{\mu}v)\delta\varepsilon + v\mu'\varepsilon_0\delta\varepsilon_\| + (\tilde{\mu}\varepsilon_0 + \tilde{D}\nabla)\delta v \tag{3.4.2e}$$

$$\delta\rho = -\frac{H(\lambda)}{G(\lambda)}\delta v - \frac{F(\lambda)}{G(\lambda)}\delta\varepsilon_\| \; . \tag{3.4.2f}$$

Here $\mu' = (d\tilde{\mu}/d\varepsilon)$, δH is the magnetic-field fluctuation normalized in units of $4\pi eN_D^* D_0/c$, and $\kappa := \epsilon_s L_D^2/(\tau_M^2 c^2)$ is the inverse squared velocity of light in the medium in units of L_D/τ_M. In practical cases (Table 3.2), κ is a very small number of the order of $10^{-8} \ldots 10^{-10}$. In (3.4.2e) we have used $\nabla v = 0$ (homogeneous steady state).

Note that (3.4.2b) is the integrated form of (3.1.21); in the absence of current vortices δj^{tot} is just the spatially constant current fluctuation in the external circuit.

Eliminating δv in (3.4.2e) by (3.4.2f, d), we find the dynamic current density-field relation

$$\delta j^{tot} = \sigma(\lambda)\delta\varepsilon \tag{3.4.3a}$$

with the dynamic differential conductivity tensor

$$\sigma(\lambda) := \lambda + \tilde{\mu}v + v\mu'\varepsilon_0 \otimes \hat{\varepsilon}_0 - \frac{G(\lambda)}{H(\lambda)}(\tilde{\mu}\varepsilon_0 + \tilde{D}\mathbf{V}) \otimes \mathbf{V} - \frac{F(\lambda)}{H(\lambda)}(\tilde{\mu}\varepsilon_0 + \tilde{D}\mathbf{V}) \otimes \hat{\varepsilon}_0 \ ,$$

$$(3.4.3b)$$

which describes the response of the current density upon a space- and time-dependent field fluctuation. Here $\hat{\varepsilon}_0$ is the unit vector in the direction of ε_0, and $\otimes$ denotes the tensor product.

In the special case of a static homogeneous longitudinal field fluctuation $\delta\varepsilon_\parallel$ (3.4.3b) simplifies to the static differential conductivity

$$\sigma_{\text{diff}} = v(\tilde{\mu} + \mu'\varepsilon_0) - \frac{F(0)}{H(0)}\tilde{\mu}\varepsilon_0 = v(\tilde{\mu} + \mu'\varepsilon_0) - \left(\frac{\partial\rho}{\partial v}\right)_{\varepsilon_0}^{-1}\left(\frac{\partial\rho}{\partial\varepsilon_0}\right)_v \tilde{\mu}\varepsilon_0 \ , \qquad (3.4.3c)$$

where we have used (3.1.38, 39). This shows that NDC ($\sigma_{\text{diff}} < 0$) can have two different causes: negative differential mobility ($d\mu/d\varepsilon < 0$) or a g-r instability ($\partial\rho/\partial v)_{\varepsilon_0} > 0$, corresponding to NNDC and SNDC, respectively.

From (3.4.2a, b, 3) we obtain the governing equation for the most general field fluctuations:

$$0 = \nabla \times \nabla \times \delta\varepsilon + \lambda\kappa\delta j^{\text{tot}} = \nabla \times \nabla \times \delta\varepsilon + \lambda\kappa\sigma(\lambda)\delta\varepsilon \ . \qquad (3.4.4)$$

It should be emphasized that (3.4.4) is more general than previously considered governing equations for field fluctuations in SNDC elements [3.19, 20] in the following respects:

(i) It allows for fluctuations of the carrier concentrations, which is essential for g-r induced SNDC.

(ii) It does not use any special assumptions about symmetry (cylindrical, etc.) or relative orientation of k, $\delta\varepsilon$, and ε_0.

(iii) It includes both NNDC and SNDC bulk mechanisms, thus allowing for the simultaneous occurrence of drift and g-r instabilities, as observed, for instance, in n-InSb [3.18].

The limiting case $F(\lambda) = 0$, $H(\lambda)/G(\lambda) = 1$ corresponds to the conventional Gunn effect with no allowance for electron trapping. The other limiting case $\tilde{\mu} = \tilde{D} = 1$, $\mu' = 0$ corresponds to a pure g-r instability with no drift nonlinearity.

For a rectangular geometry it is convenient to express (3.4.4) in terms of its components parallel and perpendicular to ε_0:

$$0 = \left\{-\nabla_\perp^2 + \alpha\left[\tilde{v} - \tilde{D}\nabla_\parallel^2 - \frac{F}{G}(\tilde{\mu}\varepsilon_0 + \tilde{D}\nabla_\parallel) - \tilde{\mu}\varepsilon_0\nabla_\parallel\right]\right\}\delta\varepsilon_\parallel$$

$$+ [\nabla_\parallel - \alpha(\tilde{\mu}\varepsilon_0 + \tilde{D}\nabla_\parallel)]\nabla_\perp\delta\varepsilon_\perp$$

$$0 = \left[\nabla_\parallel - \alpha\tilde{D}\left(\frac{F}{G} + \nabla_\parallel\right)\right]\nabla_\perp\delta\varepsilon_\parallel + [-\nabla_\parallel^2 + \alpha(\tilde{v} - \tilde{D}\nabla_\perp^2)]\delta\varepsilon_\perp \ , \qquad (3.4.5)$$

where

$$\alpha(\lambda) := \lambda\kappa\frac{G(\lambda)}{H(\lambda)} \tag{3.4.6}$$

and $\tilde{V}(\lambda) := [\lambda + v(\tilde{\mu} + \mu'\varepsilon_0)][H(\lambda)/G(\lambda)]$ is defined in analogy with (3.1.33). Here $\nabla_\perp \| \delta\varepsilon_\perp$ has been assumed for convenience.

For a cylindrical geometry, the vector components of (3.4.4) in terms of the cylindrical coordinates ζ, R are:

$$0 = \left\{-\frac{1}{R}\frac{\partial}{\partial R}R\frac{\partial}{\partial R} + \alpha\left[\tilde{V} - \tilde{D}\frac{\partial^2}{\partial\zeta^2} - \frac{F}{G}\left(\varepsilon_0\tilde{\mu} + \tilde{D}\frac{\partial}{\partial\zeta}\right) - \tilde{\mu}\varepsilon_0\frac{\partial}{\partial\zeta}\right]\right\}\delta\varepsilon_\|$$

$$+ \left[\frac{\partial}{\partial\zeta} - \alpha\left(\varepsilon_0\tilde{\mu} + \tilde{D}\frac{\partial}{\partial\zeta}\right)\right]\frac{1}{R}\frac{\partial}{\partial R}R\delta\varepsilon_\perp$$

$$0 = \left[\frac{\partial}{\partial\zeta} - \alpha\tilde{D}\left(\frac{F}{G} + \frac{\partial}{\partial\zeta}\right)\right]\frac{\partial}{\partial R}\delta\varepsilon_\|$$

$$+ \left[-\frac{\partial^2}{\partial\zeta^2} + \alpha\left(\tilde{V} - \tilde{D}\frac{\partial}{\partial R}\frac{1}{R}\frac{\partial}{\partial R}R\right)\right]\delta\varepsilon_\perp \ . \tag{3.4.7}$$

The field fluctuation $\delta\varepsilon$ can be decomposed into sources ($\nabla \times \delta\varepsilon = 0$) and vortices ($\nabla \cdot \delta\varepsilon = 0$), corresponding to longitudinal and transverse modes, respectively.

Before we proceed to discuss the possible modes in detail, it should be noted that the contact geometry induces important boundary conditions: If we assume metallic contacts in the $\zeta = 0$ and $\zeta = L_\zeta$ planes, the total voltage fluctuation

$$\delta V = \int_0^{L_\zeta} \delta\varepsilon_\| \, d\zeta \tag{3.4.8a}$$

must be independent of the transverse coordinates. Therefore $\delta\varepsilon_\|$ must either depend upon both the longitudinal and the transverse coordinates, or upon the longitudinal coordinate ζ only, but cannot depend upon the transverse coordinate alone. This condition was ignored in [3.19, 20]. A further boundary condition at metallic contacts is

$$\delta\varepsilon_\perp|_{\zeta=0} = \delta\varepsilon_\perp|_{\zeta=L_\zeta} = 0 \ , \tag{3.4.8b}$$

which states that $\delta\varepsilon_\perp$ cannot be independent of ζ unless it vanishes everywhere. Conditions (3.4.8a, b) require that both $\delta\varepsilon_\|$ and $\delta\varepsilon_\perp$ depend upon ζ as a result of the contacts. Also, our approximation of a spatially homogeneous ε_0 clearly breaks down at the contacts.

The boundary conditions of vanishing transverse current at the lateral surfaces requires

$$\delta j_\perp \equiv \tilde{D}\nabla_\perp\delta v + (\lambda + \tilde{\mu}v)\delta\varepsilon_\perp \equiv -\frac{G}{H}\tilde{D}\nabla_\perp\left(\nabla \cdot \delta\varepsilon + \frac{F}{G}\delta\varepsilon_\|\right) + (\lambda + \tilde{\mu}v)\delta\varepsilon_\perp = 0 \ . \tag{3.4.8c}$$

In conclusion, we have derived the basic eigenvalue equation (3.4.4) for the linear modes of the homogeneous steady state, where the dynamic differential conductivity tensor $\sigma(\lambda)$ defined in (3.4.3) contains all the material properties of the medium, i.e., the nonlinear drift mobility and g-r processes. More explicit forms of (3.4.4) are given in (3.4.5) for a rectangular geometry and in (3.4.7) for a cylindrical geometry. In Sects 3.4.2–4 we will adopt the view that a real wavevector k is specified by the boundary conditions, and investigate the linear modes $\sim \exp(\lambda\tau)$ with complex λ which are possible for each value of k. To this purpose we will consider all possible orientations of the applied static electric field ε_0, the electric-field fluctuation $\delta\varepsilon$, and the wavevector k relative to each other.

3.4.2 Transverse Modes ($k \perp \delta\varepsilon$)

First we derive the eigenvalue equations for purely *transverse* modes ($\nabla \cdot \delta\varepsilon = 0$), where the associated wavevector k is perpendicular to $\delta\varepsilon$. The latter property results from the Fourier transformation $\delta\varepsilon(\xi) = \int \delta\varepsilon(k)\exp(ik\xi)\,dk$. In this case, by (3.4.2) fluctuations of the charge density are excluded. Since the g-r instability manifests itself essentially in the charge density fluctuation, the transverse modes do not couple to the g-r instability, and we anticipate these modes to be always stable, whence $\lambda > 0$. We can use the vector relation

$$\nabla \times \nabla \times \delta\varepsilon = \nabla(\nabla \cdot \delta\varepsilon) - \Delta\delta\varepsilon \qquad (3.4.9)$$

to rewrite (3.4.4) as

$$[-\Delta + \kappa\lambda(\lambda + \tilde{\mu}v)]\delta\varepsilon + \kappa\lambda v\mu'\varepsilon_0\delta\varepsilon_\| - \kappa\lambda\frac{F(\lambda)}{H(\lambda)}(\tilde{\mu}\varepsilon_0 + \tilde{D}\nabla)\delta\varepsilon_\| = 0 \ . \qquad (3.4.10)$$

We can distinguish several cases according to the orientation of k and $\delta\varepsilon$ relative to ε_0:

(a) $k \| \varepsilon_0 \perp \delta\varepsilon$

For field fluctuations perpendicular to the applied field ε_0, and spatial modulation (k-vector) in the direction of ε_0, Eq. (3.4.10) can be simplified by setting $\delta\varepsilon_\| = 0$:

$$[-\Delta + \kappa\lambda(\lambda + \tilde{\mu}v)]\delta\varepsilon_\perp = 0 \ . \qquad (3.4.11)$$

We set $\delta\varepsilon_\perp \sim \sin k\zeta$, which satisfies the boundary conditions for $k = n\pi/L_\zeta$, and obtain the dispersion relation

$$k^2 + \kappa\lambda(\lambda + \tilde{\mu}v) = 0 \qquad (3.4.12)$$

which is solved by

$$\lambda = \frac{\tilde{\mu}v}{2} \pm \left[\left(\frac{\tilde{\mu}v}{2}\right)^2 - \frac{k^2}{\kappa}\right]^{1/2} \ . \qquad (3.4.13)$$

The spectrum is real for $k \leq (\tilde{\mu}v/2)\sqrt{\kappa}$, and complex for $k > (\tilde{\mu}v/2)\sqrt{\kappa}$. The latter case describes an electromagnetic wave propagating in the ε_0-direction with

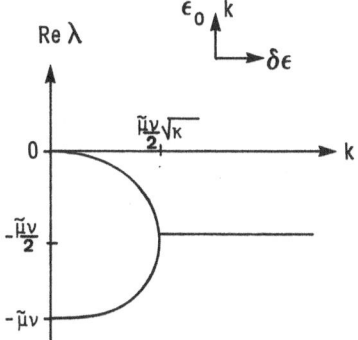

Fig. 3.9. Transverse electromagnetic modes of a single-carrier mechanism (schematic). The inset shows the orientation of the applied static electric field ε_0, the field fluctuation $\delta\varepsilon$, and the wavevector k relative to each other. All modes are damped $(\mathrm{Re}\{\lambda\} < 0)$; they are complex (propagating) for $k > \dfrac{\tilde{\mu}}{2}v\sqrt{\kappa}$

phase velocity $[1/\kappa - (\tilde{\mu}v)^2/4k]^{1/2}$ and damping constant $\tilde{\mu}v/2$. The real part of the spectrum is plotted in Fig. 3.9. It is always damped since $\mathrm{Re}\{\lambda\} \leq 0$. The damping (dielectric relaxation) is due to the conductivity ($\sigma = \tilde{\mu}v$ in our normalized units) which is, unlike the *differential* conductivity σ_{diff}, always positive.

(b) $k \perp \varepsilon_0 \parallel \delta\varepsilon$

For fluctuations $\delta\varepsilon$ parallel to the applied field ε_0, the spatial modulation must be perpendicular to ε_0 because of $\nabla \cdot \delta\varepsilon = 0$. We set $\delta\varepsilon_\perp = 0$ and $\nabla_\parallel \delta\varepsilon_\parallel = 0$ in (3.4.10):

$$\left\{-\Delta + \kappa\lambda\left[(\lambda + \tilde{\mu}v + v\mu'\varepsilon_0) - \frac{F(\lambda)}{H(\lambda)}\tilde{\mu}\varepsilon_0\right]\right\}\delta\varepsilon_\parallel = 0 \qquad (3.4.14a)$$

$$\kappa\lambda\frac{F(\lambda)}{G(\lambda)}\nabla_\perp\delta\varepsilon_\parallel = 0 \ . \qquad (3.4.14b)$$

Such fluctuations are excluded by the boundary condition (3.4.8a). In fact, from (3.4.14) it follows that either $\nabla_\perp\delta\varepsilon_\parallel = 0$, or $\lambda = 0$, or $F(\lambda) = 0$. In the first case $\delta\varepsilon_\parallel$ is spatially constant, and no transvere electromagnetic mode is obtained. In the second case only a static homogeneous mode results, which describes a uniform shift of the operating point along the current-field characteristic. The carrier concentration adjusts itself such that local neutrality is preserved. Finally, for $F(\lambda) = \mu' = 0$ (3.4.14a) is reduced to the dispersion relation (3.4.11), which requires $\lambda \leq 0$.

(c) $k \perp \varepsilon_0 \perp \delta\varepsilon$

For field fluctuations modulated perpendicular to both $\delta\varepsilon$ and ε_0, we obtain again the dispersion relation (3.4.12). These fluctuations do not, however, satisfy the boundary condition (3.4.8b).

(d) General Orientation

Assuming $k_\perp \neq 0$ and $\delta\varepsilon_\parallel \neq 0$, we obtain with the transversality condition $k_\parallel\delta\varepsilon_\parallel + k_\perp\delta\varepsilon_\perp = 0$ from the Fourier-transformed equation (3.4.10):

$$k^2 + \kappa\lambda\left[(\lambda + \nu(\tilde{\mu} + \mu'\varepsilon_0) - \frac{F(\lambda)}{H(\lambda)}(\varepsilon_0\tilde{\mu} + ik_\parallel\tilde{D})\right] = 0$$

$$k^2 + \kappa\lambda\left[(\lambda + \tilde{\mu}\nu) + \frac{F(\lambda)}{H(\lambda)}i\tilde{D}\frac{k_\perp^2}{k_\parallel}\right] = 0$$

Subtraction of these two equations yields

$$-\kappa\lambda\nu\mu'\varepsilon_0 + \kappa\lambda\frac{F(\lambda)}{H(\lambda)}[\tilde{\mu}\varepsilon_0 k_\parallel + i\tilde{D}(k_\parallel^2 + k_\perp^2)] = 0 \ ,$$

which cannot be solved with real k except for degenerate cases discussed in (b).

In conclusion, a spectrum of transverse electromagnetic modes exists only if $\delta\varepsilon \perp \varepsilon_0$; all modes are stable and do not couple with the g-r instability.

3.4.3 Longitudinal Modes ($k \parallel \delta\varepsilon$)

Next we consider *longitudinal* modes ($\nabla \times \delta\varepsilon = 0$), where k is parallel to $\delta\varepsilon$. From (3.4.2a) it follows that there are no time-dependent magnetic fields. Disregarding $\lambda = 0$, we obtain the tensor eigenvalue equation

$$\sigma(\lambda)\delta\varepsilon = 0 \tag{3.4.15}$$

Equation (3.4.15), by (3.4.9, 3b), takes the form

$$[\tilde{\bar{V}}(\lambda) - \tilde{D}\Delta]\delta\varepsilon_\parallel + [\tilde{V}(\lambda) - \tilde{D}\Delta]\delta\varepsilon_\perp - \tilde{\mu}\varepsilon_0(\nabla \cdot \delta\varepsilon) - \frac{F(\lambda)}{G(\lambda)}(\varepsilon_0\tilde{\mu} + \tilde{D}\nabla)\delta\varepsilon_\parallel = 0 \ ,$$

$$\tag{3.4.16}$$

where we have multiplied by $H(\lambda)/G(\lambda)$, assuming $G(\lambda) \neq 0$.

Again, k and $\delta\varepsilon$ can be orientated relative to ε_0 in different ways:

(a) $k \parallel \delta\varepsilon \parallel \varepsilon_0$

First we consider the case where the field fluctuation and the spatial modulation are both in the direction of ε_0. Using $\delta\varepsilon_\perp = 0$ and $\nabla_\perp \delta\varepsilon_\parallel = 0$ in (3.4.16), we obtain the one-dimensional eigenvalue problem

$$\left[\tilde{\bar{V}}(\lambda) - \tilde{\mu}\varepsilon_0\nabla_\parallel - \tilde{D}\nabla_\parallel^2 - \frac{F(\lambda)}{G(\lambda)}(\varepsilon_0\tilde{\mu} + \tilde{D}\nabla_\parallel)\right]\delta\varepsilon_\parallel = 0 \ . \tag{3.4.17}$$

If the mobility is constant ($\mu' = 0, \tilde{\mu} = \tilde{D} = 1$), this is equivalent to (3.3.2). The resulting modes become undamped on the middle branch of a tilted S-shaped current-field characteristic (Fig. 3.10), although $\sigma_{\text{diff}} > 0$, as discussed in Sect. 3.3 (domain instability). If, on the other hand, trapping is neglected ($F = 0, H/G = 1$), (3.4.17) results in the conventional dispersion relation of bulk NNDC

$$\lambda = -\nu(\tilde{\mu} + \mu'\varepsilon_0) - \tilde{D}k^2 + ik\varepsilon_0\tilde{\mu} \tag{3.4.18}$$

which gives a domain instability for $\sigma_{\text{diff}} = \nu(\tilde{\mu} + \mu'\varepsilon_0) < 0$ (NDM).

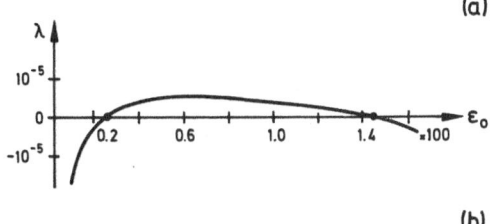

(a)

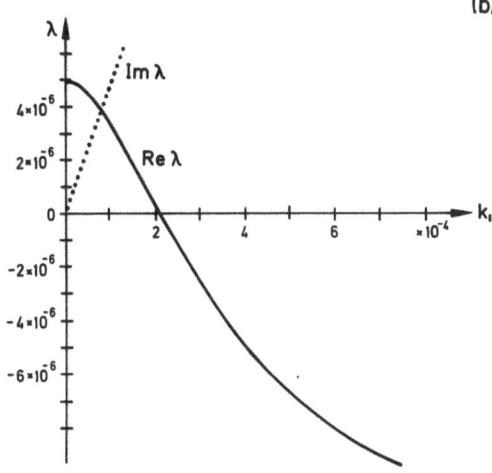

(b)

Fig. 3.10a, b. Longitudinal domain-type modes $(k \parallel \delta\varepsilon \parallel \varepsilon_0)$ for a two-level single-carrier mechanism giving anomalous-tilted SNDC (numerical parameters as in Table 3.3). (a) Largest eigenvalue λ versus control parameter ε_0 for $k_\parallel = 0$. (b) Eigenvalue λ versus the longitudinal wavevector $k_\parallel$ for $\varepsilon_0 = 50$. Solid line: $Re\{\lambda\}$, dotted line: $Im\{\lambda\}$ in units of τ_M^{-1}, $k_\parallel$ is in units of L_D^{-1}

Thus there are two different mechanisms for a domain instability: the conventional negative differential mobility known from the Gunn effect, and a novel type of g-r instability, as discussed in Sect. 3.3. Unstable complex, oscillatory modes, which may also be solutions of (3.4.17), will be discussed in Sect. 3.5.

(b) $k \parallel \delta\varepsilon \perp \varepsilon_0$

If we assume that the field fluctuation and the spatial modulation are both transverse to ε_0 ($\delta\varepsilon_\parallel = 0, \nabla_\parallel \delta\varepsilon_\perp = 0$), the transverse component of the vector equation (3.4.16) reduces to:

$$[\tilde{V}(\lambda) - \tilde{D}\nabla_\perp^2]\delta\varepsilon_\perp = 0 . \tag{3.4.19}$$

However, in general the parallel component does not vanish, but gives a contribution to the current density as defined by (3.4.3a):

$$-\varepsilon_0 \tilde{\mu}\nabla \cdot \delta\varepsilon \equiv \frac{H(\lambda)}{G(\lambda)} \delta j_\parallel^{tot} \neq 0 . \tag{3.4.20}$$

This is due to the generation of a ripple in the drift current density $j_\parallel = \tilde{\mu}\varepsilon_0 v$ by a transversally modulated carrier-concentration fluctuation $\delta v(k_\perp)$, even if $\delta\varepsilon_\parallel = 0$. The carrier-concentration fluctuation $\delta v(k_\perp)$ is in turn coupled to the transverse-field fluctuation $\delta\varepsilon_\perp$ by

Table 3.3. Numerical parameters for the two-level model used in Figs. 3.10, 11, 13, 6.11, 15 (linear modes)

Fig.	$T_1^S N_D^* \tau_M$	$T^* \tau_M$	$X_1^S \tau_M$	$X^* \tau_M$	$X_1 N_D^* \tau_M$	$X_1^* N_D^* \tau_M$	$(N_t - N_D^*)/N_D^*$	$\tilde{\mu}$	Remark
3.2, 7 ⎫⎬ 3.11 ⎭	10^{-2}	10^{-5}	5×10^{-6}	10^{-6}	$5 \times 10^{-4} \exp(-6/\varepsilon_0)$	$10^{-2} \exp(-1.5/\varepsilon_0)$	0.5	1	normal SNDC
3.10	10^{-2}	10^{-5}	10^{-6}	7.5×10^{-6}	$5 \times 10^{-4} \exp(-6/\varepsilon_0)$	$10^{-2} \exp(-1.5/\varepsilon_0)$	0.5	1	anomalous-tilted SNDC ($\varepsilon_1 \approx 25, \varepsilon_2 \approx 145$)
3.13	10^{-2}	10^{-5}	10^{-7}	10^{-7}	$5 \times 10^{-4} \exp(-6/\varepsilon_0)$	$10^{-2} \exp(-1.5/\varepsilon_0)$	0.3	$\dfrac{\arctan(0.3\varepsilon)}{0.3\varepsilon}$	Hopf bifurcation at $\varepsilon_c \approx 98$
6.11	10^{-2}	10^{-5}	5×10^{-6}	5×10^{-6}	$5 \times 10^{-4} \exp(-6/\varepsilon_0)$	$10^{-2} \exp(-1.5/\varepsilon_0)$	0.5	"	at $\varepsilon_c = 173.675$
6.15	10^{-2}	10^{-5}	5×10^{-6}	5×10^{-6}	$6 \times 10^{-4} \exp(-6/\varepsilon_0)$	$10^{-2} \exp(-1.5/\varepsilon_0)$	0.5	"	at $\varepsilon_{c_1} \approx 247$ $\varepsilon_{c_2} \approx 317$

$$\nabla_\perp \delta\varepsilon_\perp = -\frac{H(\lambda)}{G(\lambda)} \delta v \ , \tag{3.4.21}$$

see (3.4.2d, f) with $\delta\varepsilon_\| = 0$. A longitudinal current fluctuation $\delta j_\|^{tot} \neq 0$, by (3.4.4) violates the longitudinality condition $\nabla \times \delta\varepsilon = 0$ and induces either a nonzero field fluctuation $\delta\varepsilon_\|$ or a spatial dependence of $\delta\varepsilon_\perp$ in the direction of ε_0. However, since $\kappa\lambda$ is a small factor ($\lambda, \kappa \ll 1$), $\nabla \times \delta\varepsilon$ is small compared to $\delta\varepsilon_\perp$ and can be ignored. This is equivalent to neglecting time-dependent magnetic effects; it is reasonable because of the slow time-scale of g-r processes, in particular near the onset of a g-r instability.

Equation (3.4.19) is equivalent to (3.2.3) and leads to undamped recombination-diffusion modes, as discussed in Sect. 3.2, see the spectrum shown in Fig. 3.6. However, the contact boundary condition (3.4.8b) cannot be satisfied. Note that this mode does not couple to the negative differential mobility mechanism. Therefore, NDM cannot lead to filamentation.

(c) General Orientation

Considering one transversal direction only and assuming $k_\| \neq 0$, we can use the longitudinality condition $\nabla \times \delta\varepsilon = 0$ in the Fourier-transformed form

$$k_\perp \delta\varepsilon_\| = k_\| \delta\varepsilon_\perp \tag{3.4.22}$$

to eliminate $\delta\varepsilon_\perp$ from (3.4.16):

$$\left[(\tilde{V} + \tilde{D}k^2) - i\varepsilon_0\tilde{\mu}k_\| \left(1 + \frac{k_\perp^2}{k_\|^2}\right) - \frac{F}{G}(\varepsilon_0\tilde{\mu} + ik_\|\tilde{D}) \right]\delta\varepsilon_\| = 0 \tag{3.4.23}$$

$$\left[(\tilde{V} + \tilde{D}k^2) - \frac{F}{G}ik_\|\tilde{D} \right]\delta\varepsilon_\| = 0 \ . \tag{3.4.24}$$

Subtraction of (3.4.23, 24) yields

$$\left(v\mu'\varepsilon_0\frac{H}{G} - \frac{F}{G}\tilde{\mu}\varepsilon_0 \right)k_\| - i\tilde{\mu}\varepsilon_0 k^2 = 0 \ , \tag{3.4.25}$$

which does not have a real solution $k_\|, k_\perp$, hence no dispersion relation $\lambda(k_\|, k_\perp)$ is obtained.

In conclusion, for NNDC and for anomalous-tilted SNDC we have obtained unstable longitudinal modes giving rise to moving domains. In the limit of negligible retardation effects (slow g-r process: $\lambda \approx 0$) normal SNDC leads to undamped modes $k \| \delta\varepsilon \perp \varepsilon_0$ causing filamentation. Unstable oscillatory modes giving rise to self-sustained oscillations and chaos are also possible, as discussed in Sect. 3.5.

3.4.4 Mixed Modes

A rigorous theory of filamentation that is consistent with the contact boundary conditions (3.4.8) requires the solution of the full two-dimensional eigenvalue problem (3.4.5) for a rectangular geometry, or of (3.4.7) for a cylindrical geometry, since $\delta\varepsilon_\parallel$ and $\delta\varepsilon_\perp$ must depend upon both the longitudinal and the transverse coordinate in order to describe a filamentary mode which satisfies (3.4.8). We confine attention to a pure g-r instability ($\tilde{\mu} = \tilde{D} = 1, \mu' = 0$). In case of (3.4.7) we attempt a solution in the form

$$\delta\varepsilon_\parallel(R, \zeta) = \delta\tilde{\varepsilon}_\parallel J_0(k_\perp R)e^{ik_\parallel\zeta}$$

$$\delta\varepsilon_\perp(R, \zeta) = \delta\tilde{\varepsilon}_\perp J_1(k_\perp R)e^{ik_\parallel\zeta} \tag{3.4.26}$$

where J_0, J_1 are Bessel functions of the first kind of order 0 and 1, respectively, which satisfy the relations

$$\frac{1}{R}\frac{\partial}{\partial R}R\frac{\partial}{\partial R}J_0(k_\perp R) = -k_\perp^2 J_0(k_\perp R) , \qquad \frac{\partial}{\partial R}J_0 = -k_\perp J_1$$

$$\frac{\partial}{\partial R}\frac{1}{R}\frac{\partial}{\partial R}RJ_1(k_\perp R) = -k_\perp^2 J_1(k_\perp R) , \qquad \frac{1}{R}\frac{\partial}{\partial R}RJ_1 = +k_\perp J_0 . \tag{3.4.27}$$

Substitution of (3.4.26) reduces (3.4.7) to the algebraic equation

$$\begin{pmatrix} k_\perp^2 + \alpha\left[\tilde{V} + k_\parallel^2 - \dfrac{F}{G}(\varepsilon_0 + ik_\parallel) - ik_\parallel\varepsilon_0\right] & [ik_\parallel(1 - \alpha) - \alpha\varepsilon_0]k_\perp \\[2mm] -\left[ik_\parallel(1 - \alpha) - \alpha\dfrac{F}{G}\right]k_\perp & [k_\parallel^2 + \alpha(\tilde{V} + k_\perp^2)] \end{pmatrix}\begin{pmatrix} \delta\tilde{\varepsilon}_\parallel \\[2mm] \delta\tilde{\varepsilon}_\perp \end{pmatrix} = 0 . \tag{3.4.28}$$

The dispersion relation $\lambda(k_\parallel, k_\perp)$ is given by the condition that the determinant of the matrix in (3.4.28) vanishes:

$$0 = \alpha\left\{k_\parallel^4 + k_\perp^4 + 2k_\parallel^2 k_\perp^2(1 + \alpha) + k_\parallel^2\left[\tilde{V}(1 + \alpha) - \frac{F}{G}\varepsilon_0\right]\right.$$

$$\left. + k_\perp^2\tilde{V}(1 + \alpha) + \alpha\tilde{V}\left(\tilde{V} - \frac{F}{G}\varepsilon_0\right) - ik_\parallel\left(\varepsilon_0 + \frac{F}{G}\right)(k_\parallel^2 + k_\perp^2 + \alpha\tilde{V})\right\} . \tag{3.4.29}$$

Note that $\alpha, \tilde{V}, F, G$ depend upon λ, and by (3.4.11, 17) $\alpha\tilde{V} = \lambda\kappa(\lambda + \nu)$. The same dispersion relation is obtained in Cartesian coordinates by a Fourier transformation of (3.4.5).

A numerical solution of (3.4.29) is shown in Fig. 3.11 for normal SNDC; there is a domain of unstable modes (Re$\{\lambda\} > 0$) in the $(k_\parallel, k_\perp)$ plane. The boundary condition (3.4.8c) requires values of $k_\perp$ such that $J_1(k_\perp R_0) = 0$ if R_0 is the radius of a cylindrical sample, which follows from (3.4.26, 27). The permitted values of $k_\parallel$ have to satisfy the boundary conditions (3.4.8a, b), hence, for zero voltage fluctuations, $k_\parallel = n(2\pi/L_\zeta)$, where $n\in\mathbb{N}$. In order to gain some analytical insight into the dispersion relation we consider two limiting cases:

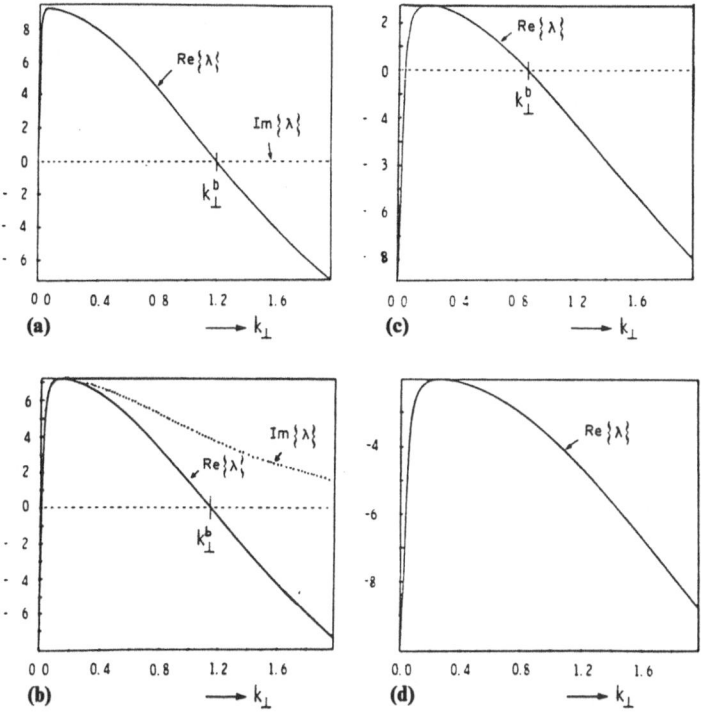

Fig. 3.11a–d. Mixed electromagnetic filamentary modes of a two-level single-carrier SNDC mechanism (numerical parameters as in Table 3.3 with $\varepsilon_0 = 8.4$). Re$\{\lambda\}$ (*solid line*) and Im$\{\lambda\}$ (*dotted line*) are plotted versus $k_\perp$ for the following values of $k_\parallel$: (**a**) 0, (**b**) $0.001/L_D$, (**c**) $0.002/L_D$, (**d**) $0.003/L_D$. λ is in units of $10^{-6}/\tau_M$, $k_\perp$ is in units of $0.1/L_D$.

(a) $k_\perp = 0$

The matrix in (3.4.28) becomes diagonal, and we obtain two decoupled modes: the longitudinal, potentially unstable domain mode $\delta\tilde{\varepsilon}_\parallel$ given by (3.4.17) in Sect. 3.4.3a, and the transverse, stable dielectric-relaxation mode $\delta\tilde{\varepsilon}_\perp$ given by (3.4.12) in Sect. 3.4.2a.

(b) $k_\parallel = 0$

Assuming large L_r, we neglect $k_\parallel$ in (3.4.29) and obtain the dispersion relation (which in general does not decouple):

$$\alpha(\lambda)\left\{[k_\perp^2 + \alpha(\lambda)\bar{V}(\lambda)][\bar{V}(\lambda) + k_\perp^2] - \frac{F(\lambda)}{G(\lambda)}\varepsilon_0\alpha(\lambda)\bar{V}(\lambda)\right\} - 0 \ . \tag{3.4.30}$$

If the explicit field dependence of φ_t is negligible, $F(\lambda) \approx 0$, then the expression enclosed in braces in (3.4.30) factorizes into

$$[k_\perp^2 + \lambda\kappa(\lambda + \nu)](\bar{V} + k_\perp^2) = 0 \ , \tag{3.4.31}$$

where the factor $[k_\perp^2 + \lambda\kappa(\lambda + v)]$ describes a stable transverse dielectric-relaxation mode $k \perp \varepsilon_0 \| \delta\varepsilon$ with a dispersion discussed in Sect. 3.4.2a. The factor $(\tilde{V} + k_\perp^2)$ gives the dispersion relation of Sect. 3.4.3b, 3.2 for a mixed mode with $k \perp \varepsilon_0$ and

$$\delta\varepsilon_\| / \delta\varepsilon_\perp = \alpha(\lambda)\varepsilon_0 ik_\perp / [k_\perp^2 + \lambda\kappa(\lambda + v)] \ . \tag{3.4.32}$$

The dispersion relation has an undamped branch, see (3.2.11).

Note that in the limit of small $\lambda\kappa$ the second mode is essentially longitudinal, since $\delta\varepsilon_\| \to 0$ for $\lambda \to 0$ by (3.4.6).

The term $F(\lambda)$ in (3.4.30) leads to the coupling of the two modes. The exact spectrum can be found analytically by solving the dispersion relation (3.4.30), which yields:

$$k_\perp^2 = -\tfrac{1}{2}(\tilde{V}(\lambda) + \lambda\kappa(\lambda + v) \pm \{[\tilde{V}(\lambda) - \lambda\kappa(\lambda + v)]^2$$
$$+ 4\lambda\kappa(\lambda + v)\varepsilon_0 F(\lambda)/G(\lambda)\}^{1/2}) \ . \tag{3.4.33}$$

We will first discuss the behavior of the spectrum in the vicinity of certain distinguished points. In the limit $k_\perp \to 0$ (3.4.30) gives

$$\alpha(\lambda)\lambda\kappa(\lambda + v)\left[\tilde{V}(\lambda) - \frac{F(\lambda)}{G(\lambda)}\varepsilon_0\right] = 0 \ . \tag{3.4.34}$$

Besides the solutions $\lambda = 0$ and $\lambda = -v$, (3.4.34) has as a solution the critical eigenvalue $\lambda_\|$, for which the square brackets vanish. It has been discussed in connection with the longitudinal instability in Sect. 3.3, see (3.3.4). We have found that $\lambda_\|$ is positive on the middle branch of the current-field characteristic if $\sigma_{\text{diff}} > 0$, i.e., $\delta\varepsilon$ is parallel to ε_0.

In order to find the behavior of the spectrum in the vicinity of the point $(\lambda = \lambda_\|, k_\perp = 0)$, we set

$$\lambda \approx \lambda_\| - \tilde{D}k_\perp^2 \tag{3.4.35}$$

and expand $\tilde{V}(\lambda)$ and $F(\lambda)/G(\lambda)$ around $\lambda_\|$. By equating coefficients up to order $O(k_\perp^2)$, we obtain from (3.4.30):

$$\tilde{D} = \left[\frac{\partial\tilde{V}}{\partial\lambda}(\lambda_\|)\right]^{-1}[1 + \tilde{V}(\lambda_\|)/(\lambda_\|\kappa(\lambda_\| + v))] \ . \tag{3.4.36}$$

The approximate form of the spectrum near $(\lambda = 0, k_\perp = 0)$ follows from an expansion of (3.4.33) in terms of λ:

$$k_\perp \approx (-\kappa\lambda\sigma_{\text{diff}})^{1/2} \qquad \text{or} \tag{3.4.37}$$

$$\lambda \approx -\frac{k_\perp^2}{\kappa\sigma_{\text{diff}}} \ , \tag{3.4.38}$$

which is damped for $\sigma_{\text{diff}} > 0$ and undamped for $\sigma_{\text{diff}} < 0$. The corresponding eigenvector is parallel to ε_0 for small λ. In order to obtain the condition for the crossing of eigenvalues from negative to positive values we set $\lambda = 0$ in (3.4.33), which yields, apart from the solution $k_\perp = 0$, which has been discussed above,

$$k_\perp = \sqrt{-\tilde{V}(0)} \ . \tag{3.4.39}$$

This wavevector is exactly the same as that of the transverse fluctuation instability studied in Sect. 3.2. It was shown to lead to the bifurcation of a family of transversally modulated filamentary structures from the middle branch of the $j - \varepsilon_0$ characteristic, cf., (3.2.11).

Here the corresponding eigenvector follows from substituting $\lambda = 0$ and $k_\perp = [-\tilde{V}(0)]^{1/2}$ into (3.4.28): $\delta\varepsilon_\parallel = 0$. Thus, at the bifurcation point the approximation of a purely longitudinal mode with $k \| \delta\varepsilon \perp \varepsilon_0$, invoked in Sect. 3.4.3b, becomes exact. Hereby we have corroborated our filamentation argument of Sect. 3.2. We can compute the behavior of the dispersion $\lambda(k_\perp)$ in the vicinity of the crossing point (3.4.39) by expanding (3.4.33) up to the lowest order in λ. The result is

$$k_\perp \cong \sqrt{-\tilde{V}(0)} - \frac{1}{\sqrt{-\tilde{V}(0)}} \left[\frac{d\tilde{V}}{d\lambda}(0) + \kappa\varepsilon_0 \frac{F(0)}{H(0)} \right] \lambda \ . \tag{3.4.40}$$

The term in parentheses in (3.4.40) is positive, and hence $dk_\perp/d\lambda < 0$.

The relevant part of the spectrum $(\lambda > \lambda_1^\infty)$ is plotted in Fig. 3.12 for the middle branch of the current density-field characteristic. Figure 3.12a shows the dispersion for a normal S-shaped characteristic with negative differential conductivity, while Fig. 3.12b corresponds to the part $\varepsilon_1 < \varepsilon_0 < \varepsilon_2$ of the tilted S-characteristic shown in Fig. 3.3b. In both cases, on the critical branch the electromagnetic mode gradually changes from transverse $(k \perp \delta\varepsilon \| \varepsilon_0)$ to longitudinal $(k \| \delta\varepsilon \perp \varepsilon_0)$ when $k_\perp$ increases from zero to the bifurcation vector $k_\perp^b = [-\tilde{V}(0)]^{1/2}$. The bifurcation wavevector itself varies with increasing ε_0 as shown in Fig. 3.7 and thus leads to the bifurcation of a family of filamentary solutions.

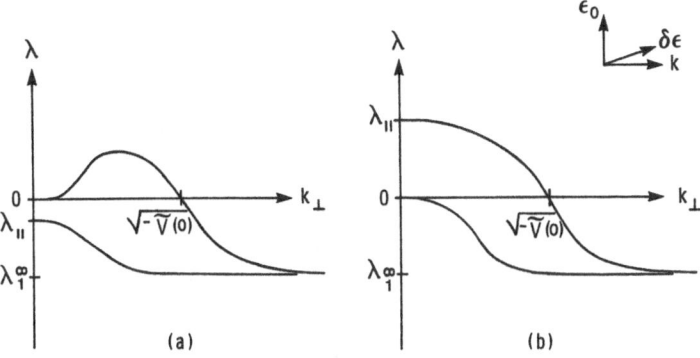

Fig. 3.12a, b. Mixed electromagnetic filamentary modes of a single-carrier SNDC mechanism (schematic). (a) Normal SNDC, as in Fig. 3.3a, (b) anomalous-tilted SNDC, as in Fig. 3.3b ($\varepsilon_1 < \varepsilon_0 < \varepsilon_2$). The inset shows the orientation of the applied static electric field ε_0, the field fluctuation $\delta\varepsilon$, and the wavevector k relative to each other

Equation (3.4.38) is the dispersion relation used by *Shaw* et al. [3.19] and *Bass* et al. [3.20] in their filamentation arguments. From our analysis which includes fluctuations of the carrier concentrations, it follows that the dielectric mode (3.4.38) approximates the spectrum only in the vicinity of ($\lambda = 0, k_\perp = 0$), and that the bifurcation of filamentary structures occurs, in fact, at a larger wavevector $k_\perp^b$ (3.4.39), which is determined by g-r processes coupled to transverse diffusion, and which is independent of the sign of the static differential conductivity σ_{diff}, as long as the middle of three branches of the current-field characteristic is considered, cf., Figs. 3.3a, b. Our bifurcation argument requires that the real part of λ crosses from negative to positive values at a finite $k_\perp$, which cannot happen with the dispersion relation (3.4.38).

In conclusion, a rigorous analysis of filamentary modes compatible with the boundary conditions has given an undamped spectrum of mixed modes $\lambda(k_\perp)$, changing from purely longitudinal at $k_\perp = 0$ to purely transverse at the bifurcation point $k_\perp = k_\perp^b$ of current filaments. In case of normal SNDC (as opposed to anomalous-tilted SNDC) the upper branch of the spectrum (Fig. 3.12a) is very similar to the spectrum $\lambda(k)$ obtained by the Cahn-Hilliard theory of *spinodal decomposition* in equilibrium phase transitions, as will be discussed in Sect. 5.4.1. This adds another important feature to the analogy of phase transitions in equilibrium and in nonequilibrium systems.

3.5 Oscillatory Instability

In the last section of this chapter we discuss the possibility of undamped oscillatory longitudinal modes, and derive analytical conditions for the bifurcation of self-sustained oscillations (limit cycles) from the homogeneous steady state [3.21]. These limit cycles will be investigated in the nonlinear regime in Sect. 6.2.2, and it will be shown in Sect. 6.3.4 that for appropriate numerical parameters the limit cycles undergo a sequence of secondary bifurcations leading eventually to chaotic oscillations.

We confine attention to longitudinal modes and longitudinal fluctuations ($k \parallel \delta\varepsilon \parallel \varepsilon_0$):

$$\delta\varepsilon_\parallel \sim \exp[\lambda(k_\parallel)\tau + ik_\parallel\zeta] \, ,$$

as in Sect. 3.4.3a. The dispersion relation is then given by (3.4.17), with the definitions

$$\tilde{V}(\lambda) := (\lambda + \tilde{v})H(\lambda)/G(\lambda), \quad \tilde{v} := v\, dv/d\varepsilon, \quad v(\varepsilon) := \tilde{\mu}(\varepsilon)\varepsilon, \quad \text{and}$$

F, G, H as in (3.1.27), (3.1.24), (3.1.26), respectively:

$$(\lambda + \tilde{v})H(\lambda) - (ik_\parallel v - \tilde{D}k_\parallel^2)G(\lambda) - (v + i\tilde{D}k_\parallel)F(\lambda) = 0 \tag{3.5.1}$$

all evaluated at the homogeneous steady state.

An oscillatory instability occurs if $\text{Re}\{\lambda\} > 0$ and $\text{Im}\{\lambda\} \neq 0$. In particular, for homogeneous fluctuations ($k_\parallel = 0$), a Hopf bifurcation of a limit cycle (Fig. 1.13) is found if two complex conjugate eigenvalues λ cross the imaginary axis.

3.5.1 Two-Level Models

An explicit form of the Hopf bifurcation condition is readily found for models with two trap levels, in which case (3.5.1) is reduced, with $k_\parallel = 0$, to a real cubic polynomial in λ:

$$\lambda^3 + g_2\lambda^2 + g_1\lambda + g_0 = 0 \ , \qquad \text{where} \tag{3.5.2}$$

$$g_0 := \tilde{v}\det\{\tilde{A}\} + v(\varepsilon_0)\alpha_2$$

$$= \tilde{v}\lambda_1\lambda_2 + v(\varepsilon_0)\left[(\tilde{A}_{22} - \tilde{A}_{21})\frac{\partial\varphi_{t_1}}{\partial\varepsilon} + (\tilde{A}_{11} - \tilde{A}_{12})\frac{\partial\varphi_{t_2}}{\partial\varepsilon}\right] \ , \tag{3.5.3}$$

$$g_1 := \det\{\tilde{A}\} - \tilde{v}\,\text{tr}\{\tilde{A}\} + v(\varepsilon_0)\alpha_1$$

$$= \lambda_1\lambda_2 - \tilde{v}(\lambda_1 + \lambda_2) + v(\varepsilon_0)\frac{\partial\varphi_0}{\partial\varepsilon} \ , \tag{3.5.4}$$

$$g_2 := \tilde{v} - \text{tr}\{\tilde{A}\} = \tilde{v} - \lambda_1 - \lambda_2 \ . \tag{3.5.5}$$

Here we have used (3.1.34, 35), and some of the expressions listed in Table 3.1. We recall that λ_1, λ_2 are the eigenvalues of the g-r matrix $\tilde{A}$. For standard mass-action kinetics we have shown in Chap. 2, (2.1.42), using a theorem by Tyson and Light, that $\text{tr}\{\tilde{A}\} < 0$ always holds. Hence for positive differential mobility $(dv/d\varepsilon > 0)$ it follows that

$$g_2 > 0 \ . \tag{3.5.6}$$

Further, we note that the static differential conductivity is given by

$$\sigma_{\text{diff}} = \tilde{v} - v(\varepsilon_0)\left(\frac{\partial\rho}{\partial\varepsilon}\right)_v\bigg/\left(\frac{\partial\rho}{\partial v}\right)_\varepsilon \ , \tag{3.5.7}$$

in analogy with (3.1.41, 42) which applies to the special case $\tilde{\mu}(\varepsilon_0) = 1$, i.e., $v(\varepsilon_0) = \varepsilon_0$. Hence g_0 can be manipulated into the form

$$g_0 = -\sigma_{\text{diff}}\left(\frac{\partial\rho}{\partial v}\right)_\varepsilon \Delta \ , \qquad \text{where} \tag{3.5.8}$$

$$H(0) = \det\{\tilde{A}\} = -\left(\frac{\partial\rho}{\partial v}\right)_\varepsilon \Delta \ ,$$

$$F(0) = -\alpha_2 = -\left(\frac{\partial\rho}{\partial c}\right)_v \Delta \ ,$$

$$G(0) = \det\{B\} = \Delta > 0 \tag{3.5.9}$$

has been used.

From (3.5.8) it follows that for normal positive differential conductivity $[\sigma_{\text{diff}} > 0, (\partial\rho/\partial v)_\varepsilon < 0]$ *or* for normal SNDC $[\sigma_{\text{diff}} < 0, (\partial\rho/\partial v)_\varepsilon > 0]$

$$g_0 > 0 \tag{3.5.10a}$$

holds, whereas for NNDC $[\sigma_{\text{diff}} < 0, (\partial\rho/\partial v)_\varepsilon < 0]$ *or* for anomalous tilted SNDC $[\sigma_{\text{diff}} > 0, (\partial\rho/\partial v)_\varepsilon > 0]$

$$g_0 < 0 \tag{3.5.10b}$$

holds.

A zero eigenvalue bifurcation of the saddle-node type (A1) in Fig. 1.12 occurs when g_0 changes sign. This happens at the peak current density of an NNDC characteristic (Fig. 1.1a), and at the field ε_1 and ε_2 of an anomalous tilted SNDC characteristic (see Fig. 3.3b). In both cases it is associated with the crossing of a real eigenvalue $\lambda = \lambda_\parallel$ of (3.5.2) from negative to positive values, and with the bifurcation of a moving domain, as discussed in Sect. 3.3, 3a.

A Hopf bifurcation (Fig. 1.13), on the other hand, occurs when $\text{Re}\{\lambda\} =: \lambda_0$ changes sign, and at the bifurcation point $\lambda = \pm i\omega \neq 0$ is a solution of (3.5.2). In the vicinity of the bifurcation point we have two complex conjugate solutions $\lambda_0 \pm i\omega$ and, in addition, one real solution $\lambda_\parallel < 0$ of (3.5.2). With this ansatz it follows that the coefficients of the cubic polynomial (3.5.2) have the form

$$g_0 = -\lambda_\parallel(\lambda_0^2 + \omega^2) > 0 \tag{3.5.11a}$$

$$g_1 = \lambda_0^2 + \omega^2 + 2\lambda_0\lambda_\parallel > 0 \qquad \begin{array}{l}(g_1, g_2 > 0 \\ \text{sufficiently close to} \\ \text{bifurcation point)}\end{array} \tag{3.5.11b}$$

$$g_2 = -\lambda_\parallel - 2\lambda_0 > 0 \tag{3.5.11c}$$

The bifurcation point is given by $\lambda_0 = 0$; by (3.5.11) we obtain for this case:

$$g_0 - g_1 g_2 = 0 \ . \tag{3.5.12}$$

The inequalities $g_0 > 0$, $g_2 > 0$ have been shown to hold for standard mass-action kinetics and positive differential mobility, and for normal SNDC in (3.5.6, 10a) above. Hence (3.5.12) represents the condition for the Hopf bifurcation of a limit cycle from the homogeneous NDC state of a normal SNDC characteristic [provided that $g_0 - g_1 g_2$ has different sign above and below the point on the SNDC characteristic given by (3.5.12)]. Using the definitions (3.5.3–5) we can recast the condition (3.5.12) as

$$0 = g_0 - g_1 g_2$$

$$= v(\varepsilon_0)\left(K_1 \frac{\partial\varphi_{t_1}}{\partial\varepsilon} + K_2 \frac{\partial\varphi_{t_2}}{\partial\varepsilon}\right) + (\lambda_1 + \lambda_2)(\tilde{v} - \lambda_1)(\tilde{v} - \lambda_2) \tag{3.5.13}$$

with

$$K_1 := \tilde{v} - \tilde{A}_{11} - \tilde{A}_{21} \ , \qquad K_2 := \tilde{v} - \tilde{A}_{22} - \tilde{A}_{12} \ . \tag{3.5.14}$$

For standard g-r kinetics, with impact-ionization coefficients monotonically increasing with field, and trapping cross sections not increasing with field, $\partial\varphi_{t_1}/\partial\varepsilon$ and

$\partial \varphi_{t_2}/\partial \varepsilon$ are negative, furthermore $\tilde{v}$, K_1, and K_2 are positive (assuming $dv/d\varepsilon > 0$), as will be shown explicitly for the g-r mechanism of Fig. 2.4 in Chap. 6, (6.2.57). Hence it follows from (3.5.13) with $\lambda_1 + \lambda_2 = \text{tr}\{\tilde{A}\} < 0$ and $\lambda_2 < 0$ that

$$\tilde{v} \equiv v(dv/d\varepsilon) < \lambda_1 \qquad (3.5.15)$$

is a necessary condition for a Hopf bifurcation. This can be achieved on the NDC branch (where $\lambda_1 > 0$!) of the static current-field characteristic if the electron concentration v is low and the differential mobility $dv/d\varepsilon$ is small (e.g., if the drift velocity is near saturation). We note that $\tilde{v}^{-1}$ is the dimensionless differential dielectric-relaxation time (in physical units: $\epsilon_s(en\,dv/d\varepsilon)^{-1}$, cf., (3.1.5), and λ_1^{-1} is a characteristic g-r lifetime essentially influenced by impact ionization. Therefore (3.5.15) singles out *relaxation semiconductors* [3.22] – as opposed to standard *lifetime* semiconductors – as candidates for oscillatory instabilites. These are typically found at low temperatures and at low doping (high-purity materials), see the experimental survey in Sect. 6.3.2. Note, however, that (3.5.15) is only a necessary, but not sufficient condition.

Figure 3.13a depicts the complex eigenvalue $\lambda = \lambda_0 + i\omega$ of (3.5.2) as a function of the static electric field ε_0 for the two-level model of Fig. 2.4. As ε_0 is increased, $\text{Re}\{\lambda\}$ changes from negative to positive values at a critical field ε_c given by the Hopf bifurcation condition (3.5.13). At ε_c spatially homogeneous self-sustained oscillations bifurcate from the NDC state. Note that the two complex conjugate eigenvalues $\lambda_0 \pm i\omega$ become real above $\varepsilon_0 > 170$. In Fig. 3.13b the solution $\lambda(k_\parallel)$ of the dispersion relation (3.5.1) is plotted for the same numerical parameters. It can be seen that $\text{Re}\{\lambda\}$ slightly increases with $k_\parallel$ up to a local maximum, and then falls off due to diffusion-type damping.

The essential difference of the oscillatory modes of this section and the domain-type instabilities of Sect. 3.3 is that

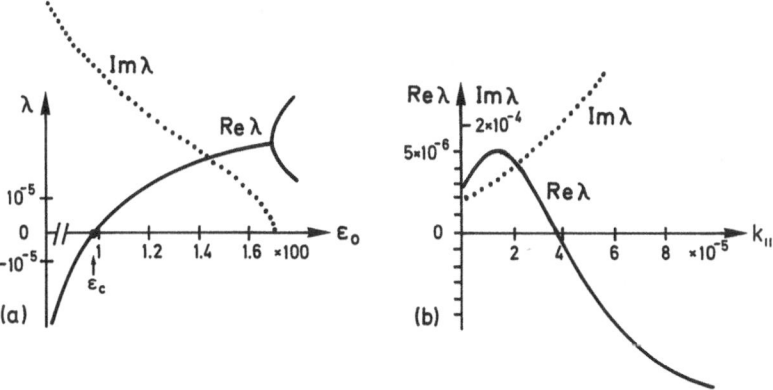

Fig. 3.13a, b. Longitudinal oscillatory modes ($k \parallel \delta\varepsilon \parallel \varepsilon_0$) of a two-level single-carrier SNDC mechanism (numerical parameters as in Table 3.3). (a) Complex eigenvalue λ versus control parameter ε_0 for $k_\parallel = 0$. (b) Complex eigenvalue λ versus longitudinal wavevector $k_\parallel$ for $\varepsilon_0 = 100$. λ is in units of τ_M^{-1}, $k_\parallel$ is in units of L_D^{-1}

$$\lim_{k_\parallel \to 0} \text{Im}\{\lambda(k_\parallel)\} = \omega \neq 0$$

holds here, which leads to *nonpropagating* spontaneous oscillations $\delta\varepsilon_\parallel \sim \exp(i\omega\tau)$. The domain instability, in contrast, is characterized by $\text{Im}\{\lambda(k_\parallel)\} \approx \tilde{v}k_\parallel + O(k_\parallel^2)$, which leads to propagating waves $\delta\varepsilon_\parallel \sim \exp[ik_\parallel(\zeta + \tilde{v}\tau)]$.

The oscillatory instability will be further elaborated in Chap. 6.

3.5.2 Single-Level Models

In this section we shall investigate the possibility of a Hopf bifurcation in single-level g-r models. In this case the matrices $\tilde{A}$ (3.1.29) and B (3.1.17), and the vector f(3.1.20) reduce to scalars, and (3.5.1) becomes quadratic in λ, with

$$H(\lambda) = \lambda - \lambda_1 = \lambda - (\partial\varphi_t/\partial v_t)_{v=1-v_t} \ ,$$

$$G(\lambda) = \lambda - \lambda_1^\infty = \lambda - (\partial\varphi_t/\partial v_t)_{v,\varepsilon} \ ,$$

$$F(\lambda) = f = (\partial\varphi_t/\partial\varepsilon)_{v,v_t} \ , \tag{3.5.16}$$

where $\lambda_1 := \tilde{A}$ is the g-r eigenvalue of homogeneous, charge-neutral carrier-density fluctuations, and $\lambda_1^\infty := B$ describes fluctuations δv_t with $\delta v = 0$. For standard mass-action kinetics the steady state $v > 0$ is always stable against homogeneous neutral fluctuations, and $\lambda_1 < 0$, $\lambda_1^\infty < 0$ holds, cf., Sect. 2.1.1. With the g-r processes X_1^S, T_1^S and X_1 (Fig. 1.10), e.g., the resulting explicit expressions are

$$\lambda_1 = -(X_1^S/v + T_1^S N_D^* v + X_1 N_D^* v)\tau_M \ ,$$

$$\lambda_1^\infty = -(X_1^S + T_1^S N_D^* v + X_1 N_D^* v)\tau_M \ ,$$

$$f = -\frac{\partial}{\partial\varepsilon}[X_1^S v_t + X_1 N_D^* v v_t - T_1^S(N_t - N_D^* v_t)]\tau_M \ . \tag{3.5.17}$$

For homogeneous fluctuations, $k_\parallel = 0$, (3.5.1) then reduces to

$$\lambda^2 + h_1\lambda + h_0 = 0 \qquad \text{where} \tag{3.5.18}$$

$$h_0 := -(\tilde{v}\lambda_1 + vf) \qquad h_1 := \tilde{v} - \lambda_1 \ . \tag{3.5.19}$$

The static differential conductivity is given by

$$\sigma_{\text{diff}} = \tilde{v} - vF(0)/H(0) = \tilde{v} + vf/\lambda_1 \ , \tag{3.5.20}$$

cf., (3.1.43), hence

$$h_0 = -\sigma_{\text{diff}}\lambda_1 \ . \tag{3.5.21}$$

The solutions λ of (3.5.18) can be completely classified by considering the signs of h_0, h_1, and $h_1^2 - 4h_0$ (see Sect. 1.2.3). Two complex eigenvalues with negative (positive) real parts, corresponding to a stable (unstable) focus, exist if $h_0 > 0$,

$h_1^2 - 4h_0 < 0$, and $h_1 > 0$ ($h_1 < 0$). A Hopf bifurcation occurs when h_1 changes sign, while $h_0 > 0$:

$$\tilde{v} = \lambda_1$$

$$-\sigma_{\text{diff}}\lambda_1 > 0 \ . \tag{3.5.22}$$

If the differential mobility $dv/d\varepsilon$ is positive, $\tilde{v} \equiv v\, dv/d\varepsilon > 0$ holds. On the other hand, $\lambda_1 < 0$ holds, and therefore a Hopf bifurcation cannot occur.

If, however, the differential *mobility* is negative, while the differential *conductivity* is positive, (3.5.22) can be satisfied, this requires a *negative* differential dielectric-relaxation time v^{-1} of the order of the g-r lifetime λ_1^{-1}. In contrast to the two-level model of Sect. 3.5.1, the destabilizing process is not g-r based, but mobility based (undamped differential dielectric "relaxation"). The requirement of positive σ_{diff} can be met by a negative f of sufficiently large magnitude, see (3.5.20). This could be due to an impact-ionization coefficient X_1 increasing with field ε, by (3.5.17), such that the rise in carrier concentration with increasing field overcompensates the decrease in mobility. An oscillatory instability can thus be induced by the combination of a drift instability (e.g., intervalley transfer of electrons) with impurity impact ionization.

In conclusion, a Hopf bifurcation at $k_\parallel = 0$ cannot occur in single-level g-r models with positive differential mobility, but may be induced by negative differential mobility in combination with impact ionization. Also, other oscillatory instabilities at non-zero wavevector $k_\parallel > 0$ are not excluded by the present analysis. Such instabilities could arise from the nonmonotonic behavior of the dispersion relation $\lambda(k_\parallel)$. Nonmonotonic dispersion relations of this type can also occur in reaction-diffusion systems with second- and fourth-order spatial derivatives, where they have been shown [3.23] to lead to the bifurcation of traveling and standing waves.

4. Stationary Transverse Spatial Structures

In the previous chapter we have shown by a linear-mode analysis that g-r instabilities of SNDC type can lead to the growth of spatial fluctuations modulated perpendicular to the homogeneous field and current flow, and to the bifurcation of stationary plane current layers (sheaths) and cylindrical current filaments. In this chapter we calculate the fully developed stationary transverse structures from the *nonlinear* transport equations [4.1–6]. In particular, we analyze the resulting current-voltage characteristics and their dependence upon boundary conditions. The first three sections deal with single-carrier g-r mechanisms, while the last two consider two-carrier mechanisms. Dissipative structures in semiconductors, like current filaments, do not only essentially influence the current-voltage characteristics, but can also be observed directly in experiments.

4.1 Plane Current Layers

We assume a single-carrier g-r mechanism for SNDC as in Sect. 2.1, and consider a semiconductor slab shaped as a parallelepiped with sides L_ξ, L_η, L_ζ [4.2]. Intuited by the linear-mode analysis of Sect. 3.2, we will analyse the transport equations for stationary spatial structures that are modulated only along one of the two transverse directions (say that ζ-direction). We anticipate a ζ-dependent current density as shown schematically in Fig. 4.1. The bifurcation argument of Sect. 3.2.2 has established this in the vicinity of the bifurcation points. The values of the applied field ε_0 at which these bifurcations occur are given by $k_\perp^b(\varepsilon_0) = 2\pi n/L_\xi$ for periodic boundary conditions, and $k_\perp^b(\varepsilon_0) = \pi n/L_\xi$ for finite boundary conditions where $n \in \mathbb{N}$, and the function $k_\perp^b(\varepsilon_0)$ is shown in Fig. 3.7.

In order to calculate explicitly the spatial profiles in the whole SNDC range $\varepsilon_h < \varepsilon_0 < \varepsilon_{th}$, we must now retain the nonlinearities in the transport equations.

Adding the nonlinear continuity equations for free (3.1.8) and trapped electrons (3.1.9) and observing the particle conservation (3.1.11), we obtain in the steady state a divergence-free conduction current j:

$$\nabla \cdot j \equiv \nabla \cdot (\varepsilon v + \nabla v) = 0 \ . \tag{4.1.1}$$

The second governing equation for the steady state is Maxwell's equation:

$$\nabla \cdot \varepsilon = \rho(v, \varepsilon) \ , \tag{4.1.2}$$

where the trapped electron densities have been eliminated by (3.1.14). Note that (4.1.1) reflects the differential form of Ampère's law for time-independent fields

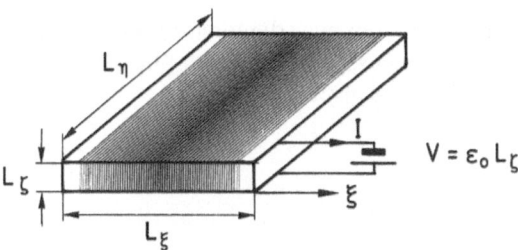

Fig. 4.1. Current sheath (*shaded dark*) in a semiconductor slab of dimensions L_ξ, L_η, L_ζ, connected to a voltage bias $V = \varepsilon_0 L_\zeta$ (schematic)

$$\nabla \times H = \frac{4\pi}{c} j \, , \tag{4.1.3}$$

where H is the magnetic field.

For plane transverse spatial structures we can divide the electric field into an externally applied homogeneous field ε_0 in the direction of the current flow, and an internal, ξ-dependent field $\varepsilon_\perp$ perpendicular to ε_0, which is generated by the lateral carrier diffusion:

$$\varepsilon(\xi) = \varepsilon_0 + \varepsilon_\perp(\xi) \, . \tag{4.1.4}$$

Substituting (4.1.4) into (4.1.1, 2), and observing the boundary condition of vanishing transverse current density, we obtain:

$$j_0(\xi) = \varepsilon_0 v(\xi) \tag{4.1.5}$$

$$\frac{d}{d\xi} v = -\varepsilon_\perp v \tag{4.1.6}$$

$$\frac{d}{d\xi} \varepsilon_\perp = \rho(v, \varepsilon_0) \, . \tag{4.1.7}$$

Here j_0 is the current density parallel to the applied field ε_0. In (4.1.7) we have assumed $|\varepsilon_\perp| \ll |\varepsilon_0|$, so that the field dependence of ρ (through the g-r coefficients) is upon the applied external field ε_0 only. Thus ρ depends upon ε_0 as an external control parameter. Equations (4.1.6, 7) are the governing equations which describe the spatial variation of the carrier density, and hence the current density, perpendicular to the current flow.

4.1.1 Phase Portraits

An important tool in the investigation of nonlinear dissipative structures, as given e.g., by (4.1.6, 7), is the phase-portrait analysis. This method was introduced by *Böer* and collaborators [4.7] in the analysis of NNDC instabilities, and has been extensively used in the theory of longitudinal domains in the Gunn effect and other NNDC instabilities [4.7–13]. Here we shall use the phase-portrait analysis to investigate *transverse* spatial structures.

Equations (4.1.6, 7) represent an autonomous nonlinear dynamic system in the two variables v, $\varepsilon_\perp$, which can be conveniently discussed in the $(v, \varepsilon_\perp)$ phase plane. The trajectories of the system in the phase plane are given by

$$\frac{d\varepsilon_\perp}{dv} = -\frac{\rho(v, \varepsilon_0)}{\varepsilon_\perp v} , \tag{4.1.8}$$

which can be solved by the exact first integral

$$\varepsilon_\perp(v) = \pm\{2[C - \phi(v, \varepsilon_0)]\}^{1/2} , \tag{4.1.9}$$

where C is a constant, and

$$\phi(v, \varepsilon_0) := \int \rho(v, \varepsilon_0) \frac{dv}{v} \tag{4.1.10}$$

is an elementary integral since the integrand is rational for all common (i.e., polynomial) g-r rates.

Different trajectories belong to different values of the integration constant $C \geq \phi(v_2, \varepsilon_0)$. The trajectories are parametrized by the transverse coordinate ζ. They are symmetric with respect to the v-axis, since (4.1.8) is invariant against inversion $\varepsilon_\perp \to -\varepsilon_\perp$. This reflects the fact that neither direction of the transverse field $\varepsilon_\perp$ is distinguished.

The null-isoclines, where either

$$d\varepsilon_\perp/dv = 0 \qquad \text{or}$$

$$dv/d\varepsilon_\perp = 0$$

are given by $\rho(v, \varepsilon_0) = 0$ and $\varepsilon_\perp v = 0$, respectively. If ε_0 is chosen in the SNDC range $(\varepsilon_h < \varepsilon_0 < \varepsilon_{th})$, $\rho(v, \varepsilon_0) = 0$ has three solutions $v_1(\varepsilon_0)$, $v_2(\varepsilon_0)$, $v_3(\varepsilon_0)$. For $\varepsilon_0 < \varepsilon_h$ and for $\varepsilon_0 > \varepsilon_{th}$ there is only one solution. Therefore the phase-plane trajectories cross the straight lines $v = v_1$, $v = v_2$, $v = v_3$ (where they exist) with zero slope $d\varepsilon_\perp/dv$. When a trajectory crosses the v-axis, its slope is infinite. The phase portraits for different values of ε_0 are shown in Fig. 4.2. Figures 4.2a–c correspond to increasing values of ε_0 within the SNDC range. From the definition (4.1.10), it follows that the extrema of ϕ as a function of v are the homogeneous steady states, where $\rho(v, \varepsilon_0) = 0$. More precisely, stable homogeneous states correspond to maxima since (Sect. 3.1.2)

$$\left[\frac{\partial^2 \phi}{\partial v^2}\right]_{v_i} = \left\{\frac{\partial}{\partial v}\left[\frac{1}{v}\rho(v, \varepsilon_0)\right]\right\}_{v_i} = \left[-\frac{1}{v^2}\rho(v, \varepsilon_0) + \frac{1}{v}\left(\frac{\partial \rho}{\partial v}\right)_{\varepsilon_0}\right]_{v_i}$$

$$= \left[\frac{1}{v}\left(\frac{\partial \rho}{\partial v}\right)_{\varepsilon_0}\right]_{v_i} < 0$$

and unstable homogeneous states correspond to minima. For $\varepsilon_0 < \varepsilon_h$ and $\varepsilon_0 > \varepsilon_{th}$, ϕ has one maximum, while for $\varepsilon_h < \varepsilon_0 < \varepsilon_{th}$ (Figs. 4.2a–c) one minimum and two maxima exist. Fig. 4.2a corresponds to the case $\phi(v_1, \varepsilon_0) > \phi(v_3, \varepsilon_0)$, which occurs

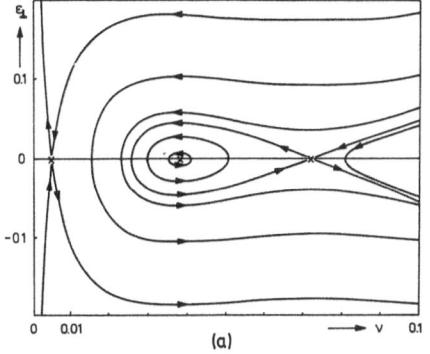

(a)

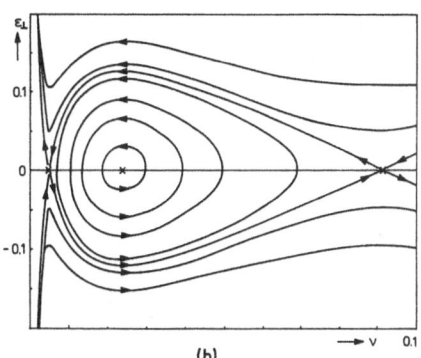

(b)

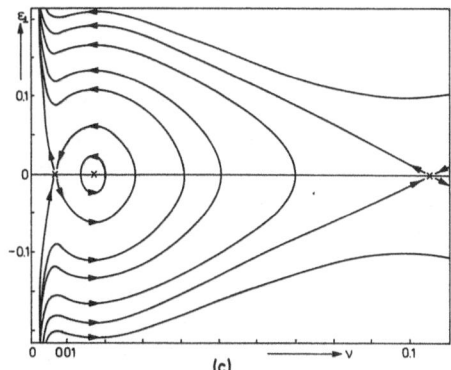

(c)

Fig. 4.2a–c. Phase portraits of the inhomogeneous steady states: transverse field $\varepsilon_\perp$ versus electron concentration v. The parameter varying along the phase trajectories is the transverse coordinate ξ. The singular points (x) correspond to the homogeneous steady states v_1, v_2, v_3. The numerical parameters are those of Table 3.2 (a) $\varepsilon_0 = 8$, (b) $\varepsilon_0 = 8.49$, (c) $\varepsilon_0 = 9$. Here and in the following figures the carrier concentrations are in units N_D^*, and $\varepsilon_\perp, \varepsilon_0$ are in units $(kT)/(eL_D)$

at lower values of ε_0, whereas Fig. 4.2c corresponds to $\phi(v_1, \varepsilon_0) < \phi(v_3, \varepsilon_0)$ which belongs to higher values of ε_0.

We shall now discuss the phase portraits in detail for ε_0-values in the SNDC range. The singular points of (4.1.6, 7) correspond to the spatially homogeneous steady states $(v = v_i, \varepsilon_\perp = 0)$ for $i = 1, 2, 3$, defined by $\rho(v_i, \varepsilon_0) = 0$. Since the system (4.1.6, 7) has a first integral, i.e., is conservative, the singular points may be saddles or centers.

Linearization of (4.1.6, 7) around $(v = v_i, \varepsilon_\perp = 0)$ yields

$$\begin{pmatrix} \dfrac{d}{d\xi} \delta v \\[2mm] \dfrac{d}{d\xi} \varepsilon_\perp \end{pmatrix} = \begin{pmatrix} 0 & -v_i \\[2mm] \left(\dfrac{\partial \rho}{\partial v} \right)_{\varepsilon_0} & 0 \end{pmatrix} \begin{pmatrix} \delta v \\[2mm] \varepsilon_\perp \end{pmatrix} \tag{4.1.11}$$

with $\delta v := v - v_i$. The eigenvalues of the matrix in (4.1.11) are

$$\kappa_i = \pm \left[-v_i \left(\frac{\partial \rho}{\partial v} \right)_{\varepsilon_0} \right]^{1/2} . \tag{4.1.12}$$

Hence (Fig. 3.1), the singular point represents a saddle point if the corresponding homogeneous steady state is stable, i.e., $(\partial\rho/\partial v)_{\varepsilon_0} < 0$, and a center if the homogeneous state is unstable, $(\partial\rho/\partial v)_{\varepsilon_0} > 0$. In case of a saddle point the separatrices become tangent to the eigenvector at the saddle point, and their slopes are thus given by

$$\frac{d\varepsilon_\perp}{dv} = \pm\left[-\left(\frac{\partial\rho}{\partial v}\right)_{\varepsilon_0}\bigg/v_i\right]^{1/2} \qquad (i = 1, 3) \ .$$

The slope is larger for v_1 than for v_3 since $v_1 < v_3$. The center is surrounded by closed trajectories corresponding to periodic oscillatory solutions. These are ellipses near the center, and represent small-amplitude harmonic spatial oscillations $\delta v(\xi) \sim \sin k_\perp\xi$ with a wavevector

$$k_\perp := i\kappa_2 = \pm\left[v_2\left(\frac{\partial\rho}{\partial v}\right)_{\varepsilon_0}\right]^{1/2} \ . \qquad (4.1.13)$$

A comparison of (4.1.13) and (3.2.11) reveals that $k_\perp$ is the bifurcation wavevector for the given value of ε_0, and the small-amplitude oscillations are the stationary solutions whose emergence from a bifurcation we have analysed in Sect. 3.2.2. As the amplitude of the periodic solutions increases, they become more and more anharmonic, and their period $\Lambda = 2\pi/k_\perp$ increases. The phase-plane region of periodic solutions is confined by separatrices originating at the saddle points. Three topologically different cases may occur (Fig. 4.2a–c). In Fig. 4.2a, two separatrices of the same saddle point at v_3 are joined together, forming a closed separatrix loop (homoclinic orbit) with a minimum electron concentration $v_{\min}$. In Fig. 4.2c which corresponds to higher values of ε_0, two separatrices of the saddle point at v_1 form a closed loop (homoclinic orbit) with the maximum $v_{\max}$. Figure 4.2b shows the limit case where the two saddle points v_1 and v_3 are connected by two symmetric separatrices (heteroclinic orbits). The corresponding inhomogeneous profiles connect the two stable steady states with a narrow interfacial layer.

All trajectories originating outside these separatrices are unbounded and represent profiles where either v or $\varepsilon_\perp$ tends to infinity for large $|\xi|$. They are associated with values of the integration constant $C > \text{Min}[\phi(v_1, \varepsilon_0), \phi(v_3, \varepsilon_0)]$.

Which of the mentioned trajectories are the physical solutions, depends upon the boundary conditions on the lateral surfaces of the SNDC elements. We shall discuss this in detail in Sect. 4.3. Here we only note that the solitary solutions are of particular interest for large systems with practically infinite L_ξ, since these solutions tend asymptotically to stable homogeneous steady states for $\xi \to \pm\infty$ and $\varepsilon_\perp(\pm\infty) = 0$, i.e., there are no surface charges. Periodic solutions of period Λ are permitted by periodic boundary conditions on the interval $L_\xi = n\Lambda, n \in \mathbb{N}$. Portions of periodic, solitary, or unbounded trajectories can be observed under finite Dirichlet or Neumann boundary conditions.

Let us now briefly comment [4.6] on the approximation $|\varepsilon_\perp| \ll |\varepsilon_0|$, which we invoked in the governing equation (4.1.7). If we extend the phase portraits to large magnitudes of $\varepsilon_\perp$, where the approximation is no longer valid, we have to replace ε_0 in (4.1.7) by $(\varepsilon_0^2 + \varepsilon_1^2)^{1/2}$. Thus the magnitude of the local field is enhanced in regions with a large carrier-density gradient. At large enough values of $|\varepsilon_\perp|$, such that $\varepsilon_0^2 + \varepsilon_1^2 > \varepsilon_{\text{th}}^2$, the effective local field exceeds the SNDC range, where

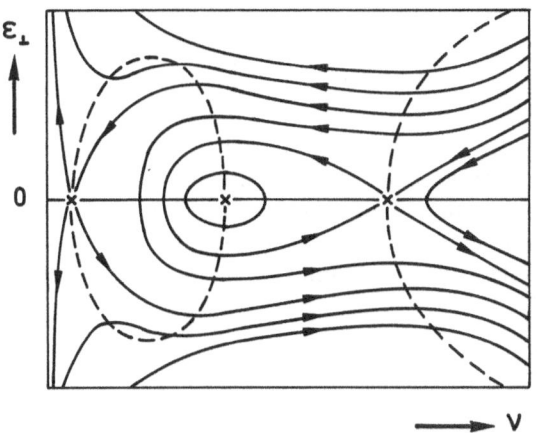

$\varepsilon_\perp$

0

v

$\rho(v, (\varepsilon_0^2 + \varepsilon_\perp^2)^{1/2}) = 0$ has three positive solutions; hence the two lowest solutions of $\rho(v, (\varepsilon_0^2 + \varepsilon_\perp^2)^{1/2}) = 0$ no longer exist; only the highest solution is still possible. Therefore the null-isoclines $d\varepsilon_\perp/dv = 0$ are no longer straight lines parallel to the $\varepsilon_\perp$-axis, but have the form indicated in Fig. 4.3. The trajectories are slightly distorted, as compared to Fig. 4.2, far from the v-axis, but the qualitative topological nature of the phase portrait, in particular the nature of the singular points and the inversion symmetry $\varepsilon_\perp \to -\varepsilon_\perp$, is not affected. This case is of relevance for SNDC mechanisms (as discussed in Sect. 2.1) which involve very shallow donors at low temperatures such that SNDC occurs at very low values of the electric field.

4.1.2 Equal-Areas Rule

We shall now show that the phase portrait of Fig. 4.2b is distinguished by a special condition which allows for spatial coexistence of the two stable homogeneous phases with a thin-plane interfacial layer in an infinitely extended medium. This condition can be cast into the form of a simple geometrical construction: an equal-areas rule.

The system (4.1.6, 7) is not a Hamiltonian system, where the first integral is given by energy conservation; it may, however, be transformed to such a form by introducing the new variable

$$\mu := \ln v \ , \tag{4.1.14}$$

which is, up to a constant, the reduced quasi-Fermi level of the electrons. Setting $\tilde{\rho}(\mu, \varepsilon_0) := \rho(v, \varepsilon_0)$, we obtain

$$\mu''(\xi) + \tilde{\rho}(\mu, \varepsilon_0) = 0 \ , \tag{4.1.15}$$

which has the first integral

$$\tfrac{1}{2}(\mu')^2 + \tilde{\phi}(\mu) = \text{const} \tag{4.1.16}$$

with the potential

$$\tilde{\phi}(\mu) := \int \tilde{\rho}(\mu, \varepsilon_0)\, d\mu \ . \tag{4.1.17}$$

(We suppress the dependence upon ε_0). Of course, (4.1.15) still describes a dissipative physical system, but it is formally "Hamiltonian" in the sense that (4.1.15) is analogous to the equation of motion of a particle of unity mass under a conservative force $-\tilde{\rho}$, if we identify μ with the spatial coordinate, and ζ with the time. The "energy conservation" (4.1.16) is equivalent to the first integral (4.1.9) in the phase plane; note that

$$\mu' = -\varepsilon_\perp \quad \text{and} \quad \tilde{\phi}(\ln v) = \phi(v, \varepsilon_0) \ . \tag{4.1.18}$$

The "potential" $\tilde{\phi}(\mu)$ has two maxima at μ_1, μ_3, and one minimum at μ_2, corresponding to the two stable and one unstable homogeneous steady states v_1, v_3, and v_2, respectively (Fig. 4.4). Spatial coexistence between v_1 and v_3 can be described by the boundary conditions

$$\lim_{\zeta \to -\infty} \mu(\zeta) = \mu_1 , \qquad \lim_{\zeta \to +\infty} \mu(\zeta) = \mu_3 \tag{4.1.19}$$

(or, equivalently, μ_1 and μ_3 interchanged). From (4.1.19) it follows that

$$\lim_{\zeta \to \pm\infty} \mu'(\zeta) = 0 \ . \tag{4.1.20}$$

Using (4.1.19, 20) as upper and lower limits of integration in (4.1.16), we find the main result of this section:

$$\tilde{\phi}(\mu_3) - \tilde{\phi}(\mu_1) \equiv \int_{\mu_1}^{\mu_3} \tilde{\rho}(\mu, \varepsilon_0)\, d\mu = 0 \ . \tag{4.1.21}$$

In terms of the electron density v, (4.1.21) may be written as

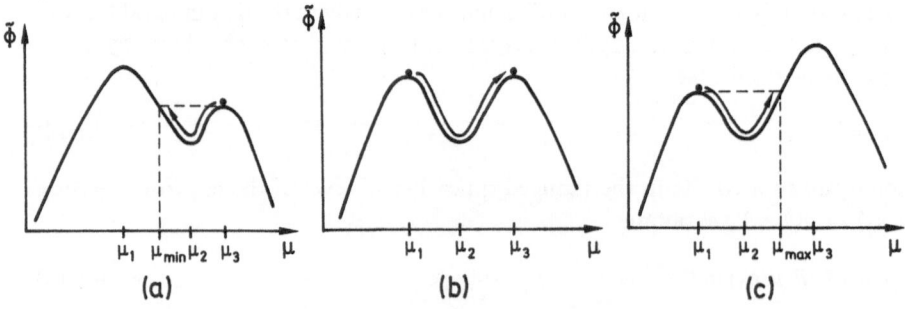

Fig. 4.4a–c. Potential $\tilde{\phi}$ versus $\mu = \ln v$ corresponding to the phase portraits in Fig. 4.2a–c, respectively (schematic). The extrema at μ_1, μ_2, μ_3 correspond to the steady states v_1, v_2, v_3. The mechanical analogy of motion of a mass point in the potential $\tilde{\phi}$ is indicated schematically by arrows

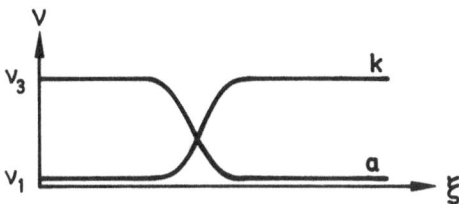

$$\int_{v_1}^{v_3} \rho(v, \varepsilon_0) \frac{dv}{v} = 0 \; . \tag{4.1.22}$$

Condition (4.1.21) singles out a specific value of the electric field ε_0: the coexistence field ε_{co} with $\varepsilon_h < \varepsilon_{co} < \varepsilon_{th}$. In terms of phase portraits this corresponds to Fig. 4.2b, and the coexistence solutions are the two heteroclinic orbits (saddle-to-saddle separatrices).

In the mechanical analogy, the boundary conditions (4.1.19, 20) correspond to a motion where the particle is initially at rest on top of one of the potential peaks, then travels through the minimum in a relatively short time, and comes asymptotically to a rest at the other potential peak (Fig. 4.4b). Therefore the two maxima of $\tilde\phi$ must be of equal height, as stated in (4.1.21). Near the peaks the particle moves slowly since the force is small. Hence the profile $v(\xi)$ rises from v_1 to v_3 (or drops from v_3 to v_1) essentially within a thin interfacial layer, and is almost constant everywhere else (Fig. 4.5).

The profiles resemble a kink or antikink, and describe phase coexistence of two phases with high (v_3) and low (v_1) conductivity, i.e., two plane layers with high- and low-current density. The position of the interface between these two layers is not fixed by the boundary conditions, but can be shifted arbitrarily in ξ-direction.

The coexistence condition (4.1.21) or (4.1.22) may be visualized as an equal-areas rule which requires that the two hatched areas in Fig. 4.6 are equal. This is analogous to Maxwell's rule in equilibrium thermodynamics that determines the pressure under which the liquid and the gas phases can coexist with a plane boundary; this is characteristic of a first-order equilibrium phase transition (Fig. 4.7). A similar construction for phase coexistence in a nonequilibrium phase transition of a chemical reaction system was first given by *Schlögl* [4.14]. *Adler* et al. [4.15] derived essentially the same equal-areas rule as in (4.1.22) for SNDC semiconductors by slightly different arguments. Other equal-areas rules are known for electron temperature induced SNDC [4.16] and for domain formation in the Gunn effect (NNDC) [4.17]. The latter will be treated in Sect. 6.1.2 in connection with transit-time oscillations. Note that in contrast to the Schlögl model [4.14] it is not the species concentration v, but $\ln v$, that is analogous to the mechanical position coordinate. As a consequence, the profiles $v(\xi)$ are not symmetric with respect to the line $v = v_2$, as it is the case in the Schlögl model, even if the coexistence condition (4.1.21) is satisfied.

We have seen that the equal-areas rule (4.1.21) determines the condition for spatial coexistence in systems infinitely extended in one direction. Apart from that,

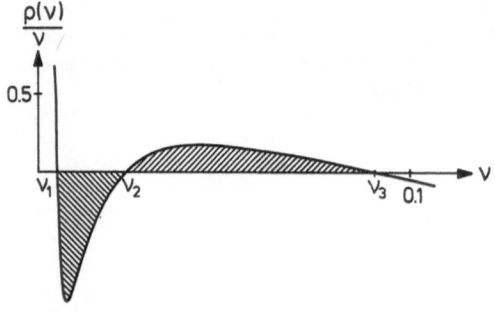

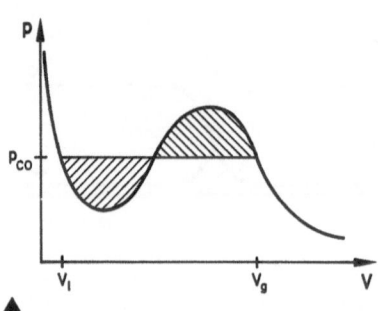

Fig. 4.7. Maxwell's construction for the Van der Waals' gas in equilibrium thermodynamics. An isotherm with $T < T_c$ (critical temperature) is plotted in the pressure (p) – volume (V) diagram. The coexistence pressure p_{co} for coexistence of the liquid phase (volume V_l) with the vapor phase (volume V_g) is determined by the equality of the two hatched areas

Fig. 4.6. Condition for coexistence of the two phases v_1 and v_3 with a plane boundary layer: The two hatched areas must be equal, in order to secure $\phi(v_1) = \phi(v_3)$. This is the equal-areas rule for plane current layers

the importance of the coexistence field ε_{co} lies in the fact that it divides the SNDC range $\varepsilon_h < \varepsilon_0 < \varepsilon_{th}$ into two regimes with topologically different phase portraits. We shall see in Sects 4.2, 3 that even for systems with a cylindrical symmetry or with a finite width L_ξ and with boundary conditions that are not compatible with the coexistence profiles of Fig. 4.5, ε_{co} divides the solution manifold into two qualitatively different classes, $\varepsilon_h < \varepsilon_0 < \varepsilon_{co}$ and $\varepsilon_{co} < \varepsilon_0 < \varepsilon_{th}$. Therefore ε_{co} is an important quantity.

4.1.3 Electron-Density Profiles

Let us now discuss various electron-density profiles in detail [4.2]. Substitution of (4.1.9) into (4.1.5) yields an integral representation of the spatial profiles:

$$\xi = \pm \int v^{-1} \{2[C - \phi(v, \varepsilon_0)]\}^{-1/2} \, dv \ . \tag{4.1.23}$$

The integral can not, in general, be expressed by an elementary function. First, we shall analyse the profiles qualitatively, using the phase portrait of Fig. 4.2a ($\varepsilon_0 < \varepsilon_{co}$). We restrict ourselves to trajectories satisfying

$$\varepsilon_\perp(0) = 0 \ . \tag{4.1.24}$$

This includes, in particular, all profiles which are symmetric with respect to $\xi = 0$.

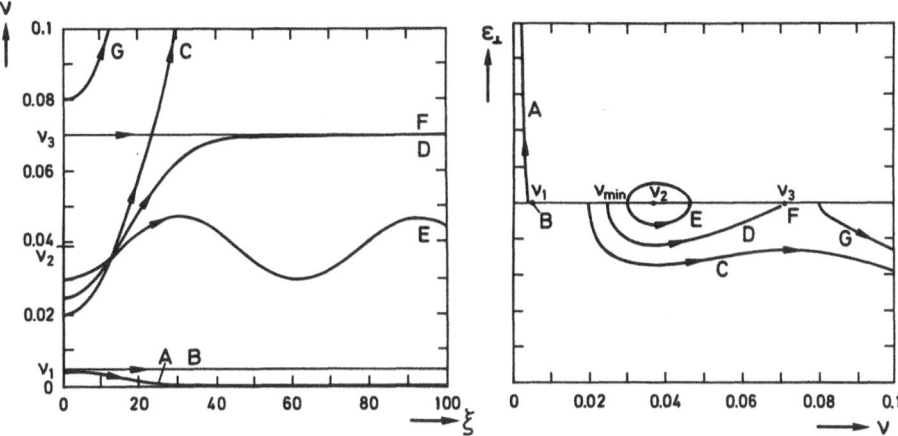

Fig. 4.8. Calculated electron concentration profiles v versus transverse coordinate ζ (*left*) and associated phase trajectories (*right*) for a plane geometry. Letters A – G denote profiles and trajectories corresponding to different boundary values $v(0)$. The numerical parameters are given in Table 3.2. The applied field is $\varepsilon_0 = 8.0$. Here and in the following figures the spatial coordinates are in units L_D

All solutions with $v(0) < v_1$ tend to $v \to 0$ and $\varepsilon_\perp \to \infty$ for $\zeta \to \infty$ (trajectory A in Fig. 4.8). For $v(0) = v_1$ the profile remains constant (B). If the electron density at $\zeta = 0$ satisfies $v_1 < v(0) < v_{min}$, $v(\zeta)$ increases monotonically (C). For $v(0) = v_{min}$ the profile corresponds to one-half of the separatrix loop and tends asymptotically to the stable homogeneous steady state v_3 for $\zeta \to \infty$ (D). For $v_{min} < v(0) < v_3$ the solutions are periodic, and the corresponding profiles oscillate around the unstable steady state (E). For $v(0) = v_3$ the profile is constant (F), and for $v(0) > v_3$ the profiles again diverge: $v \to \infty$ (G). Of all the possible profiles only the separatrix solution (D) ressembles a filamentary profile: it has a low electron-density core and a monotonic increase in density up to the stable homogeneous value v_3, away from the core.

The points of inflection of $v(\zeta)$ satisfy, by (4.1.6, 7)

$$\frac{d^2v}{d\zeta^2} = v[\varepsilon_\perp(v)^2 - \rho(v, \varepsilon_0)] = 0 \tag{4.1.25}$$

and hence lie at values $v_p > v_2$ because of

$$\left(\frac{d^2v}{d\zeta^2}\right)_{v=v_2} = v_2\varepsilon_\perp(v_2)^2 > 0 \ . \tag{4.1.26}$$

In case of the periodic solutions, the slope of $v(\zeta)$ at $v = v_2$ by (4.1.6) increases with increasing amplitude, and attains, by (4.1.9), its maximum value

$$\left|\left(\frac{dv}{d\zeta}\right)_{v=v_2}\right| = v_2 \left|2 \int_{v_2}^{v_i} \rho(v, \varepsilon_0)\frac{dv}{v}\right|^{1/2} \tag{4.1.27}$$

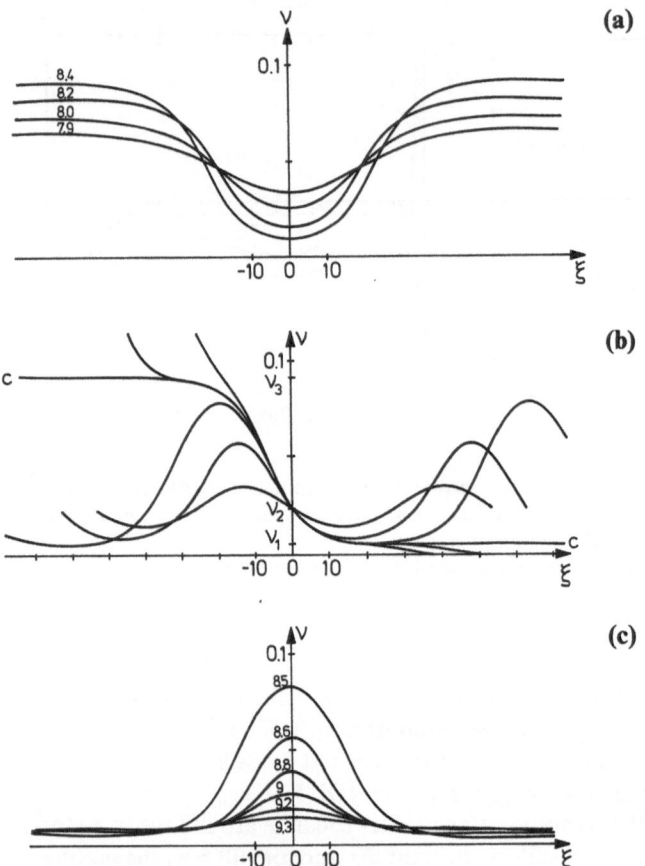

Fig. 4.9a–c. Calculated inhomogeneous concentration profiles for the numerical parameters of Table 3.2. (a) Solitary solutions (depletion layers) for $\varepsilon_0 < \varepsilon_{co}$. (b) Oscillatory, solitary (antikink), and unbounded solutions for $\varepsilon_0 = \varepsilon_{co} = 8.49$. (c) Solitary solutions (accumulation layers) for $\varepsilon_0 > \varepsilon_{co}$. The parameter denoting different curves in (a), (c) is ε_0

for the separatrix solutions (homoclinic or heteroclinic), where $i = 3$ applies to $\varepsilon_0 \leq \varepsilon_{co}$, and $i = 1$ to $\varepsilon_0 \geq \varepsilon_{co}$. The period Λ of the oscillatory solutions increases, too, and tends to infinity as the separatrix solution is approached. These features of the periodic solutions are illustrated in Fig. 4.9b.

Next, we shall discuss how the profiles change when ε_0 is varied (Fig. 4.9). For $\varepsilon_0 = \varepsilon_h$ the two singular points at v_2 and v_3 merge into one degenerate singular point in the phase plane. There are no periodic solutions or separatrix loops. The potential $\phi(v, \varepsilon_0)$ has one maximum at v_1, and a point of inflection with a horizontal tangent at $v_2 = v_3$. For $\varepsilon_h < \varepsilon_0 < \varepsilon_{co}$, the peaks of the potential satisfy $\phi(v_1, \varepsilon_0) > \phi(v_3, \varepsilon_0)$. Hence in the phase plane the homoclinic solution is a loop around v_2, starting and ending at v_3. The associated profile $v(\xi)$ represents a high-current state (v_3) with a depletion layer of minimum electron density $v_{min} < v_2$ in its center. With increasing

ε_0 also $v_3 - v_{min}$ increases, and the depletion layer becomes wider and deeper (Fig. 4.9a). Its width tends to infinity as $\varepsilon_0 \to \varepsilon_{co}$. At the coexistence field ε_{co}, the depletion-layer profile becomes a kink or antikink (Fig. 4.9b). For $\varepsilon_{co} < \varepsilon_0 < \varepsilon_{th}$ we have $\phi(v_1, \varepsilon_0) < \phi(v_3, \varepsilon_0)$, and the solitary solution represents a low-current state (v_1) with an accumulation layer $(v_{max} > v_2)$ which shrinks with increasing ε_0 and vanishes altogether at $\varepsilon_0 = \varepsilon_{th}$ (Fig. 4.9c).

In three special cases, simple analytical expressions for $v(\xi)$ can be obtained:

(a) Oscillatory Solutions of Small Amplitude a

For $a \ll v_3 - v_2, v_2 - v_1$ one finds harmonic oscillations

$$v(\xi) = v_2 + a \sin(k_\perp \xi) \tag{4.1.28}$$

around the unstable homogeneous steady state, by (4.1.13).

(b) Depletion Layers Near the Boundary of the SNDC Range

The condition $\varepsilon_0 - \varepsilon_h \ll \varepsilon_h$ implies $v_3 - v_2 \ll v_3$, and the potential ϕ can be expanded near v_3 in the universal cubic form [4.18], see Fig. 4.10,

$$\phi(v, \varepsilon_0) = -\left[\frac{c_1}{3}(v - v_c)^3 - c_2(\varepsilon_0 - \varepsilon_h)(v - v_c) \right] \tag{4.1.29}$$

where c_1, c_2 are positive constants, and

$$v_c = v_2|_{\varepsilon_0 = \varepsilon_h} = v_3|_{\varepsilon_0 = \varepsilon_h} \ .$$

For $\varepsilon_0 > \varepsilon_h$ the steady states are

$$v_2 = v_c - \delta/2 \ , \qquad v_3 = v_c + \delta/2 \tag{4.1.30}$$

with

$$\delta/2 := [(\varepsilon_0 - \varepsilon_h)c_2/c_1]^{1/2} \ . \tag{4.1.31}$$

The integral (4.1.23) is elementary for the depletion layer, where $C = \phi(v_3, \varepsilon_0)$, and yields

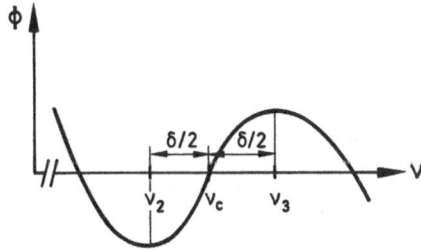

Fig. 4.10. Universal cubic form of the potential ϕ as a function of the electron concentration v near the boundary of the SNDC regime, i.e., near the holding field ε_h (schematic)

$$v(\xi) = v_3 - \frac{3}{2}\delta\left[\cosh\left(\frac{v_c}{2}\sqrt{\delta c_1}\,\xi\right)\right]^{-2} .$$
(4.1.32)

The width of the depletion layer is of the order of $[\frac{1}{2}v_c(\delta c_1)^{1/2}]^{-1}$; it tends to infinity for $\varepsilon_0 \to \varepsilon_h$, i.e., $\delta \to 0$.

An analogous profile is obtained for the accumulation layer in case of $\varepsilon_{th} - \varepsilon_0 \ll \varepsilon_{th}$.

(c) Coexistence Kinks and Antikinks Near the Critical Point

For values of the g-r constants that correspond to a critical point the homogeneous current-field characteristic has a point of inflection with infinite slope and $v_1 = v_2 = v_3$. If one varies the g-r constants slightly such that a small range of bistability is produced, and sets $\delta := v_3 - v_2 = v_2 - v_1 \ll v_3$, the potential $\phi(v, \varepsilon_{co})$ near v_2, satisfying the coexistence condition (4.1.21), is of the universal fourth-order form, shown in Fig. 4.11,

$$\phi(v, \varepsilon_{co}) = -\tfrac{1}{4}C^*(v - v_3)^2(v - v_1)^2 ,$$
(4.1.33)

which is equivalent to the Ginzburg-Landau potential in critical dynamics.

In the simple model of Sect. 2.1.2 we have

$$C^* := (T_1^S + X_1^*)X_1/[v_2\Delta(v_2 N_D^*)] .$$

The kink- and antikink-shaped coexistence profiles are, by (4.1.23),

$$v(\xi) = v_2 \pm \delta\tanh[v_2\delta(C^*/2)^{1/2}(\xi - \xi_0)] .$$
(4.1.34)

The position of the phase boundary ξ_0 is not determined and may be shifted arbitrarily in the transversal direction. The interfacial layer has a width of the order of

$$\Delta\xi \sim [v_2\delta(C^*/2)^{1/2}]^{-1} .$$
(4.1.35)

As the critical point is approached, $\delta \to 0$, and $\Delta\xi$ diverges. This reflects the universal effect that spatial correlations become infinite at the critical point. A similar behavior was found by *Schlögl* for the chemical reaction model [4.14].

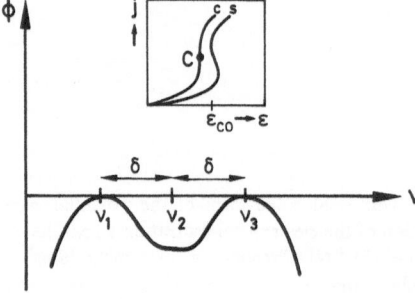

Fig. 4.11. Universal quartic form of the potential ϕ as a function of v near the critical point. The inset shows the critical (c) and a subcritical (s) static current density-field characteristic. C is the critical point. The potential $\phi(v)$ shown correponds to the coexistence field ε_{co} of the subcritical characteristic (schematic)

4.1.4 Current-Voltage Characteristics

Let us now investigate the influence of plane-transverse spatial structures upon the current-voltage relation. The total current through an SNDC element of the geometrical form shown in Fig. 4.1 is given by

$$I = I_0 \varepsilon_0 L_\eta \int_0^{L_\xi} v(\xi, \varepsilon_0) \, d\xi \, , \qquad \text{where} \tag{4.1.36}$$

$$I_0 := \mu_n N_D^* k T L_D \tag{4.1.37}$$

is the physical unit of the current.

In the approximation of a homogeneous longitudinal field ε_0, the total voltage is given by

$$V = V_0 \varepsilon_0 L_\zeta \qquad \text{where} \tag{4.1.38}$$

$$V_0 := kT/e \, . \tag{4.1.39}$$

Combination of (4.1.36, 38) gives the current $I(V)$ as a function of the voltage.

We shall analyse the I–V characteristic for current-sheath configurations such as sketched in Fig. 4.1 Strictly speaking for a finite sample width L_ξ the layered (homoclinic) electron-density profiles of Fig. 4.8a, c are not permitted as solutions. However, for sufficiently large $L_\xi \gg L_D$, the oscillatory profile with the longest possible wavelength $\Lambda = L_\xi$ may be approximated by the homoclinic solution. It follows from (4.1.36) that the total current is somewhat below, and parallel to, the high-conductivity branch of the homogeneous I–V characteristic (full line in Fig. 4.12) for $\varepsilon_h < \varepsilon_0 < \varepsilon_{co}$, and somewhat above the low-conductivity branch for $\varepsilon_{co} < \varepsilon_0 < \varepsilon_{th}$. At $\varepsilon_0 = \varepsilon_{co}$, with the coexistence solution, the current can be widely varied between the high- and low-conductivity values corresponding to v_3 and v_1, respectively, by shifting the phase boundary ξ_0; although v_3 and v_1 cannot actually be attained for finite L_ξ, due to boundary effects. If, for instance, the current is increased at fixed voltage $V_{co} = V_0 \varepsilon_{co} L_\zeta$, the high-current layer is expanded laterally:

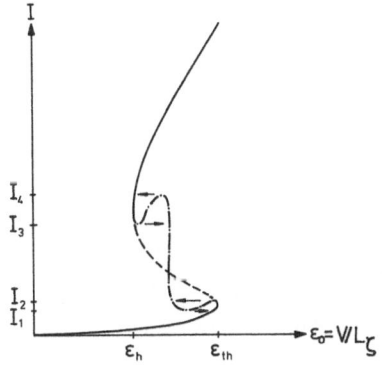

Fig. 4.12. Spatially homogeneous and inhomogeneous branches of the current-voltage characteristic for a plane geometry (schematic). Stable homogeneous steady states are represented by full lines, unstable homogeneous steady states are dashed, and solitary (depletion or accumulation layer) inhomogeneous states are dash-dotted. Nonequilibrium phase transitions between homogeneous and inhomogeneous states are indicated by arrows. The current I is plotted versus the applied voltage $V = \varepsilon_0 L_\zeta$

$$I(V_{co}; \xi_0) \approx I_0 \varepsilon_{co} L_\eta [v_3 \xi_0 + v_1 (L_\xi - \xi_0)] \ . \tag{4.1.40}$$

This means that an SNDC element containing a current layer or filament is a voltage limiter. This is analogous to the NNDC case, where excess voltage applied to the system goes into (and widens) the domain, while the current remains constant.

For $\varepsilon_0 \neq \varepsilon_{co}$, the distance between the layered and the corresponding homogeneous branches of the current-voltage characteristic tends to zero with increasing L_ξ. This shows up explicitly in the approximation (4.1.32):

$$I(\varepsilon_0) = I_0 L_\eta L_\xi \varepsilon_0 \left(v_3 - \frac{3\delta L_D}{qL_\xi} \tanh \frac{qL_\xi}{2L_D} \right) \ . \tag{4.1.41}$$

Note that the two asymptotic limits $L_\xi \to \infty$ and $q \to 0$ (i.e., $\varepsilon_0 \to \varepsilon_h$) must not be commutated.

In a more realistic analysis, the influence of the boundaries has to be taken into consideration. We shall do this in Sect. 4.3 and find that the I–V characteristic depends strongly upon the boundary conditions at the lateral surfaces and upon the lateral dimension L_ξ.

The current-voltage characteristic shown in Fig. 4.12 gives rise to new nonequilibrium phase transitions of first order between homogeneous and inhomogeneous steady states. Consider, for instance, the case of a heavily loaded external circuit, such that the load line is almost horizontal (Fig. 4.13). Then the current $I = I_{ext}$ can be varied externally thus shifting the load line upwards or downwards. With increasing I_{ext} the operating point first moves up on the homogeneous low-conductivity branch, then at I_2 jumps in a first-order phase transition to the inhomogeneous layered branch. Upon further increase of the current the high-current layer expands laterally, and at I_4 another first-order transition to the homogeneous high-conductivity branch occurs. At decreasing current I_{ext} two discontinuous nonequilibrium phase transitions occur at different values of the current, I_3 and I_1. Thus two hysteresis cycles are possible. Whether both are actually observed, depends upon the lateral dimension L_ξ, since the inhomogeneous branch moves away from the homogeneous low- and high-conductivity branches with decreasing L_ξ. As can be seen in Fig. 4.13, the $I_3 - I_4$ hysteresis cycle is usually more

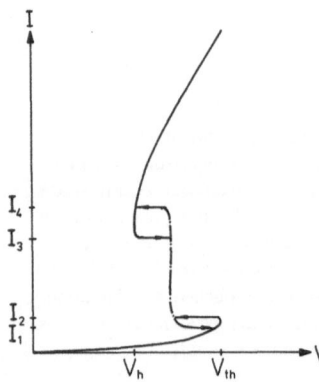

Fig. 4.13. Theoretical current-voltage characteristic for a heavily loaded external circuit, assuming a plane geometry and infinite boundary conditions. Upon variation of the external current two hysteresis cycles occur, indicated by arrows (schematic)

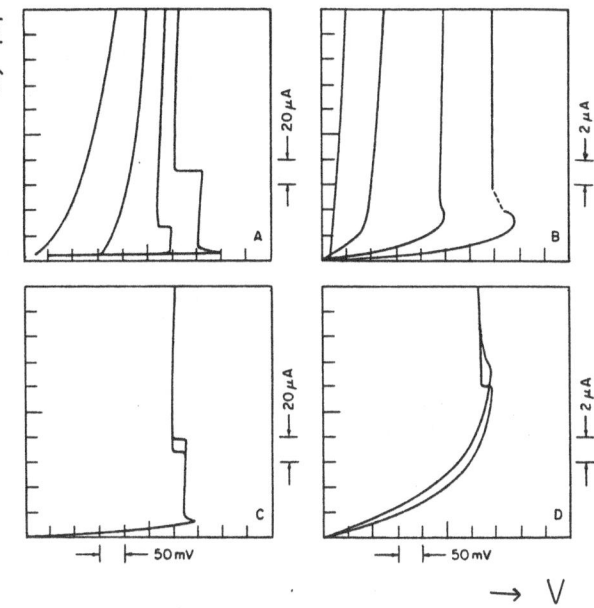

Fig. 4.14. Measured current-voltage characteristic in GaAs at 4.2 K. Curves A and C differ in wavelength and intensity of the IR irradiation. B and D represent the same measured curves as A and C, respectively, but in a different current scale [4.19]

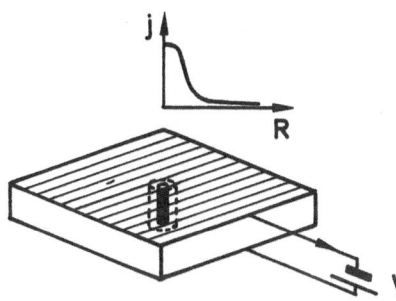

Fig. 4.15. Cylindrical current filament in the center of a semiconductor slab (schematic). The inset shows the radial current-density distribution j versus R

pronounced than the $I_1 - I_2$ cycle. This is exactly the behavior found experimentally. Figure 4.14 shows a measured current-voltage characteristic for GaAs at 4.2 K, where impact ionization of donors at 3.7 meV below the conduction band occurs [4.19]. The hysteresis cycle at $I_3 - I_4$ shows up clearly (C), while the one at $I_1 - I_2$ can be resolved with an enlarged scale only (D).

4.2 Cylindrical Current Filaments

An important geometrical configuration occurs when both transverse dimensions are comparable, and boundary effects are negligible. We can then expect a cylindrically symmetric current distribution, as sketched in Fig. 4.15. Such current filaments are widespread in SNDC elements [4.20–30]. Typical radial current-density profiles measured in a GaAs pin-diode are shown in Fig. 4.16a. The current filaments can

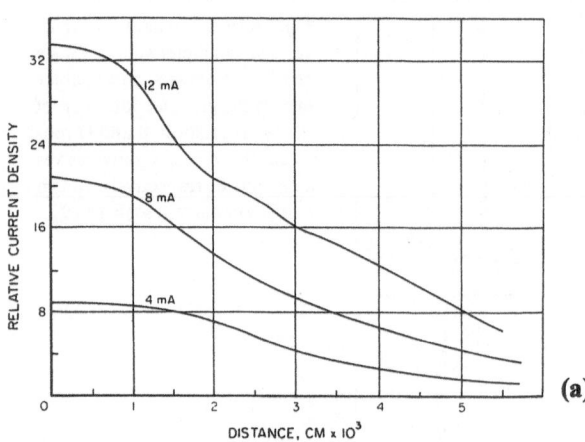

RELATIVE CURRENT DENSITY

12 mA

8 mA

4 mA

32

24

16

8

0   1   2   3   4   5

DISTANCE, CM × 10³

(a)

Fig. 4.16a, b. Observed cylindrical current filament in a GaAs pin diode at three different currents: **(a)** Measured current density j versus radius R. **(b)** Infrared photographs of light emitted from the filament at $\lambda =$ 8800 Å [4.20]

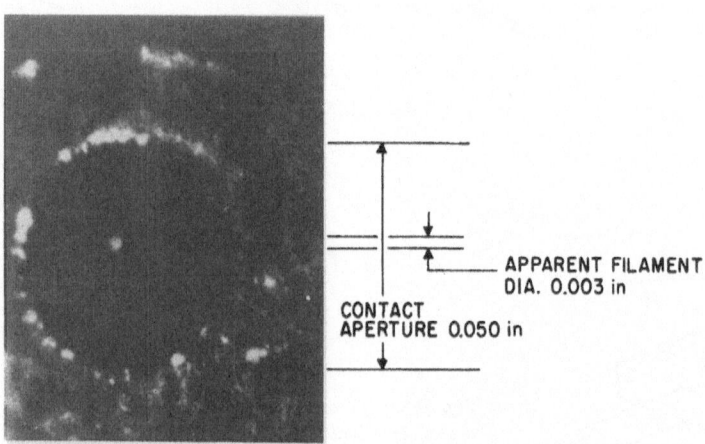

APPARENT FILAMENT
DIA. 0.003 in

CONTACT
APERTURE 0.050 in

6mA, 20-SECOND EXPOSURE

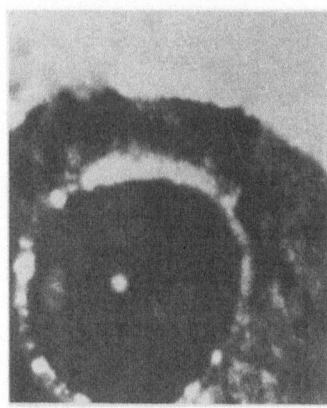

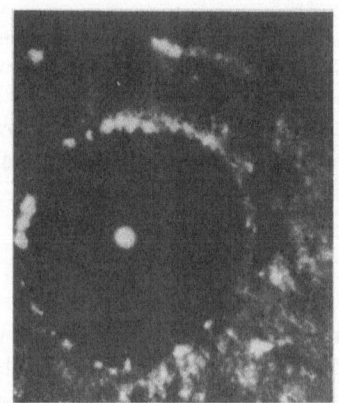

(b)

10mA, 5-SECOND EXPOSURE   15mA, 3-SECOND EXPOSURE

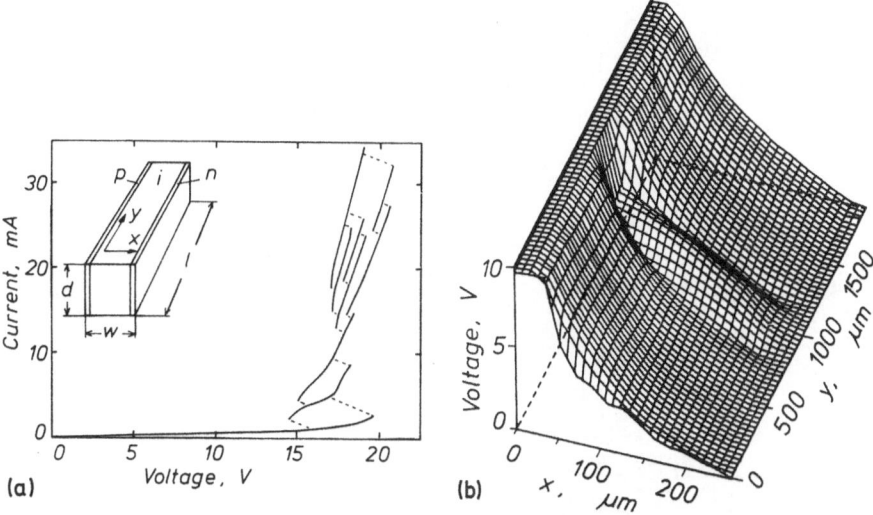

Fig. 4.17a, b. Observed current filament in a gold compensated Si pin diode at room temperature: **(a)** Measured filamentary current-voltage characteristic and schematic structure of the diodes (inset). Typical dimensions are $l = 1 \ldots 5$ mm, $w = 230 \ldots 280$ μm, $d = 200 \ldots 300$ μm. **(b)** Measured potential V at the diode versus the (x, y) coordinates of the semiconductor surface (cf., inset of Fig. 4.17a), corresponding to a point beyond the first jump in the $I-V$ characteristic shown in **(a)** [4.26b]

be observed directly through the infrared [4.20] or visible [4.59] recombination radiation which is produced by the high-carrier density in the core of the filament (Fig. 4.16b). Other direct methods of observing current filaments are: optical IR absorption measurement [4.26c]; potential probe techniques [4.26b, c]; voltage contrast [4.26a, b] or electron-beam induced voltage/current (EBIV [4.26b]/EBIC [4.30]) measurements in scanning electron microscopy [4.28]. Some experimental results are shown in Figs. 4.17, 18. Figure 4.17b depicts the potential distribution across the surface of a pin structure (inset of Fig. 4.17a) measured by pressing a tungsten probe against the semiconductor surface. While $-\partial V/\partial x$ is the longitudinal electric field in the diode, $-\partial V/\partial y$ is the transverse electric field, and $V(y)$ is thus proportional to the transverse carrier-density profiles. The potential hump indicates a current filament of width ~ 280 μm. As subsequent jumps in the $I-V$ characteristic (Fig. 4.17a) are passed, at each jump one additional filament abruptly appears in the potential distribution [4.26b, c]. Figure 4.18 shows a sequence of electron microscopy line scans of a semiconductor in the regime of low-temperature impurity breakdown. The current filament walls are marked by pronounced peaks of the EBIC signal, while the interior and the exterior of the filaments correspond to flat portions of the linescans. This behavior reflects the sensitivity of the filament walls against fluctuations, see Chap. 5 (Figs. 5.4, 5). As the bias and the current are increased (Fig. 4.18a–c) the filaments become wider. In both Figs. 4.17 and 4.18 the thickness of the samples was chosen so small that it was completely filled by the filaments. The radius R_0 of current filaments can also be measured indirectly by several independent methods as a function of the current [4.29]; for instance, by

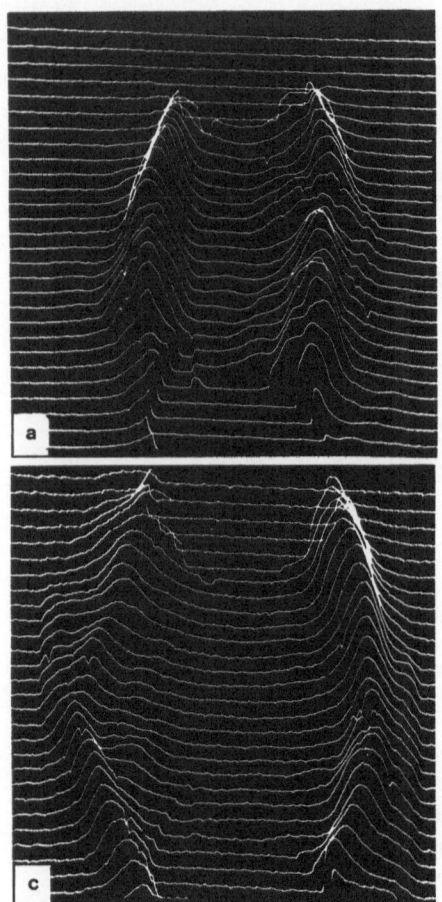

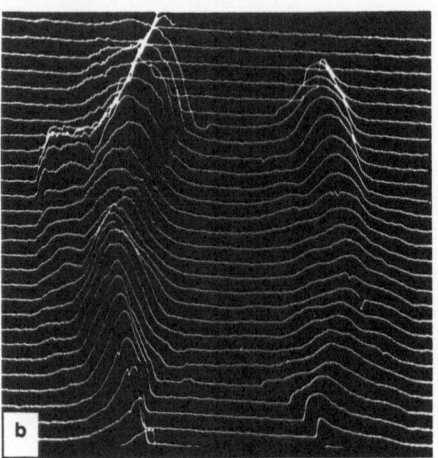

Fig. 4.18a–c. Current filament in p-Ge with $N_A = 10^{14}$ cm^{-3} at 4.2 K, observed in a scanning electron microscope via electron-beam induced current (EBIC) measurements at three different applied voltages: (a) 1.3 V; (b) 1.4 V; (c) 1.5 V. The pictures show the surface of the sample (ca., 4×4 mm^2) with Ohmic point contacts at the bottom and the top, separated by 3.14 mm. Thickness of the sample: 260 μm. [4.30]

velocity saturation in certain heterojunctions (which provides a means of determining the current density), or by transient current-voltage characteristic (TONC) techniques.

In this section we will analyse the nonlinear stationary transport equations (4.1.1, 2) for the single-carrier g-r mechanism under cylindrical symmetry, and discuss the possible current filaments [4.2, 5].

4.2.1 Electron-Density Profiles

We assume that the current density has a radial dependence $j(R)$ where $R :=$ r/L_D is the dimensionless radial coordinate. Splitting the electric field into an R-independent longitudinal component ε_0 and an R-dependent radial component $\varepsilon_\perp(R)$, we obtain from (4.1.1, 2):

$$\frac{d}{dR} v = -\varepsilon_\perp v \tag{4.2.1}$$

$$\frac{d}{dR}\varepsilon_\perp = -\frac{\varepsilon_\perp}{R} + \rho(v, \varepsilon_0) \ . \tag{4.2.2}$$

In contrast to (4.1.6, 7), the equations (4.2.1, 2) do not represent an autonomous dynamic system, due to the explicit R-dependence of the right-hand side of (4.2.2). Therefore (4.2.1, 2) do not define a unique field of directions $(dv/dR, d\varepsilon_\perp/dR)$ for each phase point $(v, \varepsilon_\perp)$ and a phase-portrait analysis cannot be conveniently performed. Equations (4.2.1, 2), however, can be reduced to a single equation, similar to (4.1.15):

$$\frac{d^2}{dR^2}\mu + \frac{1}{R}\frac{d}{dR}\mu + \tilde{\rho}(\mu, \varepsilon_0) = 0 \ , \tag{4.2.3}$$

where $\mu := \ln v$, and $\tilde{\rho}(\mu, \varepsilon_0) := \rho(v, \varepsilon_0)$. The regularity of $\varepsilon_\perp = -d\mu/dR$ at $R = 0$ imposes the boundary condition

$$\frac{d\mu}{dR}(0) = 0 \tag{4.2.4}$$

or, equivalently,

$$\frac{dv}{dR}(0) = 0 \ . \tag{4.2.5}$$

Equation (4.2.3) is the analog of the equation of motion of a particle in the potential $\tilde{\phi}(\mu)$, see (4.1.17) and Fig. 4.4, under a time-dependent friction force $R^{-1} d\mu/dR$, which goes to zero as the "time" R tends to infinity. Depending upon the "initial condition" $\mu(0)$, either damped oscillations around the potential minimum at μ_2 (corresponding to the unstable homogeneous state) or overshooting unbounded solutions are possible. As $\mu(0)$, or $v(0)$, is varied, the set of profiles shown in Fig. 4.19

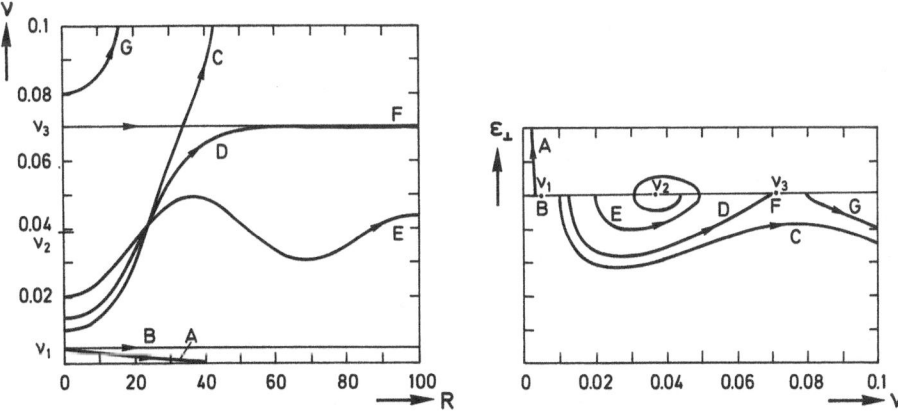

Fig. 4.19. Calculated electron concentration profiles v versus radial coordinate R (*left*) and associated phase trajectories (*right*) for a cylindrical geometry. The letters denote different boundary values, and correspond to those of Fig. 4.8 for the case of a plane geometry. The numerical parameters are given in Table 3.2. The applied field is $\varepsilon_0 = 8$

is produced. The series of solutions is qualitatively similar to that of Fig. 4.8, but the spatial variation is slower, due to the "friction force". For small amplitudes a, where $\tilde{\rho}(\mu, \varepsilon_0)$ can be linearized around μ_2, the oscillatory solutions are given by a Bessel function of zero order J_0:

$$v(R) = v_2 + a J_0(k_\perp^b R) \qquad (4.2.6)$$

as anticipated by the bifurcation analysis of Sect. 3.2.2.

In the limit case between oscillatory and unbounded solutions there is a monotonic radial density profile $v(R)$ which satisfies, in addition to (4.2.5), the boundary condition

$$\lim_{R \to \infty} v(R) = v_i \ , \qquad (4.2.7)$$

where v_i is the stable homogeneous steady state that corresponds to the lower peak of ϕ, i.e., $v_i = v_1$ for $\varepsilon_0 > \varepsilon_{co}$, and $v_i = v_3$ for $\varepsilon_0 < \varepsilon_{co}$. From the mechanical analog it follows for $\varepsilon_0 < \varepsilon_{co}$ (Fig. 4.4a) that the electron density on the cylinder axis $v_0 := v(0)$ lies in the range $v_1 < v_0 < v_{\min}$. Hence $v(R)$ represents a cylindrical filament of a low-current core, surrounded by a mantle of high-current density (curve D in Fig. 4.19). On the other hand, $\varepsilon_0 > \varepsilon_{co}$ (Fig. 4.4c) leads to $v_{\max} < v_0 < v_3$; this represents a high-current filament with a low-current mantle.

As the field is increased from $\varepsilon_0 = \varepsilon_h$ to $\varepsilon_0 = \varepsilon_{co}$, the low-current filaments change as shown in Fig. 4.20a. Since the difference in the potential peaks $\tilde{\phi}(\mu_1) - \tilde{\phi}(\mu_3)$ decreases, so does $v_{\min} - v_1$, and v_0 tends to v_1. The mechanical analog shows that in this case the initial spatial variation $v'(R)$ becomes very small, and therefore the filamentary profiles become wider. For ε_0 close to, but smaller than ε_{co}, the filamentary solution rises from a value just above v_1 to v_3 within a thin-cylindrical transition layer at $R_0 \gg 1$. Since $v(R)$ varies considerably in the transition layer only, we may approximately set $R \approx R_0$ in the second term of (4.2.3). Note that the

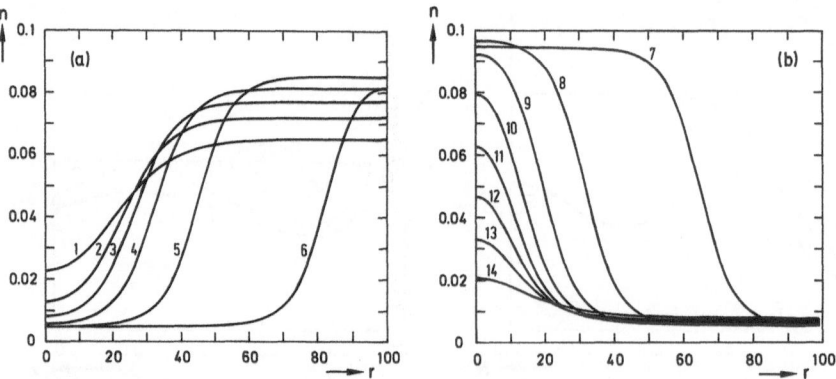

Fig. 4.20a, b. Calculated radial electron concentration profiles of cylindrical current filaments for different values of the applied electric field ε_0. (a) $\varepsilon_0 < \varepsilon_{co}$, (b) $\varepsilon_0 > \varepsilon_{co}$. The curves 1–14 correspond to increasing values of ε_0: (1) 7.9, (2) 8.0, (3) 8.1, (4) 8.2, (5) 8.3, (6) 8.4, (7) 8.6, (8) 8.7, (9) 8.8, (10) 8.9, (11) 9.0, (12) 9.1, (13) 9.2, (14) 9.3. Numerical parameters as in Table 3.2

filamentary profiles in Fig. 4.20a resemble the depletion layers in Fig. 4.9a, but at a given value of ε_0, the cylindrical filaments have a lower electron density in the center (v_0) than the plane layers (v_{min}). This is due to the curvature of the cylindrical filament walls, or, speaking in terms of the mechanical analog, the "friction force".

Decreasing the field from $\varepsilon_0 = \varepsilon_{th}$ to $\varepsilon_0 = \varepsilon_{co}$ yields an analogous behavior of the high-current filaments (Fig. 4.20b). We only have to replace v_1 with v_3, and v_{min} with v_{max}.

4.2.2 Equal-Areas Rule

In Sect. 4.1.2 we have derived an equal-areas rule for the coexistence of plane-current layers. Here we shall show that a modified equal-areas rule exists for cylindrical current filaments. The purpose of such equal-areas rules is to provide some partial information about the filaments, like the filament radius as a function of the applied field, without explicitly solving the complicated nonlinear differential equations (4.2.1, 2) that determine the filamentary carrier-density profiles, if supplemented by appropriate boundary conditions. The equal-areas rule will allow us to extract this information directly from the constitutive function $\rho(v, \varepsilon_0)$ by a simple geometrical construction. Equal-areas rules for current filaments have been derived for electron overheating instabilities [4.16], and single-carrier and two-carrier g-r instabilities [4.15], but the curvature of the filaments was neglected, i.e., the filaments were approximated by plane-current layers. Also, in the two-carrier case trapping was neglected, although this plays an essential role in two-carrier SNDC mechanisms, as we have seen in Sect. 2.2. An equal-areas rule for filamentation in single-carrier electronic mechanisms that included the effect of curvature of the filament walls has been proposed [4.1, 2], but its usefulness was limited by the fact that it involved the thickness of the filament wall ΔR as a heuristic parameter which was not specified by the theory. Recently equal-areas rules have been derived for both single-carrier and two-carrier SNDC mechanisms by which the radius and the wall thickness of filaments were calculated self-consistently, and trapping was incorporated also for two-carrier mechanisms [4.5].

Formal integration of (4.2.3) over $d\mu$ yields

$$\frac{1}{2}\left(\frac{d\mu}{dR}\right)^2 + \int \left(\frac{d\mu}{dR}\right)^2 \frac{dR}{R} + \int \tilde{\rho}(\mu, \varepsilon_0)\, d\mu = \text{const.} \tag{4.2.8}$$

A monotonic solution, matching the filamentary boundary conditions

$$\lim_{R \to 0} \frac{d\mu}{dR} = \lim_{R \to \infty} \frac{d\mu}{dR} = 0\ , \qquad \lim_{R \to \infty} \mu(R) = \mu_i \tag{4.2.9}$$

with either $i = 1$ or $i = 3$, exists if and only if the following *equal-areas rule* holds [this follows from using the boundary conditions (4.2.9) as upper and lower boundaries in the integral (4.2.8)]:

$$\tilde{\phi}(\mu(\infty)) - \tilde{\phi}(\mu(0)) \equiv \int\limits_{\mu(0)}^{\mu(\infty)} \tilde{\rho}(\mu, \varepsilon_0)\, d\mu = -\Sigma \tag{4.2.10}$$

where

$$\Sigma := \int_0^\infty \left(\frac{d\mu}{dR}\right)^2 \frac{dR}{R} = \int_0^\infty \varepsilon_1^2 \frac{dR}{R} > 0 \qquad (4.2.11)$$

represents a "surface-tension" term. Note that (4.2.10) is an exact relation, which relates the field $\varepsilon_\perp$ of the filamentary profile to the potential difference $\tilde{\phi}(\mu(\infty)) - \tilde{\phi}(\mu(0))$. Since $\varepsilon_\perp$ takes substantial values in the filament wall only, a small filament radius R_0 and a steep gradient of the filament wall correspond to large Σ. For values of ε_0 close to ε_{co} the filament has a thin wall of width $\Delta R \ll R_0$ (Fig. 4.20), and we may approximate Σ by

$$\Sigma \approx \frac{1}{R_0} \int_0^\infty \left(\frac{d\mu}{dR}\right)^2 dR = \frac{1}{R_0} \int_{\mu(0)}^{\mu(\infty)} \left(\frac{d\mu}{dR}\right) d\mu \; . \qquad (4.2.12)$$

The integral can be further evaluated by approximating the integrand by its value at μ_2:

$$\Sigma \approx \frac{1}{R_0} [\mu(\infty) - \mu(0)] \left(\frac{d\mu}{dR}\right)_{\mu=\mu_2} \; . \qquad (4.2.13)$$

This is equivalent to approximating the filament wall by a linear profile $\mu(R)$ of steepness

$$m := \left(\frac{d\mu}{dR}\right)_{\mu=\mu_2} = \frac{1}{\Delta R} [\mu(\infty) - \mu(0)] \; . \qquad (4.2.14)$$

The second equality in (4.2.14) defines the effective wall width ΔR. The steepness m can be calculated self-consistently from (4.2.8) by choosing $\mu(\infty)$ and μ_2 as the upper and the lower limits of the integrals, respectively:

$$\frac{1}{2} m^2 \approx \int_{\mu_2}^{\mu(\infty)} \tilde{\rho}(\mu, \varepsilon_0) \, d\mu + \frac{1}{R_0} [\mu(\infty) - \mu_2] m \qquad (4.2.15)$$

or

$$|m| = \left[2 \int_{\mu_2}^{\mu(\infty)} \tilde{\rho}(\mu, \varepsilon_0) \, d\mu \right]^{1/2} [\delta + (\delta^2 + 1)^{1/2}] \; , \qquad (4.2.16)$$

where

$$\delta := \frac{1}{R_0} [\mu(\infty) - \mu_2] \Big/ \left[2 \int_{\mu_2}^{\mu(\infty)} \tilde{\rho}(\mu, \varepsilon_0) \, d\mu \right]^{1/2} \qquad (4.2.17)$$

is a small parameter in case of thin filament walls; compare (4.2.14) and (4.2.17). Hence (4.2.16) can be expanded in terms of δ:

$$|m| \approx \left[2 \int_{\mu_2}^{\mu(\infty)} \tilde{\rho}(\mu, \varepsilon_0)\, d\mu \right]^{1/2} [1 + \delta + \tfrac{1}{2}\delta^2 + O(\delta^3)] \,. \qquad (4.2.18)$$

In the lowest order $\delta \approx 0$, (4.2.18) gives the result (4.1.27) of plane-current layers. The corrections in δ reflect the curvature of the cylindrical current filaments.

Equation (4.2.10) can be visualized as an equal-areas rule (Fig. 4.21). The right-hand side of (4.2.10) is always negative, therefore $\tilde{\phi}[\mu(0)] > \tilde{\phi}[\mu(\infty)]$ must hold. For $\varepsilon_0 < \varepsilon_{co}$, the absolute maximum of $\tilde{\phi}$ is at μ_1 (Fig. 4.4a); therefore $\mu(\infty)$ cannot be equal to μ_1. It follows that of the two potential boundary values μ_1 and μ_3 in (4.2.9) only $\mu(\infty) = \mu_3$ is possible. Moreover, for ε_0 sufficiently close to ε_{co}, $\mu(0) \approx \mu_1$ holds. This gives a low-current filament. Conversely, for $\varepsilon_0 > \varepsilon_{co}$ high-current filaments with $\mu(\infty) = \mu_1$, $\mu(0) \approx \mu_3$ are obtained. In both cases by (4.2.13, 14, 18), in lowest order of δ:

$$\Sigma \approx \frac{1}{R_0}(\mu_3 - \mu_1)\left[2 \int_{\mu_2}^{\mu_3} \tilde{\rho}(\mu, \varepsilon_0)\, d\mu \right]^{1/2} \,. \qquad (4.2.19)$$

For a given field ε_0, the equal-areas rule (4.2.10) determines Σ, and hence by (4.2.19) the filament radius R_0. The filament wall thickness ΔR then follows from (4.2.14). This is shown in Fig. 4.21a–d for different values of ε_0. The closer ε_0 is to ε_{co}, the thinner is the rectangle with sides $l_1 = \mu_3 - \mu_1$ and $l_2 = |m|/R_0$. It follows from

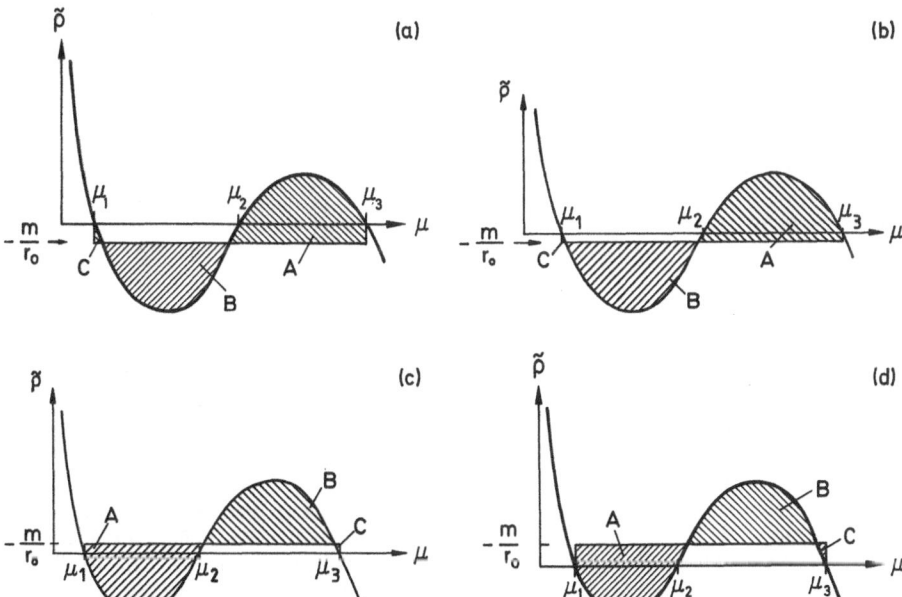

Fig. 4.21a–d. Equal-areas rule for cylindrical current filaments: The areas A + C must be equal to B (schematic). (a), (b): low-current filament ($\varepsilon_0 < \varepsilon_{co}$); (c), (d): high-current filament ($\varepsilon_0 > \varepsilon_{co}$). From (a)–(d) ε_0 increases monotonically. The local-charge density $\tilde{\rho}$ is plotted versus $\mu = \ln \nu$. The homogeneous steady states are μ_1, μ_2, μ_3

(4.2.18) with $\delta \approx 0$ that the dimensionless squared filament radius

$$R_0^2 = \int\limits_{\mu_2}^{\mu_3} \tilde{\rho}(\mu, \varepsilon_0) \, d\mu/(\tfrac{1}{2}l_2^2)$$

is of the order of the ratio of the areas B and C in Fig. 4.21.

As ε_0 approaches ε_{co}, we have in lowest order of $\varepsilon_0 - \varepsilon_{co}$

$$\tilde{\phi}(\mu_3) - \tilde{\phi}(\mu_1) \sim (\varepsilon_0 - \varepsilon_{co}) \to 0 \qquad (4.2.20)$$

and hence

$$\Sigma \sim |\varepsilon_0 - \varepsilon_{co}| \to 0 \; . \qquad (4.2.21)$$

Since m and $\mu(\infty) - \mu(0)$ are slowly varying functions of ε_0 and remain finite, R_0 tends to infinity by (4.2.13) as

$$R_0 \sim |\varepsilon_0 - \varepsilon_{co}|^{-1} \to \infty \; . \qquad (4.2.22)$$

Thus with increasing field $\varepsilon_0 < \varepsilon_{co}$, the low-current filament becomes wider, and eventually fills the whole sample homogeneously. Conversely, with decreasing field $\varepsilon_0 > \varepsilon_{co}$ the high-current filament expands according to (4.2.22). This leads to two new, filamentary branches in the current-voltage characteristic as shown in Fig. 4.22. Assuming a cylindrical sample of radius a, we can express the total current as

$$I = I_0 \varepsilon_0 2\pi \int\limits_0^a v(R, \varepsilon_0) R \, dR \; . \qquad (4.2.23)$$

For low-current filaments this may be approximated by

$$I \approx I_0 \varepsilon_0 \pi [v_1 R_0^2 + v_3(a^2 - R_0^2)] \qquad (4.2.24)$$

which gives, near ε_{co}:

$$I \approx I_3 - c(\varepsilon_0 - \varepsilon_{co})^{-2} \; , \qquad (4.2.25)$$

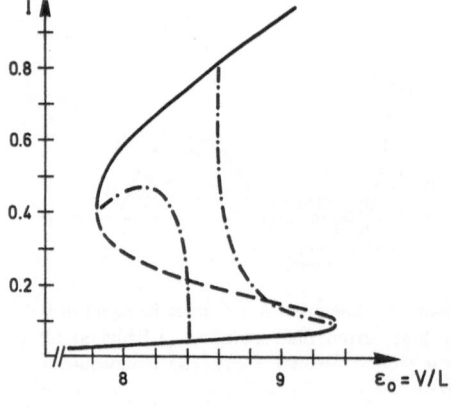

Fig. 4.22. Calculated branches of the current-voltage characteristic for a cylindrical sample of radius $a = 100$, with the numerical parameters of Fig. 4.20. Stable homogeneous steady states are represented by full lines, unstable homogeneous steady states are dashed, and filamentary states are dash-dotted (left branch: low-current filament; right branch: high-current filament)

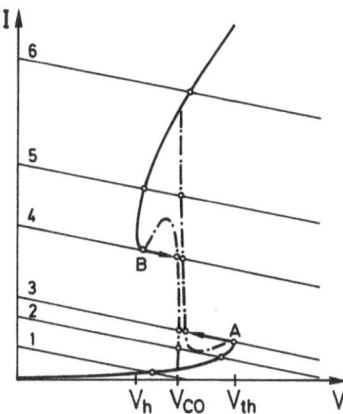

Fig. 4.23. Theoretical current-voltage characteristic for a heavily loaded circuit (schematic). The straight lines 1–6 represent the load lines for increasing values of the applied voltage bias. A and B denote operating points on the characteristic where the onset of current filamentation occurs (*arrows*)

where I_3 is the current in the homogeneous high-conductivity state v_3, and c is a positive constant. For high-current filaments we obtain, analogously,

$$I \approx I_0 \varepsilon_0 \pi [v_1(a^2 - R_0^2) + v_3 R_0^2]$$

$$\approx I_1 + c(\varepsilon_0 - \varepsilon_{co})^{-2} .$$

(4.2.26)

Obviously the current can be widely varied by shifting the filament walls, while the voltage $\varepsilon_0 L_\zeta$ changes only slightly. If a heavily loaded circuit is connected to the device, and the bias is increased, the load line changes as shown in Fig. 4.23. The operating point moves on the low-conductivity branch up to the point A, then switches to the filamentary branch in a first-order nonequilibrium phase transition. Hereby a high-current filament is spontaneously nucleated with a finite radius R_0. Upon further increase of bias, the current rises by expanding the high-current filament until the whole sample is filled homogeneously by the high-conductivity state, while the voltage $\varepsilon_0 L_\zeta$ remains almost constant. Upon further increase of bias the operating point moves up on the stable homogeneous high-conductivity branch. The reverse process of decreasing the bias can lead to another first-order phase transition at point B when a low-current filament is nucleated. It expands as the operating point travels down on the low-current filamentary branch until the homogeneous low-conductivity branch is reached. Thus an intricate hysteresis loop is generated.

We note that high-current filaments – albeit without a flat core – still exist if the g-r mechanism allows for *two* homogeneous steady states (the low-conductivity state v_1 and the NDC state v_2) only, as for instance in the two-carrier mechanism (2.2.4). But the low-current filaments do not exist in this case. Therefore there remains only one filamentary branch in the current-voltage characteristic of Fig. 4.23 (the one at $V_{co} < V < V_{th}$) and the hysteresis loop is modified.

The nucleation of current filaments is analogous to the formation and growth of liquid droplets and gas bubbles in a gas-liquid equilibrium phase transition of first order. Just as the coexistence pressure depends upon the critical radius of the droplets, so is the field ε_0 related to the filament radius R_0, as given by the equal-areas rule (4.2.10). In equilibrium the surface tension σ of spherical droplets is related to the excess pressure Δp by

$$\Delta p = \frac{2}{R_0}\sigma \tag{4.2.27}$$

while for cylindrical current filaments

$$\Sigma \approx \frac{1}{R_0}\varepsilon_\perp \ln\frac{v(0)}{v(\infty)} \tag{4.2.28}$$

holds by (4.2.13). For $R_0 \to \infty$ Maxwell's construction of phase coexistence, or the coexistence condition for plane-current layers, are regained, respectively. We shall see in Sect. 5.4 that this analogy with equilibrium phase transitions goes even deeper, and includes the dynamics of nucleation and spinodal decomposition.

In summary, we have found for cylindrical geometries qualitatively similar current-density profiles as for plane geometries. However, the coexistence condition for the high- and low-conductivity phases have to be modified by including a surface-tension term, which leads to a different equal-areas rule (4.2.10). This equal-areas rule determines the filament radius and the filament wall thickness for given values of the field ε_0. It extends previously known equal-areas rules for filamentation in SNDC elements [4.15, 16] by a self-consistent treatment of the curvature of the filament walls. The occurring first-order nonequilibrium phase transitions are analogous to nucleation phenomena in equilibrium thermodynamics.

The merit of the equal-areas rule is that it provides a simple geometrical construction for the filament radius R_0 without requiring detailed knowledge of the full spatial profiles. The integration of these profiles can in general be carried out only numerically, whereas the knowledge of R_0 as a function of ε_0 follows directly from the g-r mechanism by (4.2.10). It allows one to predict the almost vertical portion of the current-voltage characteristic of an SNDC element, by (4.2.24, 26). The approximations of Σ made in (4.2.19) apply for ε_0 close to ε_{co} only, where the filament is wide and has a flat top. However, this is not a serious restriction in practical cases, since the points of the I–V characteristic that correspond to narrow, smeared-out filaments cannot normally be reached. Rather, the system switches from the homogeneous state to a fully developed filamentary state along the load line, as shown in Fig. 4.23.

4.3 Influence of Boundaries

In the first two sections of this chapter attention has been confined to samples with large transverse dimensions, such that boundary conditions at infinity could be employed as an approximation. The current interest in submicron structures [4.31, 32] advances the need for a thorough theoretical treatment of electronic surface and interface effects in finite-size semiconductors, in particular their influence upon instabilities, switching phenomena, and filamentary transport. Whereas the influence of finite boundaries upon NNDC instabilities has been convincingly elaborated in the past [4.8, 11, 13, 33–35], no similar systematic treatment has been available for the case of SNDC except for a few special cases [4.27]. In this section we will attempt to perform such a systematic analysis of the influence of boundary conditions at the lateral surfaces of an SNDC element upon plane-transverse spatial profiles and the resulting current-voltage characteristics [4.4, 6]. We shall find that current filamentation depends sensitively upon the boundary conditions – a property which can be applied conveniently to control the switching transitions in submicron devices.

4.3.1 Lateral Boundary Conditions

We consider a semiconductor of finite transverse dimension $L_\xi = \Lambda$ as shown in Fig. 4.1. At the two lateral surfaces $\xi = 0$ and $\xi = \Lambda$ the total transverse current density vanishes, as it does in the bulk, see (4.1.6):

$$v'|_{\xi=0,\Lambda} = -v\varepsilon_\perp|_{\xi=0,\Lambda} \ . \tag{4.3.1}$$

Here $\varepsilon_\perp(0)$ and $\varepsilon_\perp(\Lambda)$ are the normal components of the electric field at the lateral surfaces. These are, in our dimensionless notation, identical with the surface charges due to the occupation of electronic surface states. Such localized surface states lying in the bandgap of the material are always present because of structural defects or impurities at semiconductor surfaces [4.36]. They act as traps or recombination centers, and may have a discrete or continuous density of states.

Their occupancy is governed by the g-r kinetics at the surface [4.37, 38], whose rate in the steady state is

$$\varphi^s(v_s) = 0 \ . \tag{4.3.2}$$

Here v_s denotes the number of electrons per unit area in the surface. In particular, for small deviations from equilibrium, the surface recombination rate [4.39–42] is given by

$$\varphi^s(v) = -r^s = -s(v - v_0) \ , \tag{4.3.3}$$

where s is the surface recombination velocity and v_0 the equilibrium electron surface density. For steady state nonequilibrium situations, as we are considering, it will in general be necessary to retain the nonlinear surface g-r rates (4.3.2).

Let us now consider two special cases of boundary conditions. First, if the surface state density is low, such that surface charges are negligible,

$$\varepsilon_\perp(0) = \varepsilon_\perp(\Lambda) = 0 \tag{4.3.4}$$

and by (4.3.1)

$$v'(0) = v'(\Lambda) = 0 \tag{4.3.5}$$

or, equivalently with $\mu = \ln v$,

$$\mu'(0) = \mu'(\Lambda) = 0 \tag{4.3.6}$$

holds. This corresponds to a flat-band condition at the surface [4.39].

Second, if the surface state density is large, and strong surface recombination, trapping, or generation occurs, then the carrier densities at the lateral surfaces are fixed to the values

$$v(0) = v_{s1} \quad \text{and} \quad v(\Lambda) = v_{s2} \tag{4.3.7}$$

determined by $\varphi^{s1}(v_{s1}) = 0$ and $\varphi^{s2}(v_{s2}) = 0$ at the respective surfaces, or equivalently

$$\mu(0) = \mu_{s1} \quad \text{and} \quad \mu(\Lambda) = \mu_{s2} \ . \tag{4.3.8}$$

The surfaces then, in general, will carry surface charges

$$\sigma_1 = -\mu'(0) \neq 0 \quad \text{and} \quad \sigma_2 = -\mu'(\Lambda) \neq 0 \ . \tag{4.3.9}$$

In the limit case of infinite transverse dimension Λ, it follows for bounded non-oscillatory profiles $v(\xi)$ that the boundary condition (4.3.5) must hold, and additionally (4.3.7) holds with v_{s1} and v_{s2} fixed to bulk steady state values. This has been the situation of Sect. 4.1.

For finite Λ, however, v_{s1} and v_{s2} are independent parameters which can often be adjusted by external controls. For example, consider strong optical generation of carriers at the surface with rate g_{opt}^s. In case of bimolecular intrinsic surface recombination $r^s = B^s v^2$, the steady-state surface carrier density is then given by

$$v_s = (g_{opt}^s / B^s)^{1/2} \ . \tag{4.3.10}$$

In the case of linear surface recombination (4.3.3) it is given by

$$v_s = v_0 + g_{opt}^s / s \ . \tag{4.3.11}$$

In both cases v_s increases with increasing optical intensity, and can be widely varied independently of the bulk g-r kinetics.

In the limit case of infinite surface recombination velocity s, and no surface generation, v_s tends to the thermal equilibrium value v_0. This is an often invoked approximation, which was used, e.g., in the investigation of steady state current filaments in semiconductors with *positive* differential conductivity and *linear* g-r kinetics in the bulk [4.41, 42].

In more general treatments [4.43–45] the carrier density v_s is given by the nonlinear-recombination kinetics at the surface, using quasi-Fermi levels. The quasi-Fermi level of the electrons E_{Fn} is *pinned* at the surface due to strong trapping into surface states in the bandgap.

We illustrate this for a set of surface states with density of states $N(E)$. If the trap states are in quasi-equilibrium with the conduction band (i.e., assuming no optical excitation), we can express the electron density by

$$v_s = N_c e^{(E_{Fn} - E_c)/(kT)} \tag{4.3.12}$$

in the standard way by the quasi-Fermi level E_{Fn} and the (normalized) effective density of states in the conduction band N_c, and the total surface trap occupancy by

$$v_t = \int_{E_v}^{E_c} (e^{(E-E_{Fn})/(kT)} + 1)^{-1} N(E)\, dE \ , \tag{4.3.13}$$

where E_v, E_c are the valence and conduction-band energies, respectively. The quasi-Fermi level E_{Fn} can be determined from (4.3.12, 13) using local charge neutrality

$$1 - v - v_t = 0 \ . \tag{4.3.14}$$

Thereby v_s is fixed. The same result can be obtained directly from the g-r rates (4.3.2) using detailed balance.

For a single discrete trap level at an energy E_t, and of density N_t, (4.3.14) yields

$$1 = N_c e^{\eta} + N_t/(e^{\varepsilon_t - \eta} + 1) \tag{4.3.15}$$

where $\eta := (E_{Fn} - E_c)/kT$; $\varepsilon_t := (E_t - E_c)/kT$. This gives an equation for $v_s = N_c \exp^{\eta}$:

$$-N + (N + N_t - 1)v_s + v_s^2 = 0 \tag{4.3.16}$$

with

$$N := N_c e^{\varepsilon_t} \ . \tag{4.3.17}$$

The solution for $N + N_t > 1$ is

$$v_s = (N + N_t - 1)\{-\tfrac{1}{2} + \tfrac{1}{2}[1 + 4N/(N + N_t - 1)^2]^{1/2}\} \tag{4.3.18}$$

which increases with increasing N from $v_s = 0$ (for $N = 0$) to $v_s = 1$ (for $N \to \infty$). The expression (4.3.17) includes thermal ionization of traps only, but can easily be extended to include optical surface excitation of traps with rate constant X_{opt}^S by redefining

$$N := N_c e^{\varepsilon_t} + \frac{X_{opt}^S}{T_1^S} \ , \tag{4.3.19}$$

where T_1^S is the capture coefficient. Again, this shows how the boundary value v_s can be controlled optically, independently of the bulk, over a wide range.

We shall now try and assess the length scale on which boundary effects become important for filamentation. The unit of length in our dimensionless notation is the effective Debye length

$$L_D = (D_0 \tau_M)^{1/2} = \left(\frac{\epsilon_s kT}{4\pi e^2 N_D^*}\right)^{1/2} \ .$$

The thickness of the walls of current filaments is of the order of the actual local Debye length

$$L_w \approx \left(\frac{\epsilon_s kT}{4\pi e^2 n}\right)^{1/2} = L_D v^{-1/2} \ ,$$

which is typically several L_D. The width of the current filaments is of the order of the period corresponding to the bifurcation wavevector $k_\perp$, see (4.1.13),

$$L_{fil} = 2\pi/k_\perp = 2\pi L_D/[v(\partial\rho/\partial v)_\varepsilon]^{1/2} = 2\pi[D_0\tau_M/(\partial\tilde{\rho}/\partial\mu)_\varepsilon]^{1/2} = 2\pi L_D(\tilde{D}_\perp/\lambda_1)^{1/2} ,$$

which is typically of the order of $10^2 L_D$. A sensitive dependence upon boundary conditions is expected if the transverse dimension is of the order of $L_{fil} \approx 10^2 L_D$, which corresponds to values in the near-micron and submicron range. This will be corroborated by numerical calculations in the following.

To this end, we want to find solutions of the governing equation (4.1.15)

$$\mu''(\xi) + \tilde{\rho}(\mu, \varepsilon_0) = 0 \qquad (4.3.20)$$

subject to either the Neumann boundary conditions (4.3.6), or the Dirichlet boundary conditions (4.3.8). These solutions, i.e., the spatial profiles $\mu(\xi)$, are implicitly given by the double integral

$$\xi(\mu) = \int_{\mu_{s1}}^{\mu} \left[\mu'(0)^2 - 2 \int_{\mu_{s1}}^{\tilde{\mu}} \tilde{\rho}(\tilde{\tilde{\mu}}, \varepsilon_0)\,d\tilde{\tilde{\mu}} \right]^{-1/2} d\tilde{\mu} , \qquad (4.3.21)$$

where in case of Dirichlet boundary conditions (4.3.8) $\mu'(0) \equiv -\varepsilon_\perp(0)$ has to be chosen such that $\xi(\mu_{s2}) = \Lambda$. In case of Neumann boundary conditions (4.3.6), μ_{s1} and μ_{s2} have to be chosen such that $\xi(\mu_{s2}) = \Lambda$ and $(d\xi/d\mu)(\mu_{s2}) = [\mu'(\Lambda)]^{-1} \to \infty$.

The qualitative nature of the solutions (4.3.21) follows from the phase portraits, and is similar for all g-r mechanisms of the considered single-carrier multilevel type. The prototype of a bistable reaction-diffusion system corresponding to this topological class of phase portraits is the Schlögl model for nonequilibrium phase transitions in one-component chemical reaction systems [4.14]. Its steady state is described by an equation of the form (4.3.20), where the variable μ corresponds to the concentration v of a chemically reacting and diffusing species (scaled such that it is zero at the critical point, and can assume positive *and* negative values), and

$$\tilde{\rho}(\mu, \varepsilon_0) \triangleq (-v^3 + \tau v + \tau + g)/D \qquad (4.3.22)$$

is a cubic polynomial in v, with constants τ, g, and D. The control parameter g corresponds to ε_0. The Schlögl model represents the generic expansion of general nonlinear bistable systems near the critical point, compare (4.1.33). We will therefore use the Schlögl model to give a systematic survey of the solution manifold and the occurring bifurcations [4.46]. The solutions for our semiconductor model look qualitatively similar, though the numerical scales are different. To distinguish the numerical values of the Schlögl model from those of the semiconductor model in the following figures, the symbols g and L will be used instead of ε_0 and Λ, respectively.

In a finite-size semiconductor, the solution manifold depends upon the control parameter ε_0, the length Λ, and the boundary values $v(0)$, $v'(0)$, $v(\Lambda)$, $v'(\Lambda)$. In general these parameters cannot be varied independently, which is evident from the phase portraits (Figs. 4.2 and 4.24). In case of homogeneous Neumann conditions $[v'(0) = v'(\Lambda) = 0]$ the starting point and the endpoint of the solution trajectories must lie

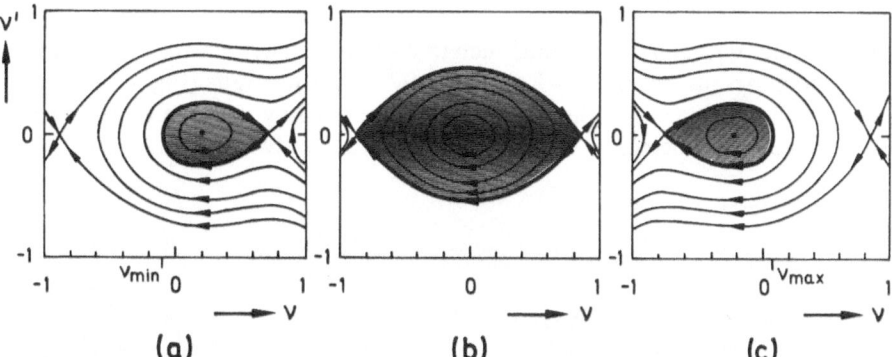

Fig. 4.24a–c. Phase portraits of v' versus v for the Schlögl model, a prototype of bistable reaction-diffusion systems. The numerical parameters are $D = 1, \tau = 0.75$ and (a) $g = -0.9$, (b) $g = g_{co} = -0.75$, (c) $g = -0.6$. The control parameter g corresponds to the suitably rescaled applied field ε_0 in semiconductor systems. The shaded areas denote the regions in phase space where solutions with Neumann boundary conditions can be found

on the line $\varepsilon_\perp = 0$. Thus for fixed ε_0, solutions exist in certain ranges of values of $v(0)$ only. If the starting point $v(0)$ and the number of extrema of the profile are suitably specified, then Λ and $v(\Lambda)$ take on fixed unique values. Similarly, for Dirichlet boundary conditions [$v(0), v(\Lambda)$ fixed], fixed ε_0, and a given number of extrema, solutions exist in a certain range of $v'(0)$ only, and Λ and $v'(\Lambda)$ are functions of $v'(0)$.

For a systematic survey of the possible solutions and their bifurcations we shall use the following diagrams:

(i)   The length Λ as a function of the starting point $v(0) \equiv v_R$ (in case of Neumann boundary conditions) or $v'(0)$ (in case of Dirichlet conditions) for fixed ε_0 and both types of boundary conditions.

(ii) The transverse carrier density profiles $v(\xi)$ for fixed Λ and different ε_0 and different boundary conditions. A sequence of such diagrams displays the bifurcations of the finite-size solutions.

(iii) The spatially averaged concentration

$$\bar{v} := \frac{1}{\Lambda} \int_0^\Lambda v(\xi)\, d\xi$$

      as a function of ε_0 for different values of Λ and different boundary conditions.

(iv) The current-voltage characteristics for different values of Λ and different boundary conditions. These diagrams are particularly useful in discussing the bistability regime and the possible bifurcations and nonequilibrium phase transitions.

The solution manifold is distinct for $\varepsilon_0 < \varepsilon_{co}$ and $\varepsilon_0 > \varepsilon_{co}$, where ε_{co} is defined by the equal-areas rule (4.1.21) for spatial coexistence between the bulk low- and high-conductivity states v_1 and v_3 in an infinitely extended system.

4.3.2 Neumann Boundary Conditions

First we consider boundary conditions such that the spatial derivatives of v vanish at both boundaries, as given by (4.3.5). From the phase portraits (Fig. 4.24) it is evident that all solutions lie in the shaded region enclosed by the saddle-to-saddle separatrices, i.e., the concentrations v_R at both boundaries always satisfy

$$v_1 < v_{min} < v_R < v_3 \qquad \text{if } \varepsilon_0 < \varepsilon_{co}$$

and

$$v_1 < v_R < v_{max} < v_3 \qquad \text{if } \varepsilon_0 > \varepsilon_{co} \ .$$

The relation between the length Λ and the boundary concentration $v_R = v(0)$ is plotted in Fig. 4.25a for different numbers of extrema and fixed $\varepsilon_0 > \varepsilon_{co}$. There are profiles with m extrema ($m = 0, 1, 2, ...$) if Λ exceeds $\Lambda_h(m + 1)/2$, where $\Lambda_h = 2\pi/k_\perp$, by (4.1.13), is the period of small harmonic spatial oscillations around the homogeneous NDC steady state $v = v_2$. At values of the length $\Lambda_m := \Lambda_h(m + 1)/2$ ($m = 0, 1, 2, ...$), successive bifurcations of pairs of inhomogeneous solutions (symmetric to one another) occur from the homogeneous steady state v_2. The homogeneous and the two inhomogeneous profiles merge at the bifurcation points. For $\varepsilon_0 < \varepsilon_{co}$ an analogous diagram as Fig. 4.25 is obtained, with v_1, v_{max} replaced by v_3, v_{min}, respectively.

If the control parameter ε_0 approaches either boundary of the bistability domain, then $k_\perp \to 0$ and $\Lambda_h \to \infty$. Hence, for fixed Λ, with varying ε_0 also a series of bifurcations occurs, whenever $\Lambda_h(m + 1)/2$ passes Λ for an integer m. The num-

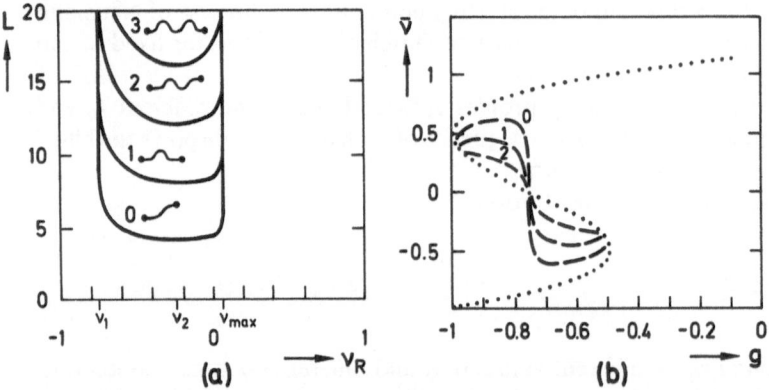

Fig. 4.25. (a) Length L (corresponding to Λ in the semiconductor model) versus concentration $v_R = v(0)$ for Neumann boundary conditions $v'(0) = v'(L) = 0$. The integers 0, 1, 2, 3 denote the number of extrema of the corresponding transverse carrier density profiles. These profiles are shown schematically. (Numerical parameters for the Schlögl model as in Fig. 4.24c). (b) Mean concentration $\bar{v}$ versus the control parameter g ($\hat{=} \varepsilon_0$), corresponding to the Neumann boundary conditions of Fig. 4.25a. The curves 0, 1, 2 correspond to $L = 15$ and 0, 1, 2 extrema, or, equivalently, $L = 15, 7.5, 5$ and no extrema. Homogeneous steady states are dotted. (Numerical parameters for the Schlögl model as in Fig. 4.24)

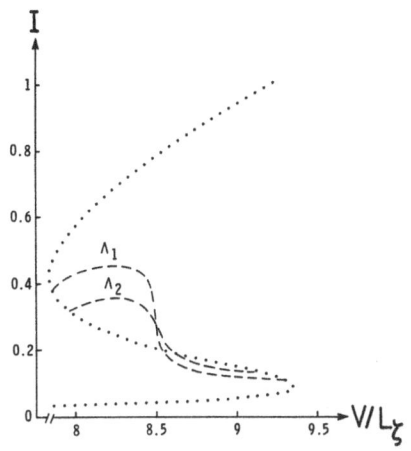

Fig. 4.26. Current I versus voltage $V = \varepsilon_0 L_\zeta$ for Neumann boundary conditions. Homogeneous steady states are dotted: the dashed curves correspond to monotonic inhomogeneous profiles with transverse dimensions $\Lambda_1 = 50$, $\Lambda_2 = 37.5$. Equivalently, the dashed curves correspond to inhomogeneous symmetric profiles of lengths $\Lambda_1 = 100$, $\Lambda_2 = 75$ with one extremum. (Calculated with the numerical parameters for the g-r model of Table 3.2)

ber and location of the bifurcation points depends upon Λ. For $\Lambda < \Lambda_h/2$ there are no inhomogeneous solutions. This is illustrated in Fig. 4.25b for the Schlögl model.

The current-voltage characteristic calculated from (4.3.20) with (4.1.36) for the two-level SNDC mechanism of Sect. 2.1 is shown in Fig. 4.26 for two values of Λ. The bifurcations of these inhomogeneous branches from the homogeneous states occur at different values of $V = \varepsilon_0 L_\zeta$, depending upon Λ and the number of extrema, as discussed above.

4.3.3 Dirichlet Boundary Conditions

The influence of Dirichlet boundary conditions upon dissipative structures has been studied in general nonlinear reaction-diffusion systems by various authors [4.46–54]. Here we shall describe systematically the effects of such boundaries, in particular the shift of the SNDC range in the current-voltage characteristics [4.4, 46]. If the concentration is held fixed at the boundaries, $v(0) = v_{s1}$, $v(\Lambda) = v_{s2}$, as given by (4.3.7), there can be 0, 1, 2, or 3 different spatial profiles depending upon the values of v_{s1}, v_{s2}. The profiles can be classified into monotonic, returning (one extremum), and oscillatory (more than one extremum).

For any value of Λ and given ε_0 there is one *monotonic* solution connecting two stable homogeneous steady states (Fig. 4.27) since Λ is a monotonic function of $v'(0)$. Some profiles are shown in Fig. 4.28.

Oscillatory profiles (with at least two extrema) exist, e.g., for $v_{min} < v_{s1} = v_{s2} < v_3$ (if $\varepsilon_0 < \varepsilon_{co}$) or for $v_1 < v_{s1} = v_{s2} < v_{max}$ (if $\varepsilon_0 > \varepsilon_{co}$).

A plot of Λ as a function of $v'(0)$ for hump-shaped (i.e., *filamentary*) symmetric profiles ($v_{s1} = v_{s2} =: v_R$) with one extremum is given in Fig. 4.29 for different values of ε_0 ($\hat{=} g$ in the Schlögl model) and v_R. It can be seen in Fig. 4.29c that in a range of values of Λ and v_R there are three inhomogeneous steady states. We shall show in Sect. 5.3.2 that the middle one of these, corresponding to dashed lines in Fig. 4.29,

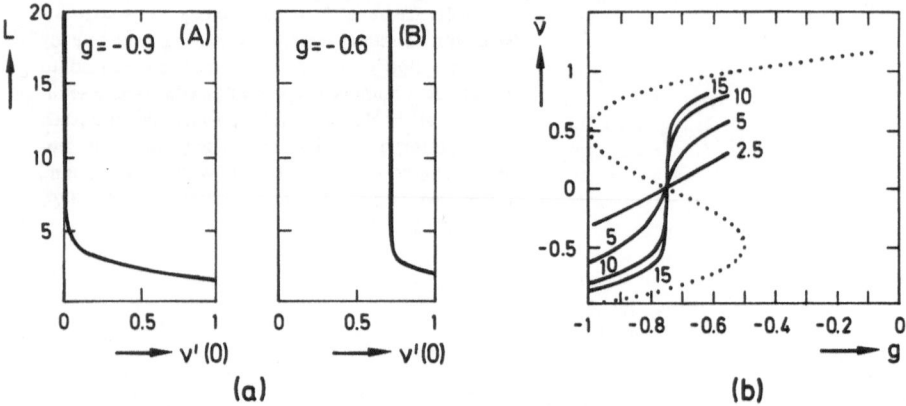

Fig. 4.27. (a) Length L ($\sim \Lambda$) versus $v'(0)$ for monotonic profiles with Dirichlet boundary conditions $v(0) = v_1$, $v(L) = v_3$, and $D = 1$, $\tau = 0.75$, (A) $g = 0.9 < g_{co}$, (B) $g = 0.6 > g_{co}$. (b) Mean concentration $\bar{v}$ versus the control parameter g ($\hat{=} \varepsilon_0$), corresponding to the boundary conditions of (a) and $L = 2.5, 5, 10, 15$. (Calculated with numerical parameters of the Schlögl model as in Fig. 4.24)

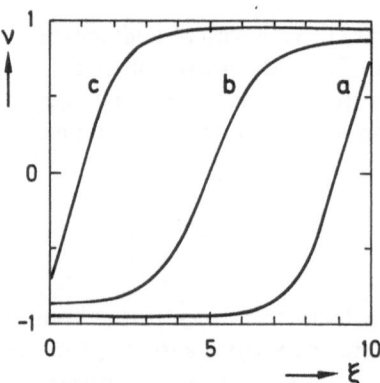

Fig. 4.28. Concentration profiles $v(\xi)$ for Dirichlet boundary conditions $v(0) = v_1$, $v(L) = v_3$ and $L = 10$. (a) $g = -0.9 < g_{co}$, (b) $g = -0.75 = g_{co}$, (c) $g = -0.6 > g_{co}$. (Calculated with numerical parameters of the Schlögl model as in Fig. 4.24)

is always unstable, while the upper and the lower profile is stable. This situation represents *filamentary bistability*.

The coexistence condition of the infinite system ε_{co} divides the solution manifold into two different classes ($\varepsilon_0 < \varepsilon_{co}$ and $\varepsilon_0 > \varepsilon_{co}$), which are symmetric to each other with respect to replacing v_1 with v_3, and $v'(0)$ with $-v'(0)$. The graphs in Fig. 4.29 can be interpreted as bifurcation diagrams, if $\Lambda(\sim L)$ is considered as a control parameter. The diagrams for $v_R = v_2$ in Fig. 4.29a, b for instance, represent a transcritical bifurcation of inhomogeneous profiles from the homogeneous solution. The bifurcation diagrams in Fig. 4.29c change in a generic way as v_R is raised: the regime of Λ in which three solution branches exist decreases and eventually ceases to exist; all branches are shifted towards smaller, i.e., more negative, values of $v'(0)$. The shift of the regime of multiple steady states is best visualized by plotting $\bar{v} := \int_0^\Lambda v(\xi)\, d\xi / \Lambda$ as a function of the control parameter ε_0 for various values of Λ

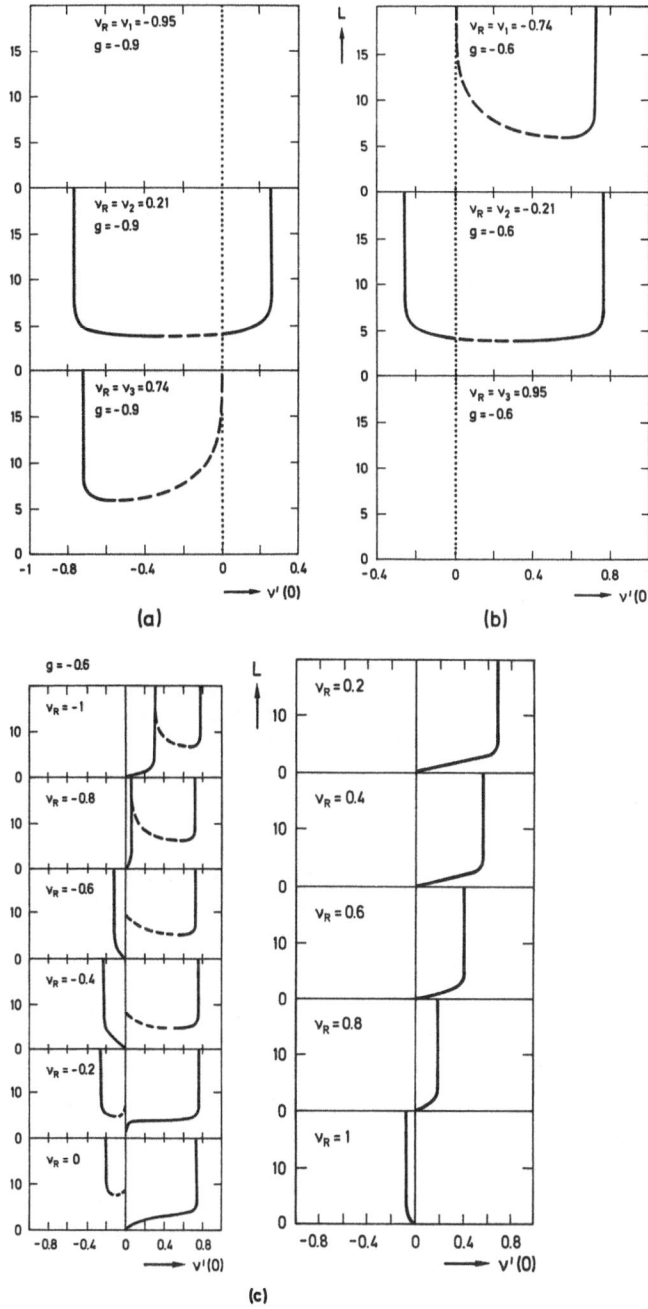

Fig. 4.29a–c. Length L ($\sim \Lambda$) versus $v'(0)$ for profiles with one extremum and symmetric Dirichlet boundary conditions $v(0) = v(L) = v_R$. Unstable state are dashed. The dotted line represents homogeneous steady states. (a) $g = -0.9 < g_{co}$; $v_R \triangleq$ homogeneous steady states; (b) $g = -0.6 > g_{co}$; $v_R \triangleq$ homogeneous steady states; (c) $g = -0.6 > g_{co}$. (Calculated with numerical parameters of the Schlögl model as in Fig. 4.24)

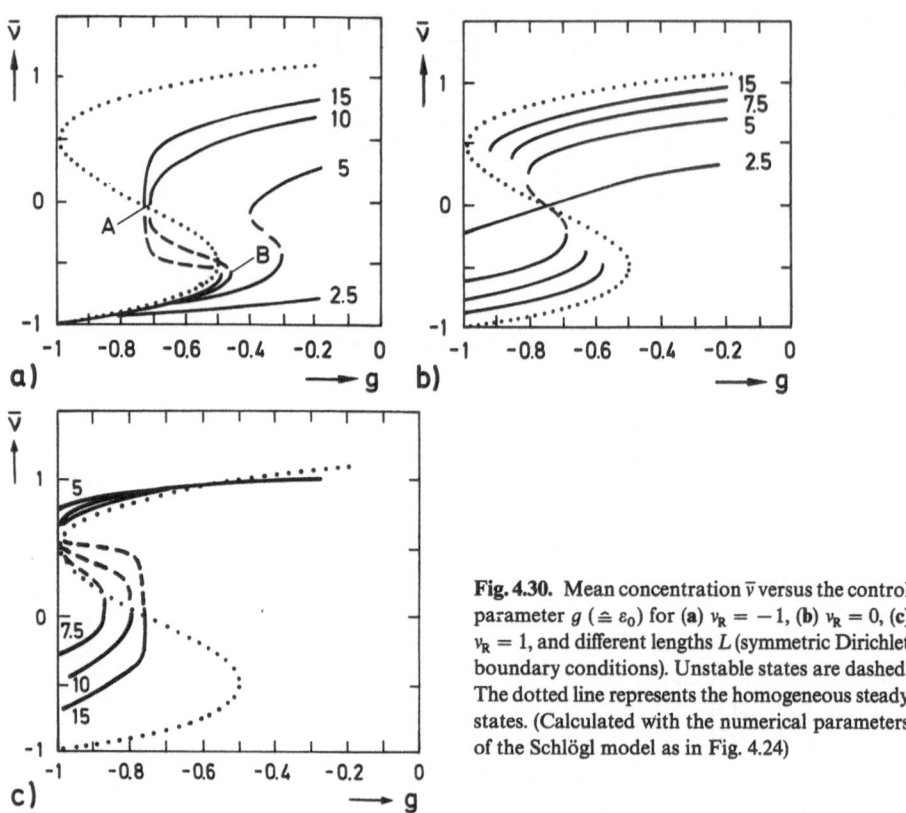

Fig. 4.30. Mean concentration $\bar{v}$ versus the control parameter g ($\hat{=}\ \varepsilon_0$) for (**a**) $v_R = -1$, (**b**) $v_R = 0$, (**c**) $v_R = 1$, and different lengths L (symmetric Dirichlet boundary conditions). Unstable states are dashed. The dotted line represents the homogeneous steady states. (Calculated with the numerical parameters of the Schlögl model as in Fig. 4.24)

(Fig. 4.30). The dotted line corresponds to the spatially homogeneous states of the infinite system. Note that the finite system can possess three solutions in regimes of ε_0 where the infinite system does not. In case of low v_R (Fig. 4.30a) the bistability regime is shifted to higher values of ε_0, and the shift is most pronounced for small Λ (corresponding to L in the Schlögl model): for $L = 5$, e.g., the whole bistability range lies completely outside the homogeneous bistability regime. For very small Λ ($L = 2.5$, e.g.,) however, the inhomogeneous profiles are quenched, since the spatial gradients are limited by diffusion: the bistability regime vanishes altogether. For large Λ the central pieces of the two stable profiles approach the two stable homogeneous states v_1 and v_3, and the bistability regime becomes $\varepsilon_{co} \leq \varepsilon_0 \leq \varepsilon_{th}$ i.e., the upper half of that of the infinite system. For intermediate values of v_R (Fig. 4.30b)the bistability regime shrinks from both sides with decreasing Λ. For high values of v_R (Fig. 4.30c) the bistability regime is shifted to lower values of ε_0. At the endpoints of the bistability regimes (e.g., A, B for the curve with $L = 10$ in Fig. 4.30a), bifurcations between inhomogenous profiles occur with ε_0 as a control parameter. At A, for instance, the upper and the lower profiles merge and annihilate upon decrease of ε_0. At B, the lower and the middle profiles merge and disappear with increasing ε_0. This is illustrated by showing the corresponding concentration profiles for a series of increasingly larger values of ε_0 in Fig. 4.31.

The corresponding current-voltage characteristics are shown in Fig. 4.32. They depend crucially upon the transverse dimension and on the boundary values of the carrier density at the lateral surfaces v_s. It is remarkable that filamentary bistability can exist for values of ε_0 above the threshold field ε_{th} of homogeneous SNDC, i.e., bistability of homogeneous steady states (Fig. 4.32a), and below the holding field ε_h of homogeneous SNDC (Fig. 4.32b). This means that appropriate boundary conditions can induce filamentary SNDC in the regime of positive differential conductivity of the homogeneous steady state [4.4]. Analogous behavior has been pointed out in chemical reaction systems [4.48]. Such behavior can be understood from the phase portrait (Fig. 4.33). For $\varepsilon_h < \varepsilon_0 < \varepsilon_{co}$ (a) there is only one phase trajectory at a given width Λ. For $\varepsilon_{co} < \varepsilon_0 < \varepsilon_{th}$ (b) there is one trajectory (denoted "a") with $v(\xi) < v_1$ throughout, and, for large enough Λ, two more trajectories (denoted "b", "c") in the shaded annular area around v_2. As ε_0 is increased beyond ε_{th}, the saddle point at v_1 and the center at v_2 merge and disappear, but – with the foregoing exception – the global picture of the trajectories changes smoothly and continuously (Fig. 4.33c). Therefore the three trajectories (a, b, c) corresponding to a given length Λ in Fig. 4.33 (b) can still be found in Fig. 4.33 (c) provided that ε_0 is not too far beyond ε_{th}. The corresponding diagrams of $\varepsilon_{\perp s} := \varepsilon_\perp(0) = -v'(0)/v(0)$ vs. Λ change with ε_0 as shown in Fig. 4.34.

Approximate analytical expressions for the profiles $\mu(\xi)$ and for the $\varepsilon_{\perp s}(\Lambda)$ relations can be obtained from a piecewise linear approximation of the local charge density $\tilde{\rho}$ as a function of μ [4.6]:

$$\tilde{\rho}(\mu, \varepsilon_0) = \begin{cases} -\alpha_1^2(\mu - \mu_1) & \text{if } \mu < \mu_2 \\ -\alpha_3^2(\mu - \mu_3) & \text{if } \mu > \mu_2 \end{cases}$$

$$\alpha_i := \left[-\frac{\partial \tilde{\rho}}{\partial \mu}(\mu_i) \right]^{1/2} \quad i = 1, 3 \ . \tag{4.3.23}$$

This approximation represents a linearization of $\tilde{\rho}$ around the two stable homogeneous steady states μ_1 and μ_3, respectively. Equation (4.3.20) can be solved analytically with (4.3.23):

$$\mu(\xi) = \mu_i + c_i e^{\alpha_i \xi} + d_i e^{-\alpha_i \xi} \ , \tag{4.3.24}$$

where $i = 1$ for $\mu < \mu_2$, $i = 3$ for $\mu > \mu_2$, and μ and μ' must be joined continuously at the boundary of the two regimes ($i = 1$ and $i = 3$).

For example, in this approximation the infinite boundary coexistence kink becomes

$$\mu(\xi) = \begin{cases} \mu_1 + c e^{\alpha_1 \xi} & (\xi < 0) \\ \mu_3 + d e^{-\alpha_1 \xi} & (\xi > 0) \end{cases} \tag{4.3.25}$$

with $c := \alpha_3 A$, $d := -\alpha_1 A$, $A := (\mu_3 - \mu_1)/(\alpha_1 + \alpha_3)$ and the coexistence condition (equal-areas rule) is

$$\alpha_1(\mu_2 - \mu_1) = \alpha_3(\mu_3 - \mu_2) \ .$$

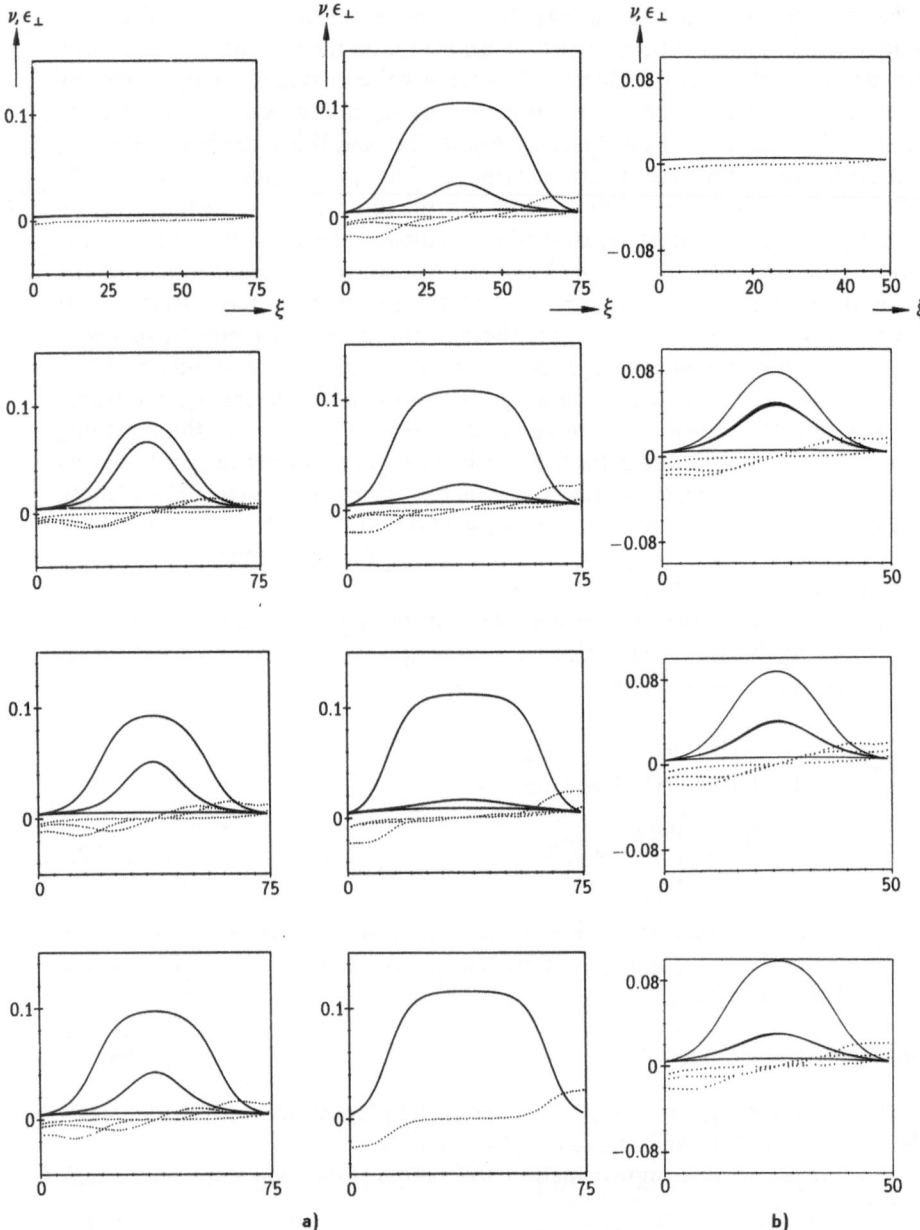

Fig. 4.31. Transverse concentration profiles v versus ξ for symmetric Dirichlet boundary conditions with (a) $v_s = 0.004 < v_1$, $\Lambda = 75$, (b) $v_s = 0.004$, $\Lambda = 50$, (c) $v_s = v_3$, $\Lambda = 50$. Sequences of profiles are shown with (a) $\varepsilon_0 = 8.5$, 8.6, 8.7, 8.8, (*left, from top to bottom*), 9.0, 9.2, 9.4, 9.6 (*right, from top to bottom*); (b) $\varepsilon_0 = 8.9$, 9.0, 9.1, 9.3, (*from top to bottom*). The dotted lines represent the respective transverse field profiles $\varepsilon_\perp$ versus ξ. In (c) a sequence of profiles is shown with $\varepsilon_0 = 7.9$ (*top*), $\varepsilon_0 = 8.0$ (*center*), $\varepsilon_0 = 8.2$ (*bottom*). For $\varepsilon_0 > 8.2$ the two lowest profiles merge and annihilate. (Calculated with the numerical parameters for the g-r model of Table 3.2)

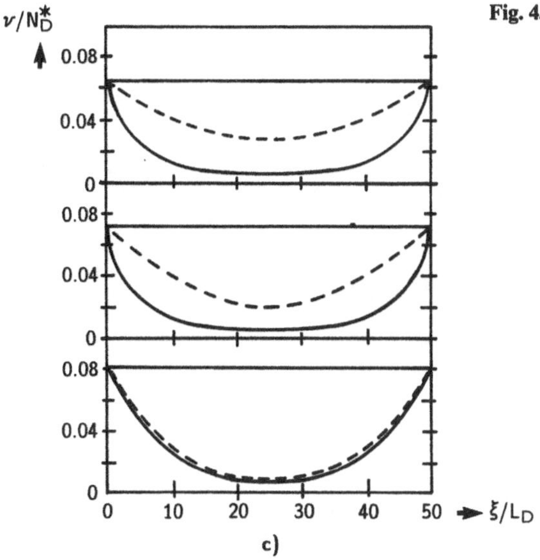

v/N_D^*

Fig. 4.31c. Caption see opposite page

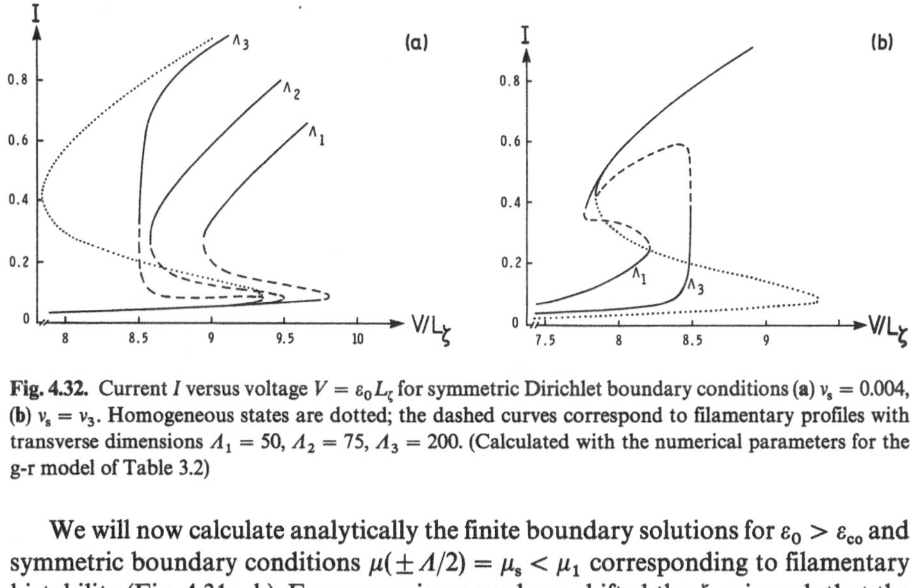

0.08

0.04

0

0.08

0.04

0

0.08

0.04

0

0   10   20   30   40   50 ➤ ξ/L_D

c)

I

(a)

0.8

0.6

0.4

0.2

0

8   8.5   9   9.5   10 ➤ V/L_ζ

Λ_3

Λ_2

Λ_1

I

(b)

0.8

0.6

0.4

0.2

0

7.5   8   8.5   9 ➤ V/L_ζ

Λ_1

Λ_3

Fig. 4.32. Current I versus voltage $V = \varepsilon_0 L_\zeta$ for symmetric Dirichlet boundary conditions (a) $v_s = 0.004$, (b) $v_s = v_3$. Homogeneous states are dotted; the dashed curves correspond to filamentary profiles with transverse dimensions $\Lambda_1 = 50$, $\Lambda_2 = 75$, $\Lambda_3 = 200$. (Calculated with the numerical parameters for the g-r model of Table 3.2)

We will now calculate analytically the finite boundary solutions for $\varepsilon_0 > \varepsilon_{co}$ and symmetric boundary conditions $\mu(\pm \Lambda/2) = \mu_s < \mu_1$ corresponding to filamentary bistability (Fig. 4.31a, b). For convenience we have shifted the ζ-axis such that the boundaries are at $\zeta = \pm \Lambda/2$, and the maximum of the profile is at $\zeta = 0$. The three profiles shown in Fig. 4.31a for fixed ε_0-values are given by:

(a) $\mu(\xi) = \mu_1 - \dfrac{(\mu_1 - \mu_s)\cosh(\alpha_1 \xi)}{\cosh(\alpha_1 \Lambda/2)}$ (4.3.26a)

(which is always below μ_1)

(b, c) $\mu(\xi) = \begin{cases} \mu_3 - e\cosh(\alpha_3 \xi), & 0 \le |\xi| \le \xi_2 \\ \mu_1 + c_1 e^{\alpha_1|\xi|} + d_1 e^{-\alpha_1|\xi|}, & \xi_2 \le |\xi| \le \Lambda/2 \end{cases}$

(4.3.26b)

(4.3.26c)

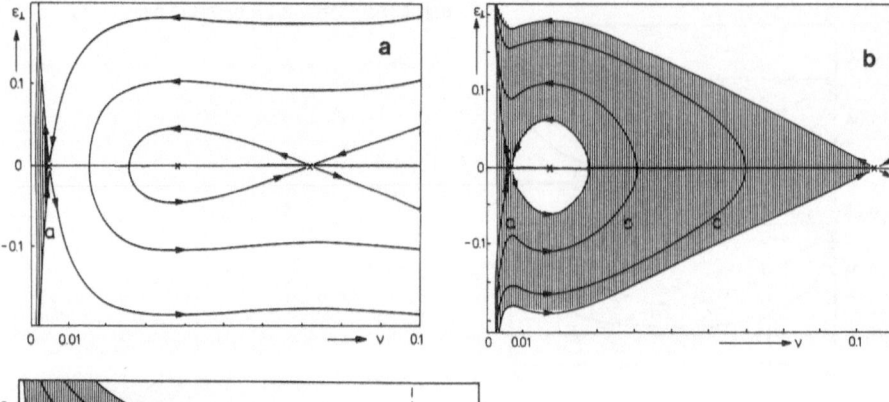

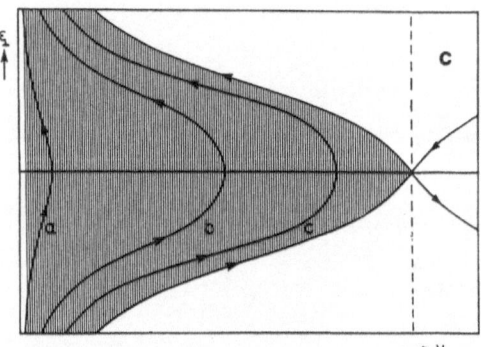

Fig. 4.33. Phase portraits of $\varepsilon_\perp$ versus v for (a) $\varepsilon_h < \varepsilon_0 < \varepsilon_{co}$; (b) $\varepsilon_{co} < \varepsilon_0 < \varepsilon_{th}$; (c) $\varepsilon_0 > \varepsilon_{th}$ (schematic). The shaded areas denote the regions in phase space where solutions with Dirichlet boundary conditions $v_{s1} = v_{s2} < v_1$ can be found. The letters a–c mark phase trajectories that correspond to profiles of a given width Λ

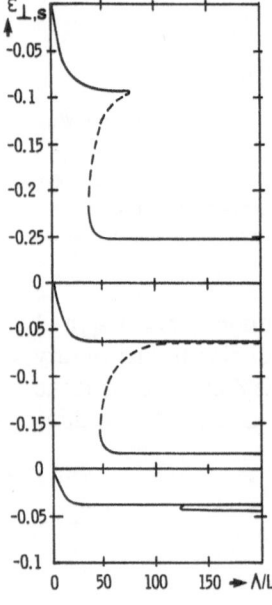

Fig. 4.34. Surface field $\varepsilon_{\perp s} := \varepsilon_\perp(0) = -\varepsilon_\perp(\Lambda)$ versus Λ for $v_s = 0.004$ and $\varepsilon_0 = 9.5 > \varepsilon_{th}$ (*top*), $\varepsilon_0 = 9.0 < \varepsilon_{th}$ (*center*), $\varepsilon_0 = 8.5 < \varepsilon_{th}$ (*bottom*). The lower two plots correspond qualitatively to the top of Fig. 4.29c. (Calculated with the numerical parameters for the g-r model of Table 3.2)

where c_1, d_1, e are determined from the continuity conditions at $\xi_2 > 0$ [ξ_2 is defined by $\mu(\xi_2) = \mu_2$]

$$\mu_3 - e\cosh(\alpha_3\xi_2) = \mu_1 + c_1 e^{\alpha_1\xi_2} + d_1 e^{-\alpha_1\xi_2} \tag{4.3.27a}$$

$$-e\alpha_3\sinh(\alpha_3\xi_2) = c_1\alpha_1 e^{\alpha_1\xi_2} - d_1\alpha_1 e^{-\alpha_1\xi_2} \tag{4.3.27b}$$

and the boundary condition

$$\mu_1 + c_1 e^{\alpha_1 A/2} + d_1 e^{-\alpha_1 A/2} = \mu_s \ . \tag{4.3.27c}$$

Note that the equation which determines ξ_2,

$$\mu(\xi_2) = \mu_2 \ , \tag{4.3.28}$$

gives two values of ξ_2, corresponding to the two solutions b, c.

The relation of $\varepsilon_{\perp s} = -\mu'(-A/2) = \mu'(A/2)$ vs. A can now be obtained from the profiles (4.3.26). The branch that corresponds to profile (a) is readily found from (4.3.26a):

$$\varepsilon_{\perp s} = -(\mu_1 - \mu_s)\alpha_1 \tanh(\alpha_1 A/2) \ . \tag{4.3.29}$$

As A tends to infinity, $\varepsilon_{\perp s} \to -(\mu_1 - \mu_s)\alpha_1$. Rather than calculate the other two branches (b, c) from (4.3.26b, c) in the same way (after eliminating ξ_2, c_1, d_1, e), we shall compute these directly from the double integral (4.3.21):

$$A/2 = \int_{\mu_s}^{\mu(0)} [\varepsilon_{\perp s}^2 - 2\tilde{\phi}(\mu, \varepsilon_0)]^{-1/2} d\mu \tag{4.3.30}$$

where

$$\tilde{\phi}(\mu, \varepsilon_0) = \int_{\mu_s}^{\mu} \tilde{\rho}(\tilde{\mu}, \varepsilon_0) d\tilde{\mu}$$

$$= \begin{cases} -\dfrac{\alpha_1^2}{2}[(\mu - \mu_1)^2 - (\mu_1 - \mu_s)^2] & \text{if } \mu < \mu_2 \\[3mm] -\dfrac{C}{2} - \dfrac{\alpha_3^2}{2}[(\mu - \mu_3)^2 - (\mu_3 - \mu_2)^2] & \text{if } \mu > \mu_2 \end{cases}$$

$$C := \alpha_1^2[(\mu_2 - \mu_1)^2 - (\mu_1 - \mu_s)^2] \ . \tag{4.3.31}$$

Hence for $\mu(0) > \mu_2$ (this holds for solutions b and c) we obtain

$$A(\varepsilon_{\perp s})$$

$$= 2\left\{\frac{1}{\alpha_1}\ln\frac{[(\mu_2 - \mu_1)^2 - (\mu_1 - \mu_s)^2 + \varepsilon_{\perp s}^2/\alpha_1^2]^{1/2} + (\mu_2 - \mu_1)}{-\varepsilon_{\perp s}/\alpha_1 - (\mu_1 - \mu_s)}\right.$$

$$\left. + \frac{1}{\alpha_3}\ln\frac{\{[\mu_3 - \mu(0)]^2 - (\mu_3 - \mu_2)^2 + (\varepsilon_{\perp s}^2 + C)/\alpha_3^2\}^{1/2} - [\mu_3 - \mu(0)]}{\sqrt{\varepsilon_{\perp s}^2 + C/\alpha_3} - (\mu_3 - \mu_2)}\right\} \ . \tag{4.3.32}$$

The first term in (4.3.32) diverges for $\varepsilon_{\perp s} \rightarrow -(\mu_1 - \mu_s)\alpha_1$, the second term for $\varepsilon_{\perp s} \rightarrow -[(\mu_3 - \mu_2)^2\alpha_3^2 - (\mu_2 - \mu_1)^2\alpha_1^2 + (\mu_1 - \mu_s)^2\alpha_1^2]^{1/2}$. Therefore these two terms correspond asymptotically, i.e., for large $\varLambda$, to the solution branches (b) and (c), respectively.

For the simple two-level g-r model of Sect. 2.1.2 the piecewise linear charge density (4.3.23) is given by the following explicit expresions for α_1 and α_3: In the low-conductivity (OFF) state μ_1, impact ionization can be neglected, and practically all available electrons are trapped. We have $\rho \approx 1 - v/v_1$ in that case. In the high-conductivity (ON) state μ_3, thermal generation is negligible and the trap ground state is depleted by impact ionization, therefore $\rho \approx v_3 - v$. This gives

$$\tilde{\rho}(\mu, \varepsilon_0) \approx 1 - e^{\mu - \mu_1} \approx -(\mu - \mu_1) \qquad \text{for } \mu \approx \mu_1 \tag{4.3.33a}$$

and

$$\tilde{\rho}(\mu, \varepsilon_0) \approx v_3\left(1 - \frac{v}{v_3}\right) = v_3(1 - e^{\mu - \mu_3}) \approx -v_3(\mu - \mu_3) \tag{4.3.33b}$$

for $\mu \approx \mu_3$, and hence, by (4.3.23)

$$\alpha_1 = 1, \qquad \alpha_3 = \sqrt{v_3} . \tag{4.3.34}$$

In Figs. 4.26, 31, 32, 34 all concentrations, lengths, and fields have been given in units of N_D^*, L_D, and $(kTE_t)/(eL_D)$, respectively. All dimensions can be scaled by choosing different temperatures T, doping concentrations N_D^*, and trap energies E_t. In particular, since $L_D \sim (kT/N_D^*)^{1/2}$, the spatial dimension can be scaled to smaller values by decreasing the temperature and increasing the doping. Similarly, the applied field ε_0 can be scaled to larger values by choosing energetically deeper traps. The dimensional units are given for a selection of typical input data in Table 4.1. As the figures show, the current-voltage characteristics of filamentary transport depend crucially upon the lateral dimension $\varLambda$ in the range $\varLambda \approx 20 \dots 100 L_D$, which corresponds to values in the near-micron and submicron range.

In conclusion, the SNDC characteristics of the infinite system (dotted lines in Figs. 4.26, 32) are significantly changed by finite boundaries: SNDC is completely suppressed for monotonic profiles; for symmetric filamentary profiles the SNDC range of fields ε_0 is shifted. As the density v_s of electrons at the surface is increased externally (this corresponds to Dirichlet boundary conditions with varying v_s), ε_h

Table 4.1. Scaling of length scale for GaAs ($\epsilon_s = 12.5$)

	T [K]	N_D^* [cm^{-3}]	L_D [µm]
Shallow donors	4.2	10^{15}	0.01
Deep traps, – low doping	300	10^{15}	0.1
– high doping	300	10^{17}	0.01

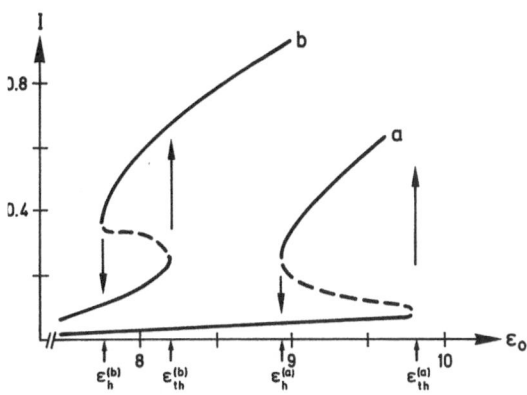

Fig. 4.35. Current I versus voltage $V = \varepsilon_0 L_\zeta$ for $\Lambda = 50$ and symmetric Dirichlet boundary conditions with two different values of v_s (corresponding to increasing optical surface excitation intensity from (a) to (b)). (a) $v_s = 0.004$, (b) $v_s = v_3 \approx 0.08$. (Calculated with the numerical parameters for the g-r model of Table 3.2)

and ε_{th} shift towards lower values, see Fig. 4.35. This provides a convenient control of the threshold and holding fields of the switching transition. The control could be achieved, for example, by optical surface generation, or by electrically adjusting the barrier height through an additional electrode at the lateral surfaces (a "gate" similar to an MOS transistor configuration). In a complementary mode of operation one could use a fixed voltage, and switch the device by applying illumination at a lateral surface [4.55]. Such optical-threshold switches are needed as logic elements in optical or hybrid-optical computers [4.56]. By changing the bias voltage, the optical-threshold intensity at which switching between the low- and the high-conductivity branch occurs can be varied: it is obvious from Fig. 4.35 that smaller voltages correspond to larger v_s, i.e., higher optical threshold intensities. In a typical numerical example, with $N_D^* \approx 10^{15}$ cm^{-3}, $v_3/v_1 \approx 10$, $v_3 \approx 0.1$, $L_D \approx 0.1$ μm at $T = 300$ K, and with a maximum optical power of 0.25 μW (bandgap light in GaAs, $\hbar\omega = 1.4$ eV, absorption efficiency 20%) available in a 2×2 μm^2 channel, a surface recombination velocity $s \approx 5 \times 10^4$ cm/s is required.

It seems, therefore, that the investigated mechanism of filamentation controlled by surface g-r, lends itself to a variety of device applications, such as optically controllable threshold switches, electrically controllable optical threshold switches, or three-terminal electronic switches with adjustable threshold voltage, especially in the near-micron and submicron range.

4.4 Filamentation in Two-Carrier Models

In Sects. 4.1–3 we have dealt with single-carrier models. The stationary transverse spatial structures were characterized by the balance of unipolar diffusion and drift currents (4.1.6). The transverse field counteracts the carrier-density gradients. The inhomogeneous carrier distributions lead to space charges ρ which couple to the transverse field $\varepsilon_\perp$ via Poisson's equation, and thus limit the drift current.

We shall now consider two-carrier models, where both electrons and holes contribute to the current [4.5]. Therefore the transverse electron and hole currents can compensate each other, and the ambipolar currents are generally much higher than the unipolar ones. Electrons and holes tend to neutralize each other approximately in a very short time of the order of the dielectric relaxation time τ_M (3.1.5), so that no space charges limit the currents. Instead, the loss of electrons and holes through recombination limits the ambipolar currents in semiconductors [4.57]. There may still be a small charge density but this space charge is accommodated by slight perturbations of the recombination determined electron and hole distributions.

4.4.1 Ambipolar Diffusion and Trapping

In the following we shall use, as in Chaps. 1 and 2, unscaled variables (time t, space r, carrier densities n, p, electric field E). The continuity equations for free electrons and holes, and trapped electrons, respectively, are given by, see (1.1.5–7),

$$\dot{n} - \frac{1}{e}\nabla\cdot j_n = \varphi_n(n, p, n_t, E) \tag{4.4.1}$$

$$\dot{p} + \frac{1}{e}\nabla\cdot j_p = \varphi_p(n, p, n_t, E) \tag{4.4.2}$$

$$\dot{n}_t = \varphi_t(n, p, n_t, E) \ , \tag{4.4.3}$$

where

$$j_n = e(\mu_n nE + D_n\nabla n) \tag{4.4.4}$$

$$j_p = e(\mu_p pE - D_p\nabla p) \tag{4.4.5}$$

are the respective current densities, $\mu_{n,p}$ and $D_{n,p}$ are the mobilities and diffusion constants, and φ_n, φ_p, φ_t are the g-r rates. Since the g-r kinetics conserve the total number of carriers,

$$\varphi_n - \varphi_p + \varphi_t = 0 \ . \tag{4.4.6}$$

As discussed above, we assume approximate local-charge neutrality

$$N_D^* - n + p - n_t \approx 0 \ . \tag{4.4.7}$$

For plane transverse stationary spatial structures (4.4.1–3) reduce to

$$-j_n' = e\varphi_n \tag{4.4.8}$$

$$j_p' = e\varphi_p \tag{4.4.9}$$

$$\varphi_t \equiv \varphi_p - \varphi_n = 0 \qquad \text{where} \tag{4.4.10}$$

$$j_n = e(\mu_n nE_\perp + D_n n') \tag{4.4.11}$$

$$j_p = e(\mu_p pE_\perp - D_p n') \tag{4.4.12}$$

are the transverse current densities, and the prime denotes the derivative with respect to the transverse coordinate. In case of cylindrical transverse structures (4.4.8, 9) become

$$-\frac{1}{r}\frac{\partial}{\partial r}(rj_\mathrm{n}) = e\varphi_\mathrm{n} \tag{4.4.13}$$

$$\frac{1}{r}\frac{\partial}{\partial r}(rj_\mathrm{p}) = e\varphi_\mathrm{p} \ . \tag{4.4.14}$$

Subtracting (4.4.8, 9), using (4.4.10) and the boundary condition of vanishing transverse current density at the lateral surfaces we find the conservation law

$$j_\mathrm{n} + j_\mathrm{p} = 0 \ . \tag{4.4.15}$$

Substituting (4.4.11, 12) into (4.4.15) yields the transverse field

$$E_\perp = \frac{-D_\mathrm{n}n' + D_\mathrm{p}p'}{\mu_\mathrm{n}n + \mu_\mathrm{p}p} \ . \tag{4.4.16}$$

We conclude from (4.4.16) that $E'_\perp$ will in general not vanish; but it must be small so that the local-charge neutrality condition (4.4.7) holds approximately:

$$E'_\perp = \frac{4\pi e}{\epsilon_\mathrm{s}}(N_\mathrm{D}^* - n + p - n_\mathrm{t}) \approx 0 \ . \tag{4.4.17}$$

A self-consistent treatment requires that we first calculate the profiles $n(x)$, $p(x)$ using (4.4.7), then determine $E_\perp$ from (4.4.16), and finally check the consistency of our assumption of quasi neutrality by substituting n, p, and $E_\perp$ into (4.4.17) and verifying that

$$E'_\perp \ll \frac{4\pi e}{\epsilon_\mathrm{s}}(N_\mathrm{D}^* + p)$$

holds.

We can eliminate the field by inserting (4.4.16) into (4.4.12) and using the Einstein relation $eD_\mathrm{n,p} = kT\mu_\mathrm{n,p}$:

$$\frac{1}{e}j_\mathrm{p} = -\frac{pn' + np'}{n/D_\mathrm{p} + p/D_\mathrm{n}} \ . \tag{4.4.18}$$

If we *neglect* the trapped electron density n_t, (4.4.7) gives $n' = p'$ and (4.4.18) can be expressed as

$$j_\mathrm{p} = -eD_\mathrm{a}p' \ , \qquad \text{where} \tag{4.4.19}$$

$$D_\mathrm{a} = \frac{n + p}{n/D_\mathrm{p} + p/D_\mathrm{n}} \tag{4.4.20}$$

is the usual ambipolar diffusion coefficient [4.15]. In two-carrier SNDC mechanisms, however, trapping plays a crucial role, as we have seen in Sect. 2.2 and therefore we cannot neglect n_t. In fact, we shall see that the presence of trapped electrons and holes or recombination centers leads to an essential modification of the ambipolar transport properties. There are two equivalent ways of eliminating the traps and deriving a renormalized ambipolar diffusion coefficient. First, we can use local charge neutrality (4.4.7) to eliminate n_t (4.4.10):

$$\psi_t(n,p) := \varphi_t(n, p, n_t = N_D^* - n + p) = 0 \ . \tag{4.4.21}$$

Hence

$$n' = -\left[\left(\frac{\partial \psi_t}{\partial p}\right)_n \bigg/ \left(\frac{\partial \psi_t}{\partial n}\right)_p\right] p' =: \alpha(n, p)p' \tag{4.4.22}$$

and, by (4.4.18),

$$j_p = -e\tilde{D}_a p' \tag{4.4.23}$$

with an effective ambipolar diffusion coefficient

$$\tilde{D}_a(n, p) = \frac{n + p\alpha(n, p)}{n/D_p + p/D_n} \tag{4.4.24}$$

$$= \frac{n\left(\frac{\partial \psi_t}{\partial n}\right) - p\left(\frac{\partial \psi_t}{\partial p}\right)}{n/D_p + p/D_n}\left(\frac{\partial \psi_t}{\partial n}\right)^{-1} \ . \tag{4.4.25}$$

By (4.4.10) we can express α and $\tilde{D}_a$ in terms of the Jacobian matrix A,

$$(A_{ij}) := \begin{pmatrix} \dfrac{\partial \psi_n}{\partial n} & \dfrac{\partial \psi_n}{\partial p} \\[2mm] \dfrac{\partial \psi_p}{\partial n} & \dfrac{\partial \psi_p}{\partial p} \end{pmatrix}$$

$$\psi_{n,p}(n, p) := \varphi_{n,p}(n, p, n_t = N_D^* - n + p) \ , \tag{4.4.26}$$

which we used in Sect. 2.2 to discuss the stability of spatially homogeneous steady states:

$$\alpha(n, p) = \frac{A_{22} - A_{12}}{A_{11} - A_{21}} \tag{4.4.27}$$

$$\tilde{D}_a(n, p) = \frac{-nA_{11} + pA_{12} + nA_{21} - pA_{22}}{(n/D_p + p/D_n)(A_{21} - A_{11})} \ . \tag{4.4.28}$$

We can use the implicit relation between n and p given by the steady state condition (4.4.21) to express $\tilde{D}_a$ and φ_p as functions of p (or n) alone so that the governing

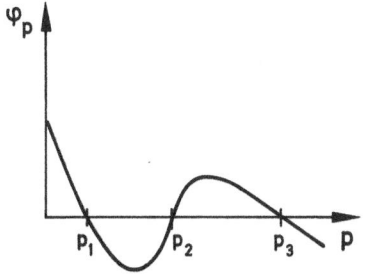

Fig. 4.36. Hole generation-recombination rate versus hole density p for a two-carrier SNDC mechanism (schematic). The homogeneous, charge-neutral steady states are p_1, p_2, p_3. The free and trapped electron densities n and n_t have been eliminated in the g-r rate by the charge neutrality and the steady state conditions

equation for the spatial profiles becomes, by (4.4.9, 23)

$$[\tilde{D}_a(p)p']' + \varphi_p(p) = 0 \tag{4.4.29}$$

for plane current layers, and

$$\frac{1}{r}[r\tilde{D}_a(p)p']' + \varphi_p(p) = 0 \tag{4.4.30}$$

for cylindrical current filaments. (For simplicity we use the same symbols $\tilde{D}_a$ and φ_p for the functions after elimination of n and n_t.)

The g-r rate φ_p, expressed as a function of p alone by (4.4.21), has three non-negative zeros in the SNDC range, and is shown schematically in Fig. 4.36. Eq. (4.4.29) reflects the fact that recombination determines the spatial profiles, and thus the concentration gradients and the current density $j_p = -j_n$, see (4.4.19). There is a net recombination traffic

$$\varphi_n = \varphi_p \neq 0$$

even in the steady state, while the charge density ρ is zero. In single-carrier stationary spatial structures, on the other hand, the recombination rate φ vanishes, but ρ does not.

An alternative derivation of (4.4.29) proceeds by first calculating $n_t(n, p)$ from $\varphi_t(n, p, n_t) = 0$ (4.4.10), and then substituting into the charge neutrality condition (4.4.7):

$$N_D^* - n + p - n_t(n, p) = 0 \ . \tag{4.4.31}$$

This yields

$$-n' + p' - \frac{\partial n_t}{\partial n} n' - \frac{\partial n_t}{\partial p} p' = 0 \ . \tag{4.4.32}$$

Using (4.4.32), we can write (4.4.18) in the form (4.4.23, 24) with

$$\alpha(n, p) = \left(1 - \frac{\partial n_t}{\partial p}\right) \Big/ \left(1 + \frac{\partial n_t}{\partial n}\right) \tag{4.4.33}$$

Table 4.2. Filamentation in the two-carrier mechanism of Sect. 2.2.2 (Fig. 2.24)

Defining equation	Expression
	$\varphi_n(n, p, n_t) = n[X_1 n_t - T_1^S(N_t - n_t) - B^S p]$
	$\varphi_p(n, p, n_t) = p[X_4(N_t - n_t) - T_2^S n_t - B^S n]$
(4.4.21)	$\psi_t(n, p) \quad = p[F_b + (X_4 + T_2^S)(n - p)] - n[F_a + (X_1 + T_1^S)(p - n)]$
(4.4.22)	$\alpha(n, p) \quad = \dfrac{F_b + n(X_4 + T_2^S - X_1 - T_1^S) - 2p(X_4 + T_2^S)}{F_a + p(X_1 + T_1^S - X_4 - T_2^S) - 2n(X_1 + T_1^S)}$
(4.4.26)	$\psi_p(n, p) \quad = p[F_b - (X_4 + T_2^S)p - (B^S - X_4 - T_2^S)np]$
(4.4.10)	$n_t(n, p) \quad = (n T_1^S + p X_4)N_t/Z$
	$Z := n(X_1 + T_1^S) + p(X_4 + T_2^S)$
(4.4.33)	$\alpha(n, p) \quad = \dfrac{1 - nC/Z^2}{1 - pC/Z^2} \qquad (C > 0 \text{ in bistability domain})$
(4.4.34)	$\tilde{D}_a(n, p) \quad = \dfrac{(n + p)Z^2 - 2npC}{(n/D_p + p/D_n)(Z^2 - pC)}$
(4.4.31)	$n(p) \quad = \dfrac{F_a' + (A - B)p}{2A} \pm \dfrac{p}{2A}\left[\left(\dfrac{F_a}{p} + A - B\right)^2 + 4A\left(B - \dfrac{F_b}{p}\right)\right]^{1/2}$

$$\tilde{D}_a(n, p) = \frac{n + p + n(\partial n_t/\partial n) - p(\partial n_t/\partial p)}{(n/D_p + p/D_n)(1 + \partial n_t/\partial n)} \ . \tag{4.4.34}$$

Finally, n is eliminated by (4.4.31). Of course, (4.4.25) and (4.4.34) are equivalent. The explicit expressions for α, $\tilde{D}_a$, and φ_p are listed in Table 4.2 for the band-trap impact-ionization mechanism analysed in Sect. 2.2.2.

For standard Shockley-Read-Hall kinetics [4.58] without impact ionization and Auger recombination, the respective expressions are:

$$\varphi_t = T_1^S[n(N_t - n_t) - n_1 n_t] - T_2^S[(pn_t - p_1(N_t - n_t)] \tag{4.4.35}$$

and

$$n_t(n, p) = (T_1^S n + T_2^S p_1)N_t/Z_0$$

$$Z_0 := T_1^S(n + n_1) + T_2^S(p + p_1) \ , \tag{4.4.36}$$

where T_1^S and T_2^S are the electron and hole capture coefficents, respectively, N_t is the total trap density, and $n_1 := n_0(N_t - n_{t0})/n_{t0}$, $p_1 := p_0 n_{t0}/(N_t - n_{t0})$ are the electron and hole concentrations for which the respective quasi-Fermi levels coincide with the trapping level (the subscript 0 denotes thermal equilibrium).

From (4.4.36) we obtain

$$\frac{\partial n_t}{\partial n} = T_1^S(T_1^S n_1 + T_2^S p)N_t/Z_0^2 > 0 \tag{4.4.37}$$

and

$$\frac{\partial n_t}{\partial p} = -T_2^S(T_1^S n + T_2^S p_1) N_t / Z_0^2 < 0 , \tag{4.4.38}$$

hence

$$\alpha(n, p) = \frac{Z_0^2 + T_2^S(T_1^S n + T_2^S p_1) N_t}{Z_0^2 + T_1^S(T_1^S n_1 + T_2^S p) N_t} > 0 \tag{4.4.39}$$

and

$$\tilde{D}_a(n, p) = \frac{(n + p) Z_0^2 + [(T_1^S)^2 nn_1 + (T_2^S)^2 pp_1 + 2 T_1^S T_2^S np] N_t}{(n/D_p + p/D_n) [Z_0^2 + T_1^S (T_1^S n_1 + T_2^S p) N_t]} . \tag{4.4.40}$$

In this example it can easily be seen that $\tilde{D}_a > 0$ holds. We assume in the following that in general $\tilde{D}_a > 0$. Without trapping, D_a by (4.4.20) always lies between D_n and D_p and is close to the minority carrier diffusion constant. A trapping factor $\alpha > 1$ enhances the hole diffusion, while $\alpha < 1$ slows it down.

4.4.2 Equal-Areas Rules for Current Layers and Filaments

Equation (4.4.23, 9) are the analog of (4.1.6) and (4.1.7), respectively, where j_p/e is replaced with $\varepsilon_\perp$, p with v, $\tilde{D}_a(p)$ with $1/v$, and φ_p with ρ. Identical arguments as those following (4.1.7) apply. In particular, a similar phase-portrait analysis can be performed. The homogeneous steady states are determined by $j_p = j_n = 0$, and $\varphi_p(p) = 0$. The governing equation (4.4.29) can be cast into the form (4.1.15) by the transformation

$$\mu(p) := \int_{p_0}^{p} \tilde{D}_a(\bar{p}) d\bar{p} . \tag{4.4.41}$$

The coexistence condition (equal-areas rule) for plane-current layers (4.1.21) is then reproduced with

$$\tilde{\phi}(\mu) := \int \varphi_p(p) \tilde{D}_a(p) dp . \tag{4.4.42}$$

In terms of the variable p, (4.1.21) reads:

$$\int_{p_1}^{p_3} \varphi_p(p, E_0) \tilde{D}_a(p, E_0) dp = 0 . \tag{4.4.43}$$

This is the equal-areas rule for two-carrier filamentation. Note that in (4.4.43) we have explicitly marked the dependence of φ_p and $\tilde{D}_a$ upon the applied field E_0 through the g-r coefficients. Therefore (4.4.43) singles out a coexistence field $E_0 = E_{co}$, just as the equal-areas rule in case of single-carrier filamentation does. The transverse carrier-density profiles have qualitatively the same behavior as those for single-carrier models.

For cylindrical current filaments the equations (4.4.23, 14) are the analog of (4.2.1, 2), and again a similar equal-areas rule as the one in (4.2.10) is obtained:

$$\int_{p(0)}^{p(\infty)} \varphi_p(p, E_0) \tilde{D}_a(p, E_0)\, dp = -\Sigma \;,$$

$$(4.4.44)$$

where in lowest order for E_0 close to E_{co}

$$\Sigma \equiv \frac{1}{e^2} \int_0^\infty j_p^2 \frac{dr}{r}$$

$$\approx \frac{1}{r_0} \left[\int_{p_1}^{p_3} \tilde{D}_a(p, E_0)\, dp \right] \left[2 \int_{p_2}^{p_3} \varphi_p(p, E_0) \tilde{D}_a(p, E_0)\, dp \right]^{1/2}$$

$$(4.4.45)$$

and $p(0) \approx p_1, p(\infty) = p_3$ for $E_0 < E_{co}$ (low-current filaments), $p(0) \approx p_3, p(\infty) = p_1$ for $E_0 > E_{co}$ (high-current filaments).

The major difference of the current layers and filaments in systems with one or two types of carriers lies in the concentration profile within the interfacial layer or wall of the filament. A dimensional analysis of (4.4.29) shows that in the two-carrier case the width of the filament wall is of the order of an effective ambipolar diffusion length

$$L_a = (\tilde{D}_a \tau)^{1/2}|_{p=p_2} \;,$$

$$(4.4.46)$$

where $\tau = (d\varphi_p/dp)^{-1}$ is an effective recombination lifetime. The width of single-carrier filament walls, see (4.1.15), is of the order of several effective Debye lengths

$$L_D = (D_0 \tau_M)^{1/2}$$

$$(4.4.47)$$

which is generally smaller than L_a since the dielectric-relaxation time $\tau_M = \epsilon_s/(4\pi e \mu_0 N_D^*)$ is in normal semiconductors ("lifetime semiconductors") much smaller than the recombination lifetime τ. Thus the filament walls in the ambipolar case are wider than in the unipolar case.

The longitudinal current density is given by

$$j_0 = e(\mu_n n + \mu_p p) E_0 \;,$$

$$(4.4.48)$$

where E_0 is the longitudinal field. For a rectangular slab of dimensions L_x, L_y with a current layer of width x_0 the total current is approximately

$$I \approx e\varepsilon_0 L_y [(\mu_n n_1 + \mu_p p_1)(L_x - x_0) + (\mu_n n_3 + \mu_p p_3)x_0] \;.$$

$$(4.4.49)$$

For a cylindrical sample of radius a with a high-current filament of radius r_0 the current is approximately

$$I \approx e\varepsilon_0 \pi [(\mu_n n_1 + \mu_p p_1)(a^2 - r_0^2) + (\mu_n n_3 + \mu_p p_3)r_0^2]$$

$$(4.4.50)$$

in analogy with (4.2.26).

In summary, by the approximation of local quasi-neutrality, we have reduced the problem of stationary transverse spatial structures in two-carrier models to the same mathematical form as the corresponding single-carrier problem. Hereby the results of Sects. 4.1–3 can readily be carried over.The equal-areas rule (4.4.44) extends previous equal-areas rules for two-carrier filamentation [4.15] in two respects: first, the curvature of the filament walls is self-consistently accounted for by the Σ-term and second, the presence of impurities leads to a dressed ambipolar diffusion coefficient $\tilde{D}_a$ and hence to a modified spatial scale.

4.5 Multiple Filaments

Multiple current filaments have been observed in pin-diodes under double injection [4.20, 26c] and in hot electron-hole plasmas in GaAs produced by impact ionization [4.59]. In Fig. 4.37 the potential probe measurement of a typical multiple filament is shown. If the curvature of the filaments is neglected, multiple filaments can be described in a very crude one-dimensional approximation as a sequence of kinks and antikinks, alternating between the high- and the low-conductivity phases, as sketched in Fig. 4.38. For sufficiently wide filaments the condition for their existence is given approximately by the equal areas rule (4.1.21). We shall see in Chap. 5 that such configurations are generally unstable against fluctuations of the carrier densities. In practical cases, however, they may be metastable on a very long time-scale, and thus observable, for instance, if the filaments are pinned to inhomogeneities or defects of the material. A phenomenological two-layer model for (multiple) current filamentation in pin diodes has been proposed by *Radehaus* et al. [4.60]. It is based

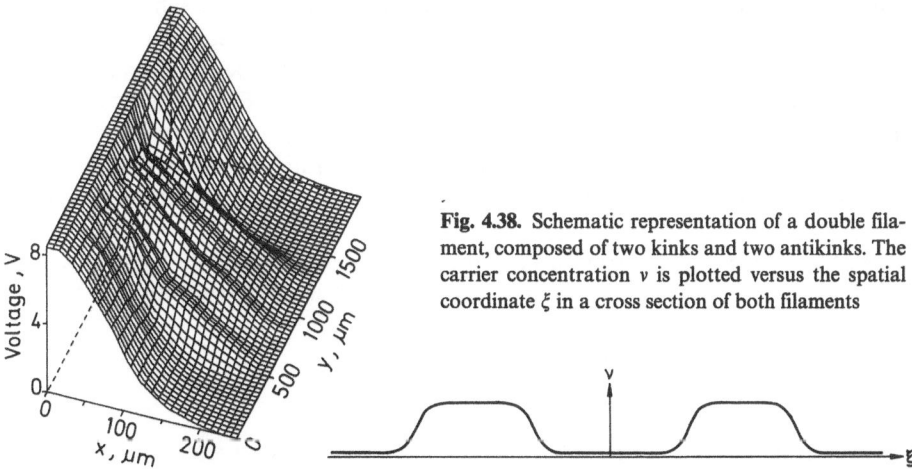

Fig. 4.38. Schematic representation of a double filament, composed of two kinks and two antikinks. The carrier concentration v is plotted versus the spatial coordinate ξ in a cross section of both filaments

Fig. 4.37. Observed multiple current filaments in a gold compensated Si pin diode at room temperature. The potential V at the surface of the diode, measured by a tungsten probe, is plotted as a function of the longitudinal (x) and transverse (y) coordinates. The three humps correspond to three current filaments occurring after three jumps in the $I-V$ characteristic. The experimental configuration is the same as in Fig. 4.17. [4.26c]

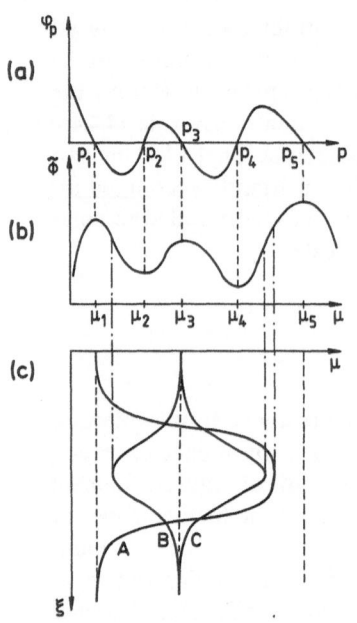

Fig. 4.39. (a) Hole generation-recombination rate versus hole density for a two-carrier g-r mechanism with tristability (schematic). The homogeneous stable (unstable) steady states are p_1, p_3, p_5 (p_2, p_4). (b) The associated potential $\tilde{\phi}$ versus $\mu = \int \tilde{D}_a(p)\,dp$ (c) Transverse spatial profiles $\mu(\xi)$ corresponding to accumulation (A, C) and depletion layers (B)

upon Drude-type intraband dynamics of the current density, similar to the momentum balance equation (1.1.17), albeit with a non-standard diffusive term, and disregards the g-r kinetics and the dynamics of the carrier densities. The resulting nonlinear evolution equations are of activator-inhibitor type [4.61] and allow for stable spatial patterns with one or more extrema.

Another kind of multiple filament formation [4.5, 6] can arise when the g-r kinetics are such that they allow for five homogeneous steady states, three of which are stable. An example of such a mechanism involving two carriers is discussed in Sect. 2.2.4. The resulting g-r rate $\varphi_p(p)$ is shown schematically in Fig. 4.39a for values of the applied field E_0 in the tristability regime. The potential $\tilde{\phi}(\mu)$ which is associated with $\varphi_p(p)$ by (4.4.42) has three maxima and two minima and is shown in Fig. 4.39b. In the case of a single-carrier g-r mechanism, $\tilde{\phi}$ and μ are associated with the charge density $\tilde{\rho}$ and the electron density ν by (4.1.14, 17).

First, we discuss plane-current layers (sheaths) in infinitely extended systems.

There are a variety of depletion- and accumulation-layer-type solutions, satisfying the boundary conditions

$$\lim_{\xi \to \pm \infty} \mu(\xi) = \mu_i \tag{4.5.1}$$

where μ_i corresponds to any one of the stable homogeneous steady states (μ_1, μ_3, μ_5). For the potential configuration of Fig. 4.39b these profiles are shown in Fig. 4.39c. Note that in contrast to the case of homogeneous bistability both depletion and accumulation layers can occur at the same value of E_0. Of particular interest are potential configurations were at least two maxima are equal. These are shown in Fig. 4.40 together with the corresponding spatial profiles.

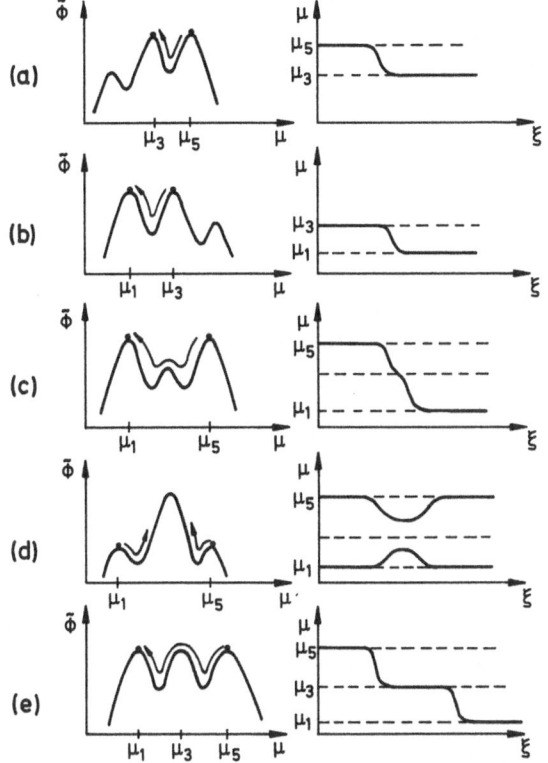

Thus coexistence* between two stable homogeneous states is possible in the following cases (Fig. 4.40a–c):

(a) $\int_{\mu_3}^{\mu_5} \tilde{\rho}\, d\mu = 0$ \hfill (4.5.2a)

(b) $\int_{\mu_1}^{\mu_3} \tilde{\rho}\, d\mu = 0$ \hfill (4.5.2b)

(c) $\int_{\mu_1}^{\mu_5} \tilde{\rho}\, d\mu = 0$, $\quad \int_{\mu_1}^{\mu_3} \tilde{\rho}\, d\mu < 0$. \hfill (4.5.2c)

Equations (4.5.2a–c) are equal-areas rules similar to (4.1.21). Note that no coexistence profiles exist in the case (Fig. 4.40d):

(d) $\int_{\mu_1}^{\mu_5} \tilde{\rho}\, d\mu = 0$, $\quad \int_{\mu_1}^{\mu_3} \tilde{\rho}\, d\mu < 0$. \hfill (4.5.2d)

Double coexistence between all three stable homogeneous steady states is possible if the following double equal-areas rule is satisfied:

(e) $\int_{\mu_1}^{\mu_3} \tilde{\rho}\, d\mu = \int_{\mu_3}^{\mu_5} \tilde{\rho}\, d\mu = 0$. \hfill (4.5.2e)

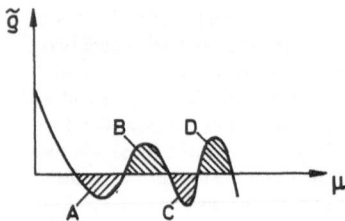

Fig. 4.41. Double equal-areas rule for concentric double filaments as shown in Fig. 4.40e. All four hatched areas A, B, C, D must be equal (schematic)

It requires that all four hatched areas in Fig. 4.41 are equal. This situation is the analog of a triple point in equilibrium thermodynamics. An analogous case occurring in nonequilibrium chemical reaction systems was investigated by *Czajkowski* [4.62].

Cylindrical current filaments with radial concentration profiles similar to those shown in Figs. 4.40a–e are possible if the equal-areas rules (4.5.2a–e) are modified by appropriate surface-tension terms Σ, as in (4.2.10). The double equal-areas rule (4.5.2e) then corresponds to two concentric cylindrical current filaments: a high-current density core with a medium current density mantle, embedded in a low-current density phase. Thus another kind of double filament formation – as opposed to the one sketched in Fig. 4.38 – is possible in tristable systems.

The current-voltage characteristics associated with double filaments like those shown in Fig. 4.40e are quite intricate and depend on the details of the path by which the "triple point" is approached in an appropriate two-dimensional control parameter space. This path may or may not touch some of the other points distinguished by equal-areas rules (4.5.2a–d). Several almost vertical sections may occur in the current-voltage characteristics, in conjunction with several voltage switch-back transitions. Although I–V characteristics of the described type have been observed experimentally, the possibility of explaining them by double filamentation in tristable g-r systems has not yet been considered.

5. Stability of Transverse Spatial Structures

In the preceding chapter we have described a variety of possible transverse stationary spatial structures induced by g-r nonlinearities. The question whether these can
actually be observed depends upon their stability character. In this chapter we shall
analyse the stability of plane current layers and cylindrical current filaments against
small fluctuations under various boundary and circuit conditions, and discuss the
associated transient behavior, in particular switching via spontaneous localized
fluctuations [5.1]. The analysis is mathematically more complicated than the treatment of the stability of spatially homogeneous steady states in Chap. 3, because the
normal mode ansatz of a plane or cylindrical wave which yielded a dispersion
relation $\lambda(k)$ does not work for small fluctuations from a spatially *in*homogeneous
steady state. However, we will be able to prove general qualitative results on
stability, and compute the linear modes analytically in some important special
cases. Throughout the present chapter we restrict ourselves to single-carrier g-r
mechanisms.

5.1 Plane Current Layers

First we analyse the stability of stationary plane current layer configurations, like
those studied in Sect. 4.1, against small fluctuations of the field and carrier densities
[5.1]. We assume that the width of the sample is sufficiently large ($L_\xi > 200 L_D$) so
that the approximation of infinite boundary conditions can be used.

The stability of similar one-dimensional spatial structures has been investigated
in various other synergetic systems, e.g., chemical reaction systems [5.2–5], combustion systems [5.6], semiconductors exhibiting current instabilities [5.7–13], or
ferromagnets [5.14, 15].

In infinitely extended media the stability of these structures can often be determined by using the analogy of the governing eigenvalue equation (which arises from
linearization of the transport equations around the steady state) with a Schrödinger
equation, and the existence of a Goldstone mode [5.5] (see also [5.16] for a
discussion of the subtle problems associated with the degeneracy of the Goldstone
mode). This has been done, e.g., for temperature profiles in flat flame fronts [5.6]
and in electron overheating instabilities [5.7, 8], for electric-field profiles in Gunn
diodes [5.9] and in other voltage-controlled current instabilities [5.13]. Note,
however, that finite boundary conditions and the influence of an external circuit
may change the stability character sensitively, as we shall see in the following
sections.

For the g-r mechanism considered here, the analogy of the stability problem
with a Schrödinger equation is not so obvious, since our model involves several

dynamic variables, namely the concentrations of free and bound electrons. Hence a more general multimode eigenvalue equation is obtained. Moreover, our model differs from conventional reaction-diffusion systems [5.2–5] by the presence of an electric field resulting in a drift term, which breaks the spherical symmetry. This term can, however, be removed from both the steady state equation and the eigenvalue equation by transforming to logarithmic concentration variables, as we have seen in earlier chapters. The role of the reaction term in the steady state equation is taken by the static charge density given as a function of electron concentration by the g-r mechanism.

In Sect. 5.1.1 we shall investigate the stability of plane spatial structures against small fluctuations using general properties of the eigenvalue equation which may be visualized as a generalized Schrödinger equation. The eigenfunctions and eigenvalues of the unstable mode of hump-shaped concentration profiles are computed explicitly in Sect. 5.1.2. Section 5.2 deals with the stability of cylindrical structures, and Sect. 5.3 with the influence of finite boundaries. Finally, in Sect. 5.4, the results are applied to discuss nucleation phenomena and nonequilibrium phase transitions induced by spontaneous localized fluctuations.

In Sect. 3.1 we linearized the transport equations for small fluctuations

$$\delta\boldsymbol{\phi}(\xi, \tau) = (\delta\varepsilon(\xi), \delta v(\xi), \delta v_t(\xi))e^{\lambda\tau}$$

around the homogeneous steady state, and, after elimination of $\delta\varepsilon$ and δv_t, obtained the multimode eigenvalue equation (3.1.32). Similarly, we shall now linearize the transport equations for small fluctuations around the *inhomogeneous* steady state corresponding to a transverse spatial structure. Throughout the following we will restrict ourselves to the "most dangerous" *transverse* fluctuations $\delta\boldsymbol{\varepsilon} \perp \boldsymbol{\varepsilon}_0$, $\nabla\delta v \perp \boldsymbol{\varepsilon}_0$.

Analogous to Sect. 3.1, we obtain the linearized Poisson equation, (3.1.25) with $\delta\varepsilon_\| = 0$

$$\nabla_\perp \delta\varepsilon_\perp = -\frac{H(\lambda)}{G(\lambda)} \delta v \tag{5.1.1}$$

and the linearized current-density equation, (3.1.22) with $\delta j_\perp^{\text{tot}} = 0$:

$$0 = (\lambda + v)\delta\varepsilon_\perp + (\varepsilon_\perp + \nabla_\perp)\delta v \ . \tag{5.1.2}$$

Since the dielectric-relaxation frequency v is in conventional lifetime semiconductors much faster than the time constant of the g-r processes, we neglect the displacement current in (5.1.2): $\lambda \ll v$. Transforming to the variable $\mu = \ln v$, $\delta\mu = v^{-1}\delta v$, and using the steady state condition $v\varepsilon_\perp = -\nabla_\perp v$ (4.1.6) for transverse spatial structures, we obtain from (5.1.2):

$$0 = e^\mu(\delta\varepsilon_\perp + \nabla_\perp\delta\mu) \ . \tag{5.1.3}$$

This can be combined with (5.1.1) to give

$$\nabla_\perp^2 \delta\mu - \frac{H(\lambda)}{G(\lambda)} e^\mu \delta\mu = 0 \ . \tag{5.1.4}$$

The functions $G(\lambda)$, $H(\lambda)$ were defined in (3.1.24, 28), and have to be evaluated at the steady state. For a homogeneous steady state, $\exp(\mu)H(\lambda)/G(\lambda)$ does not depend upon the spatial coordinates; hence (5.1.3) can be solved by the ansatz $\delta\mu(\xi) \sim \exp(ik_\perp \xi)$ or $\delta\mu(R) \sim J_0(k_\perp R)$, and is equivalent to the eigenvalue equation (3.2.3) studied in the context of the filamentary instability in Sect. 3.2. (Here we use the additional approximation $\lambda \ll v$.)

For fluctuations from an inhomogeneous steady state, however, the quantities $\exp(\mu)$, $H(\lambda)$, $G(\lambda)$ depend upon the transverse coordinate ξ via $\mu(\xi)$ and (5.1.4) represents a generalized eigenvalue problem if supplemented by suitable boundary conditions.

5.1.1 General Results

In the following we study plane stationary dissipative structures $\mu(\xi)$, given by the transport equations (4.1.15):

$$\mu''(\xi) + \tilde{\rho}(\mu, \varepsilon_0) = 0 \ . \tag{5.1.5}$$

The term $\exp(\mu)H(\lambda)/G(\lambda)$ then depends only upon the transverse coordinate ξ. We assume that the fluctuations $\delta\mu$ about $\mu(\xi)$ vanish at plus and minus infinity. The eigenvalue problem (5.1.4) is separable in the two transverse directions (ξ and η). Since a fluctuation $\delta\mu(\eta) \sim \exp(ik_\eta\eta)$ perpendicular to both ε_0 and $\varepsilon_\perp$ only adds a stabilizing constant term $-k_\eta^2$ to the left-hand side of (5.1.4), we consider $k_\eta = 0$ for simplicity. Equation (5.1.4) may then be put in the form of a generalized one-dimensional Schrödinger equation:

$$-\delta\mu''(\xi) + V(\xi, \lambda)\delta\mu(\xi) = 0 \tag{5.1.6}$$

with a generalized "potential"

$$V(\xi, \lambda) := e^\mu H(\lambda)/G(\lambda) \ . \tag{5.1.7}$$

The inhomogeneous steady states $\mu(\xi)$ are states of broken translation symmetry. A set of equivalent states $\mu(\xi + \Delta\xi)$ is generated by an infinitesimal translation $\Delta\xi$. Therefore there is a *Goldstone mode*

$$\delta\mu_G(\xi) := \frac{d\mu}{d\xi}\Delta\xi \tag{5.1.8}$$

corresponding to the eigenvalue $\lambda = 0$ [5.5]. This can be verified explicitly by differentiating (5.1.5) with respect to ξ:

$$\mu''' + (\partial\tilde{\rho}/\partial\mu)\mu' = 0 \ . \tag{5.1.9}$$

We compare this with (5.1.6), setting $\lambda = 0$, and using $H(0)/G(0) = \det \tilde{A}/\det B = -\partial\rho/\partial v$, by (3.1.34, 35, 38):

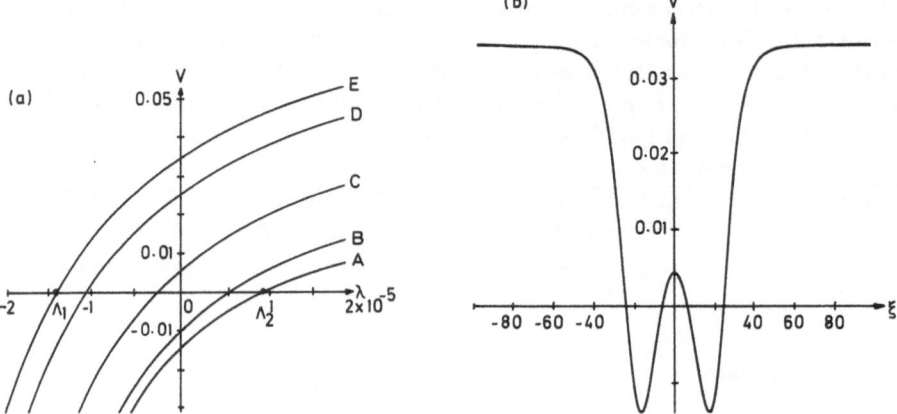

Fig. 5.1. Potential $V(\xi, \lambda)$ of the generalized eigenvalue equation versus (**a**) λ, (**b**) ξ for the depletion layer shown in Fig. 5.3a ($\varepsilon_0 = 8.4$). In (**a**) the different curves correspond to A: $v = v_2 = 0.0255$ ($\xi = \pm 16$); B: $v = 0.04$ ($\xi = \pm 21$); C: $v = 0.06$ ($\xi = \pm 26$); D: $v = 0.08$ ($\xi = \pm 34$); E: $v = v_3 = 0.0888$ ($\xi = \pm \infty$). In (**b**) $\lambda = 0$ is used. The numerical parameters are given in Table 3.2

$$\delta\mu_G'' - V(\xi, 0)\delta\mu_G \overset{\cdot}{=} \delta\mu_G'' + v(\partial\rho/\partial v)\delta\mu_G$$

$$= \delta\mu_G'' + (\partial\tilde{\rho}/\partial\mu)\delta\mu_G = 0 \ . \tag{5.1.10}$$

It follows that $\delta\mu_G \sim \mu'$.

A further conclusion about the spectrum of (5.1.6) can be obtained by a generalization of a quantum mechanical Oscillation Theorem [Ref. 5.17, Ch. III § 12]. For fixed ξ, the zeros of the function $V(\xi, \lambda)$ are the eigenvalues $\lambda_1(\xi) > \cdots > \lambda_M(\xi)$ of $\tilde{A}$, and its poles are the eigenvalues $\lambda_1^\infty(\xi) > \cdots > \lambda_M^\infty(\xi)$ of B. Negative or positive values of $\lambda_1(\xi)$ correspond to concentrations $v(\xi)$ such that charge neutral fluctuations around v are stable or unstable, respectively. In Fig. 5.1 $V(\xi, \lambda)$ is plotted for a typical stationary profile with one minimum (depletion layer). Figure 5.2 shows how $V(\xi, 0)$ changes when the applied field is varied. Figures 5.2a–d correspond to depletion layers, Fig. 5.2e to a kink, and Figs. 5.2f–j to accumulation layers. If

$$\lambda_1^\infty(\xi) < \Lambda_1 \qquad \text{for } \xi \in (-\infty, \infty)$$

$$\frac{\partial}{\partial\lambda} V(\xi, \lambda) > 0 \qquad \text{for } \xi \in (-\infty, \infty) \quad \text{and} \quad \lambda \in (\Lambda_1, \Lambda_2) \tag{5.1.11}$$

with

$$\Lambda_1 := \text{Max} \ \{\lambda_1(+\infty), \lambda_1(-\infty)\} \qquad \text{and}$$

$$\Lambda_2 := \underset{\xi \in \mathbb{R}}{\text{Max}} \ \{\lambda_1(\xi)\}$$

Fig. 5.2a–j. Potential $V(\xi, 0)$ vs. ξ for a sequence of applied fields ε_0: (a) 7.9, (b) 8.0, (c) 8.2, (d) 8.4, (e) 8.494 ($\hat{=} \varepsilon_{oo}$), (f) 8.6, (g) 8.8, (h) 9.0, (i) 9.1, (j) 9.2. The other numerical parameters are those of Fig. 5.1

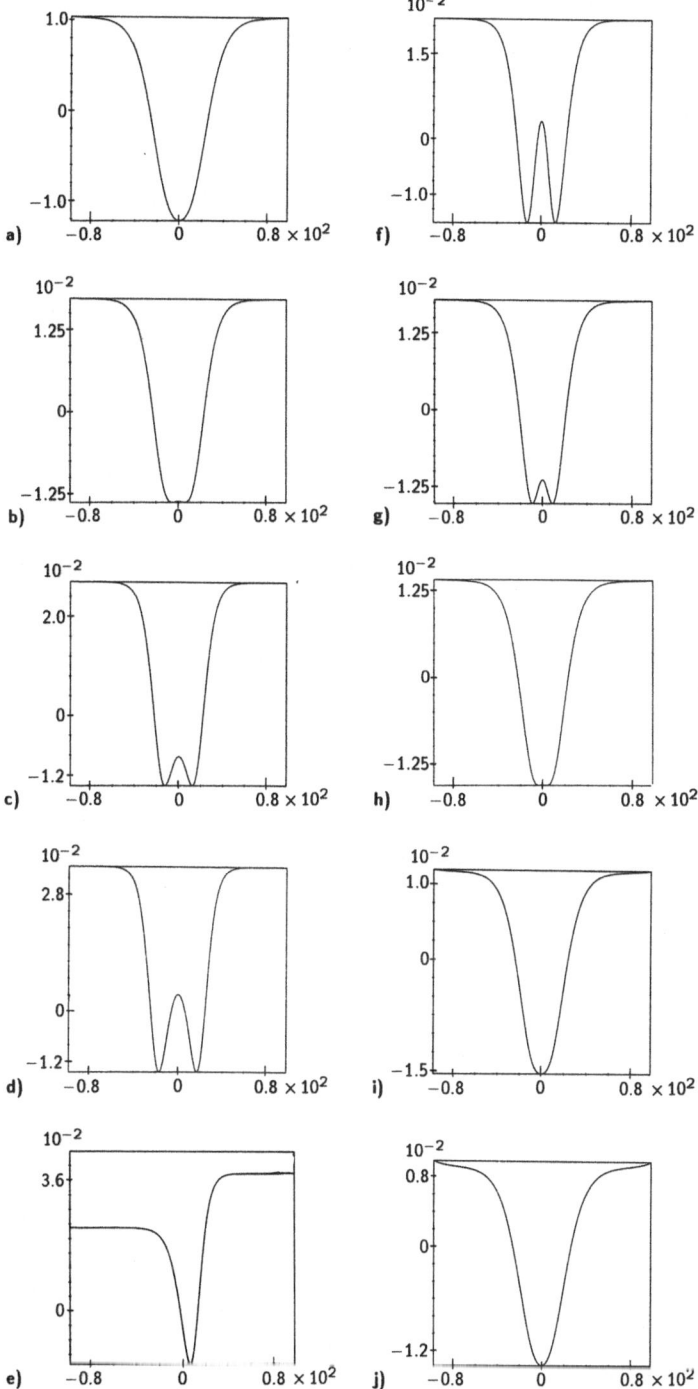

Fig. 5.2a–j. Caption see opposite page

holds, then one can prove the following:

Oscillation Theorem. The discrete eigenvalues of (5.1.6) lie in the interval (Λ_1, Λ_2). If the eigenstates are ordered by decreasing eigenvalues $\lambda_0 > \lambda_1 > \cdots$, then the nth eigenfunction has exactly n zeros in the open interval $-\infty < \xi < \infty$ ("nodes").

Proof

For $\lambda_1^\infty(\xi) < \Lambda_1$ and continuous dissipative structures $v(\xi)$ the potential $V(\xi, \lambda)$ is bounded from below and continuous with respect to ξ for $\xi \in (-\infty, \infty)$ and $\lambda \in (\Lambda_1, \Lambda_2)$.

Lemma. For any $\lambda \in (\Lambda_1, \Lambda_2)$ with $\Lambda_1 < \Lambda_2$ there is one and only one solution y_λ of (5.1.6) with $\lim_{\xi \to \infty} y_\lambda(\xi) = 0$ (up to an arbitrary factor).

This lemma follows from the general asymptotic properties of the solution of the one-dimensional Schrödinger equation [5.17, Ch. III §9] by identifying

$$V(\xi, \lambda) = U(\xi) - E \ , \tag{5.1.12}$$

where $U(\xi)$ is the Schrödinger potential and E the energy. Note that increasing λ corresponds to decreasing energy by (5.1.11). It follows that $V^{\min}(\Lambda_2) = 0$ with $V^{\min}(\lambda) := \text{Min}_{\xi \in \mathbb{R}} V(\xi, \lambda)$ defines the upper bound $\Lambda_2 = \text{Max}_{\xi \in \mathbb{R}} \lambda_1(\xi)$ of the discrete spectrum of (5.1.6), and $V^\infty(\Lambda_1) = 0$ with $V^\infty(\lambda) := \text{Min} \{V(+\infty, \lambda), V(-\infty, \lambda)\}$ defines its lower bound Λ_1 i.e., $\Lambda_1 = \text{Max} \{\lambda_1(+\infty), \lambda_1(-\infty)\}$. Now let $\xi = \xi_0(\lambda)$ be a zero of $y_\lambda(\xi)$ for arbitrary λ. From (5.1.6) it follows that $z := \partial y_\lambda / \partial \lambda$ is a solution of

$$-z''(\xi) + V(\xi, \lambda) z(\xi) = -\frac{\partial V}{\partial \lambda} y_\lambda(\xi) \ . \tag{5.1.13}$$

Multiplying (5.1.6) for y_λ by z and (5.1.13) by y_λ, subtracting and integrating from $\xi_0(\lambda)$ to infinity, one obtains the Wronskian

$$y_\lambda'(\xi_0) z(\xi_0) = \int_{\xi_0}^\infty \frac{\partial V(\xi, \lambda)}{\partial \lambda} y_\lambda^2 \, d\xi \tag{5.1.14}$$

where $y_\lambda(\xi_0) = y_\lambda(\infty) = y_\lambda'(\infty) = 0$ has been used. With

$$0 = \frac{\partial y_\lambda}{\partial \lambda}(\xi_0) + y_\lambda'(\xi_0) \frac{d\xi_0}{d\lambda} \tag{5.1.15}$$

it follows, by (5.1.11),

$$\frac{d\xi_0}{d\lambda} = -y_\lambda'(\xi_0)^{-2} \int_{\xi_0}^\infty \frac{\partial V}{\partial \lambda} y_\lambda^2 \, d\xi < 0 \ . \tag{5.1.16}$$

Hence with increasing λ the zero $\xi_0(\lambda)$ of y_λ moves towards more negative values. Whenever λ passes an eigenvalue of (5.1.6), a zero of y_λ disappears at minus infinity. Therefore, with increasing λ, each eigenfunction has one node less than the preceding one. When λ reaches Λ_2, all eigenvalues have been passed, and y_{Λ_2} has no finite zeros. To prove this by contradiction, assume that y_{Λ_2} has a zero at $\xi_0 \in (-\infty, \infty)$. Then by (5.1.6)

$$\int_{\xi_0}^{\infty} y_{\Lambda_2}'' y_{\Lambda_2} \, d\xi = -\int_{\xi_0}^{\infty} (y_{\Lambda_2})'^2 \, d\xi$$

$$= \int_{\xi_0}^{\infty} V(\Lambda_2, \xi) y_{\Lambda_2}^2 \, d\xi > 0 , \tag{5.1.17}$$

which is contradictory. It follows that the eigenfunction corresponding to the largest eigenvalue has no node. This completes the proof of the Oscillation Theorem.

Note that in general there are only a finite number of eigenvalues. The usual Oscillation Theorems of Sturmian Theory [5.18] which give an infinite number of eigenvalues with one limit point, do not hold for (5.1.6). Note also that (5.1.11) is a sufficient but not necessary condition for the validity of the Oscillation Theorem.

We shall now use the preceding results to determine quite generally the stability of dissipative structures. Three typical stationary profiles (full lines) and the corresponding Goldstone modes (dotted lines) are shown in Fig. 5.3 for $\varepsilon_0 < \varepsilon_{co}$ (a), $\varepsilon_0 = \varepsilon_{co}$ (b), and $\varepsilon_0 > \varepsilon_{co}$ (c). The following conclusions about the stability of the

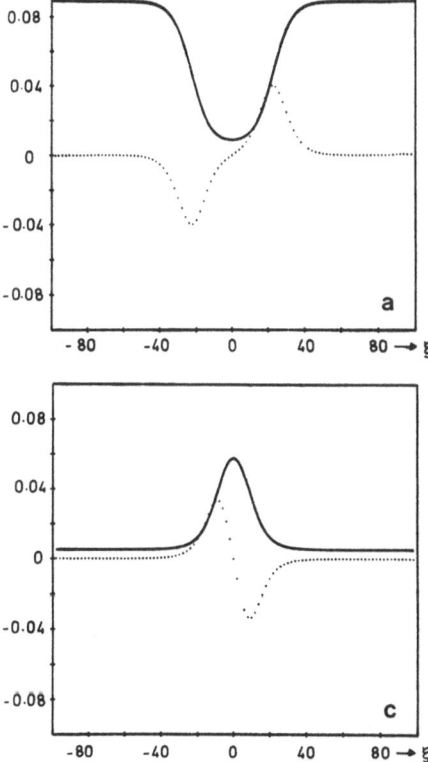

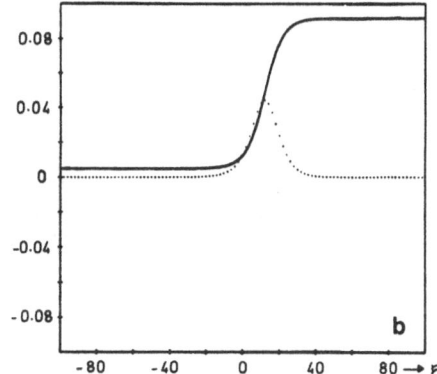

Fig. 5.3. Plane stationary concentration profiles v (———) and Goldstone mode $dv/d\xi \cdot 10$ (······) versus the transverse coordinate ξ for (a) depletion layer ($\varepsilon_0 = 8.4$), (b) kink ($\varepsilon_0 = 8.494296$), (c) accumulation layer ($\varepsilon_0 = 8.6$). The other numerical parameters are those of Fig. 5.1

plane dissipative structures can be drawn from the Goldstone Theorem and the Oscillation Theorem:

(i) For a *kink-shaped* (i.e., monotonic) concentration profile $v(\xi)$ (Fig. 5.3b) the Goldstone mode $\sim dv/d\xi$ has no node, hence by the Oscillation Theorem it corresponds to the "quantum mechanical ground state", i.e., the largest eigenvalue ($\lambda_0 = 0$), and all other eigenvalues $\lambda_1, \lambda_2, \ldots$ of (5.1.6) are smaller, i.e., negative. This proves the stability of the spatial coexistence solution $v(\xi)$.

(ii) For a *hump-shaped* (i.e., with one extremum) concentration profile $v(\xi)$ (Fig. 5.3a, c) the Goldstone mode $\sim dv/d\xi$ has one node and is the "first excited state", hence there is a "ground state" associated with the largest eigenvalue $\lambda_0 > 0$. Therefore depletion and accumulation layers are unstable with respect to one and only one discrete mode λ_0. However, since this mode everywhere either increases or decreases the concentration $v(\xi)$, it changes the total current $I \sim \varepsilon_0 \int v\, d\xi$ through the sample. Thus under constant current conditions this mode is forbidden, and $v(\xi)$ can be stabilized. A similar effect of the external current has been found in other types of SNDC instabilities in semiconductors [5.11, 25], while dissipative structures associated with NNDC can be stabilized by imposing a constant voltage [5.9, 13].

(iii) For *oscillatory* concentration profiles with n periods spaced within the width of the sample the Goldstone mode has $2n$ nodes, and hence there are $2n$ unstable modes with positive eigenvalues. Not that in this case periodic boundary conditions should be employed.

5.1.2 Unstable Modes of Depletion Layers

The unstable mode and the associated eigenvalue of the depletion (or accumulation) layer will be calculated explicitly in the following for some special cases.

Since in standard "lifetime" semiconductors, the time-scale of the g-r processes, which drive the instability, is much slower than the dielectric-relaxation time defining the unit of time, we can assume $|\lambda|, |\lambda_1(\xi)| \ll 1$, and linearize $V(\xi, \lambda)$ around $\lambda_1(\xi)$ in (5.1.6):

$$-\delta\mu''(\xi) + \frac{\lambda - \lambda_1(\xi)}{D(\xi)}\delta\mu(\xi) = 0 \qquad \text{with} \tag{5.1.18}$$

$$D(\xi)^{-1} := v\frac{\partial}{\partial\lambda}\left(\frac{H}{G}\right)\Bigg|_{\lambda=\lambda_1(\xi)} = v(\xi)\frac{\displaystyle\prod_{i=2}^{M}(\lambda_1(\xi) - \lambda_i(\xi))}{\displaystyle\prod_{i=1}^{M}(\lambda_1(\xi) - \lambda_i^\infty(\xi))}. \tag{5.1.19}$$

In the special case of a spatially homogeneous steady state the approximation (5.1.18) corresponds just to the "hydrodynamic limit" $k_\perp \ll 1$ discussed in Sect. 3.2, and $D(\xi)$ becomes the dressed diffusion constant $\tilde{D}_\perp$. [In (5.1.19) we have employed the additional approximation $\lambda \ll v$.] Since $\lambda_1(\xi) > \lambda_1^\infty(\xi)$ for standard g-r kinetics, $D(\xi)$ is positive. In Fig. 5.4 $D(\xi)$ and $\lambda_1(\xi)$ are plotted for a depletion layer at $\varepsilon_0 = 8.4$. Figure 5.5 depicts $D(\xi)$ and $\lambda_1(\xi)$ for a sequence of applied electric fields ε_0. Figures 5.5a–d correspond to a depletion layer ($\varepsilon_0 < \varepsilon_{co}$), Fig. 5.5e to a kink ($\varepsilon_0 = \varepsilon_{co}$), and Figs. 5.5f–j to an accumulation layer ($\varepsilon_0 > \varepsilon_{co}$). As ε_0 comes close to ε_{co}, the layer

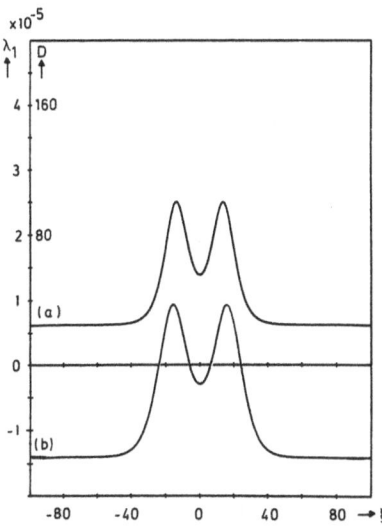

becomes wide (cf., Fig. 4.9) and the two-peak structure of $D(\xi)$ and $\lambda_1(\xi)$ becomes more pronounced, reflecting the sharp walls of the stationary profiles. At both ends of the SNDC range $[\varepsilon_0 \to \varepsilon_h(a)$ and $\varepsilon_0 \to \varepsilon_{th}(j)]$ the walls of the profile become diffuse, and the two peaks merge. At the coexistence field ε_{co} the profile has a single wall, and consequently D and λ_1 have a single peak.

If the "critical" eigenvalue $\lambda_1(\xi)$ is well separated from all other ("stable") eigenvalues of the g-r matrices $\tilde{A}$ and B, i.e., if $|\lambda_1(\xi)| \ll |\lambda_i(\xi)|, |\lambda_j^\infty(\xi)|$ $(i = 2, \ldots, M;$ $j = 1, \ldots, M)$, then (5.1.19) can be approximated with (3.1.35, 38) by

$$D(\xi) := \lambda_1(\xi)(\partial\tilde{\rho}/\partial\mu)^{-1} \ . \tag{5.1.20}$$

The unstable "ground state" eigenfunction $\delta\mu_0$ and the Goldstone mode $\delta\mu_G$ satisfy the eigenvalue equation (5.1.18) with eigenvalues $\lambda_0 > 0$ and $\lambda_1 = 0$, respectively. Without loss of generality we assume $\delta\mu_G(0) = 0$. We multiply the equation (5.1.18) for $\delta\mu_0$ by $\delta\mu_G$ and the equation for $\delta\mu_G$ by $\delta\mu_0$, and subtract one from another. After integration over ξ from zero to infinity we obtain

$$\lambda_0 = \frac{\delta\mu_0(0)\delta\mu_G'(0)}{\displaystyle\int_0^\infty \delta\mu_0(\xi)\delta\mu_G(\xi)\frac{d\xi}{D(\xi)}} \ , \tag{5.1.21}$$

where we have used $\delta\mu_0(\infty) = \delta\mu_G(\infty) = \delta\mu_G(0) = 0$. From (5.1.21) it can be seen explicitly that $\lambda_0 > 0$, because $\delta\mu_0(\xi) > 0$ for $-\infty < \xi < \infty$, $\delta\mu_G(\xi) > 0$ for $0 < \xi < \infty$ and $\delta\mu_G'(0) > 0$.

(i) *Wide Depletion Layer*

With increasing external field ε_0 the depletion layer profiles change as shown in Fig. 4.9a. For values of ε_0 slightly below the coexistence value ε_{co} the steady state

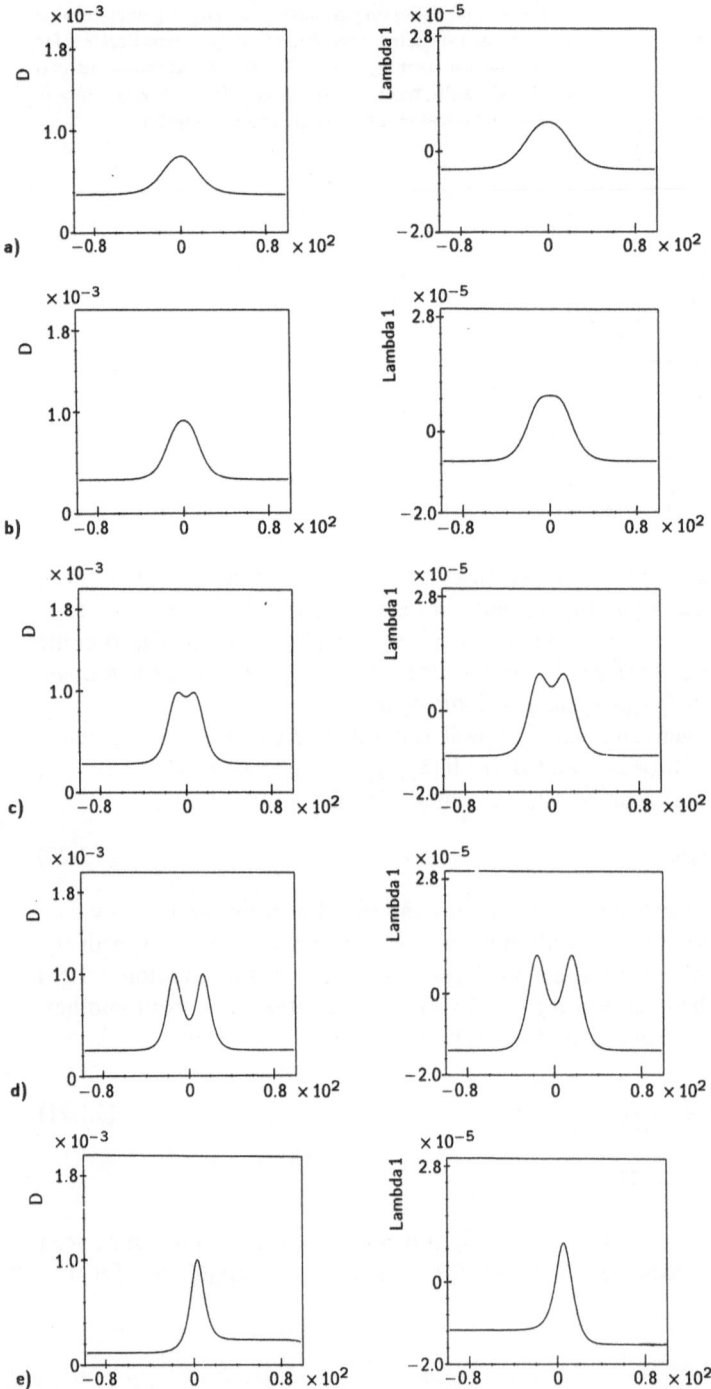

Fig. 5.5a–j. Diffusion parameter D (*left column*) and g-r eigenvalue λ_1 (*right column*) versus ζ for a sequence of applied fields ε_0: (**a**) 7.9, (**b**) 8.0, (**c**) 8.2, (**d**) 8.4, (**e**) 8.494 ($\cong \varepsilon_{co}$)

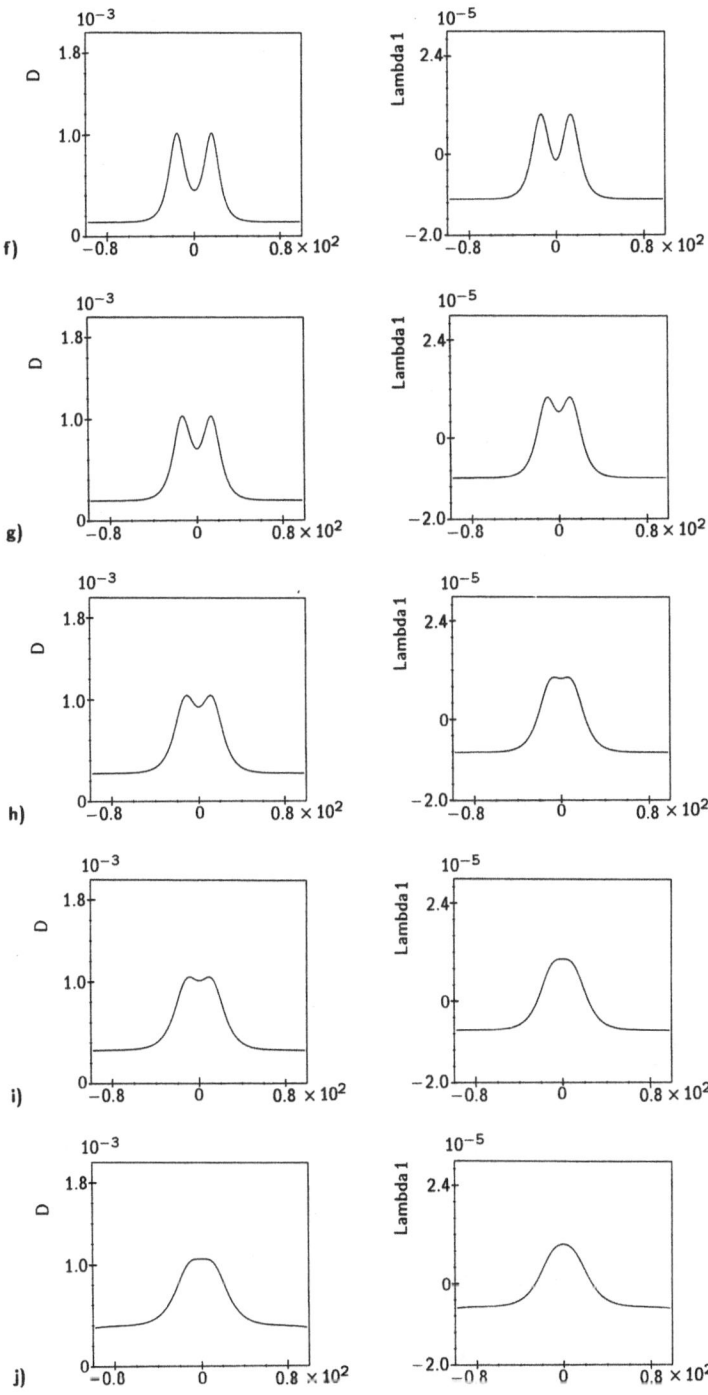

Fig. 5.5 (cont.). (f) 8.6, (g) 8.8, (h) 9.0, (i) 9.1, (j) 9.2

concentration $v_d(\xi)$ drops from the asymptotic value v_3 to its minimum $v_{min} \geq v_1$ within two thin transition layers at $\pm\xi_0$ and remains at an almost constant value v_{min} in the interior of a wide central depletion layer.

The Sturm-Liouville problem (5.1.18) can be transformed into Liouville's normal form, i.e., a "Schrödinger equation",

$$-\psi''(x) + [U(x) + \lambda]\psi(x) = 0 \tag{5.1.22}$$

by a standard Liouville transformation [5.19]

$$x := \int_0^\xi D(\xi')^{-1/2}\, d\xi' \, ,$$

$$\psi(x) := \delta\mu(\xi)D(\xi)^{-1/4} \, ,$$

$$U(x) := -\lambda_1(\xi) + \frac{3}{16D}\left(\frac{d}{d\xi}D\right)^2 - \frac{1}{4}\frac{d^2}{d\xi^2}D \, . \tag{4.1.23}$$

For a wide depletion layer the Schrödinger potential U represents two potential wells located in the transition layers ("domain walls") at $\pm x_0$. Hence the ground state ψ_0 and the first excited state ψ_G can be approximated by the bonding and the antibonding linear combinations of the single-well ground state wavefunctions $\psi^{(1)}$, $\psi^{(2)}$

$$\psi_0 = \psi^{(1)} + \psi^{(2)}$$

$$\psi_G = \psi^{(1)} - \psi^{(2)} \, , \tag{5.1.24}$$

where $\psi^{(1)}$ and $\psi^{(2)}$ are localized in the two potential wells at $+x_0$ and $-x_0$, respectively, and overlap very little. Therefore, up to exponentially small quantities, for large enough $|x|$

$$\psi_0(x) \approx \pm\psi_G(x) \qquad \text{for } x \gtrless 0 \tag{5.1.25}$$

holds. Note that (other than in the Gunn effect [5.9]) the depletion layer $v_d(\xi)$ is symmetric with respect to inversion $\xi \to -\xi$. Hence $U(x)$ is also symmetric and $\psi^{(1)}(x) = \psi^{(2)}(-x)$, so that the wavefunctions ψ_0 and ψ_G are already orthogonal as required.

The temporal development of the wide depletion layer during the early stage [i.e., while linearization around $v_d(\xi)$ is valid] is by (5.1.25, 23, 8) for $|\xi| \gg |\Delta\xi|$ given by

$$v(\xi,\tau) = v_d(\xi) + \delta v(\xi)e^{\lambda_0\tau} \approx v_d(\xi) \pm \Delta\xi\frac{dv_d}{d\xi}e^{\lambda_0\tau}$$

$$\approx v_d(\xi \pm \Delta\xi e^{\lambda_0\tau}) \, , \tag{5.1.26}$$

where $+$ and $-$ correspond to $\xi > 0$ and $\xi < 0$, respectively. Thus the unstable mode of the wide depletion layer describes two wavefronts traveling with increasing phase velocity

$$v(\tau) = \pm \lambda_0 \Delta \xi e^{\lambda_0 \tau}$$

laterally in opposite directions, after an initial perturbation $\Delta \xi \cdot dv_d/d\xi$ has shifted the walls from $\pm \xi_0$ to $\pm(\xi_0 + \Delta \xi)$; $\Delta \xi > 0$ corresponds to lateral growth of the depletion layer, $\Delta \xi < 0$ to shrinking.

Next we evaluate the eigenvalue λ_0 from (5.1.21). Since only values of ξ near ξ_0 contribute to the integral in the denominator, we approximately set $\delta \mu_0 \approx \delta \mu_G$ in the integrand and use (5.1.8) where $\Delta \xi = 1$ is assumed without loss of generality, together with the first integral of the steady state (5.1.5)

$$\tfrac{1}{2}(\mu')^2 + \int \tilde{\rho}(\mu, \varepsilon_0)\, d\mu = \text{const.} \tag{5.1.27}$$

This gives

$$\int\limits_0^\infty \delta\mu_0 \delta\mu_G \frac{d\xi}{D(\xi)} \approx \int\limits_0^\infty \left(\frac{d\mu}{d\xi}\right)^2 \frac{d\xi}{D(\xi)} = \int\limits_{\mu_{min}}^{\mu_3} \left[2 \int\limits_\mu^{\mu_3} \tilde{\rho}(\tilde{\mu})\, d\tilde{\mu} \right]^{1/2} \frac{d\mu}{D(\mu)}, \tag{5.1.28}$$

and further, by (5.1.5, 8)

$$\delta\mu_G'(0) = -\tilde{\rho}(\mu_{min}) \ . \tag{5.1.29}$$

By a WKB approximation, $\delta\mu_0(\xi)$ can be calculated in the interior of the depletion layer, where the potential U is slowly varying. From (5.1.22) one obtains with $|\lambda_0| \ll |U(0)|$, using the asymptotic condition (5.1.25),

$$\psi_0(x) = C_0 \cosh \int\limits_0^x \sqrt{U(x')}\, dx' \tag{5.1.30}$$

$$\psi_G(x) = C_0 \sinh \int\limits_0^x \sqrt{U(x')}\, dx' \ , \tag{5.1.31}$$

C_0 can be determined by comparison of (5.1.29) with the expression for $\delta\mu_G'(0)$ derived from (5.1.31) with (5.1.23):

$$C_0 = -\tilde{\rho}(\mu_{min})D(0)^{1/4}U(0)^{-1/2} \ . \tag{5.1.32}$$

Substitution into (5.1.30) yields

$$\delta\mu_0(0) = -\tilde{\rho}(\mu_{min})\sqrt{D(0)/U(0)} \ . \tag{5.1.33}$$

Inserting (5.1.28, 29, 33) into (5.1.21), one obtains λ_0. In particular, the approximation (5.1.20) gives with $U(0) \approx -\lambda_1(0)$ and extraction of D from the integral (5.1.28).

$$\lambda_0 \approx \tilde{\rho}(\mu_{min})^2 \lambda_1(\xi_0) \left\{ \sqrt{-\frac{\partial\tilde{\rho}}{\partial\mu}(\mu_{min})} \frac{\partial\tilde{\rho}}{\partial\mu}(\mu_2) \cdot \int\limits_{\mu_{min}}^{\mu_3} \left[2 \int\limits_\mu^{\mu_3} \tilde{\rho}(\mu')\, d\mu' \right]^{1/2} d\mu \right\}^{-1} \ . \tag{5.1.34}$$

The eigenvalue λ_0 is proportional to the unstable homogeneous eigenvalue taken at the wall, $\lambda_1(\xi_0)$, times the square of the central charge density $\tilde{\rho}(\mu_{min})$, which is a small factor.

We shall now discuss the critical behavior of the eigenvalue λ_0. As ε_0 approaches the coexistence field ε_{co}, the width of the depletion layer goes to infinity, and μ_{min} tends to μ_1. Hence

$$\lambda_0 \sim \tilde{\rho}(\mu_{min})^2 \sim (\mu_{min} - \mu_1)^2 \xrightarrow{\varepsilon_0 \to \varepsilon_{co}} 0 \tag{5.1.35}$$

i.e., the moving walls are slowed down critically.

Note that from (5.1.31) an analytical approximation of the wide depletion layer in its interior can be obtained by (5.1.23, 20, 8):

$$\mu_d(\xi) \approx \mu_{min} + \tilde{\rho}(\mu_{min}) \frac{\partial \tilde{\rho}}{\partial \mu}(\mu_{min})^{-1} \left[\cosh\left(\xi \sqrt{-\frac{\partial \tilde{\rho}}{\partial \mu}(\mu_{min})} \right) - 1 \right]. \tag{5.1.36}$$

In the exterior $\mu_d(\xi)$ tends asymptotically to μ_3,

$$\mu_d(\xi) - \mu_3 \sim \exp\left[\mp \xi \sqrt{-\frac{\partial \tilde{\rho}}{\partial \mu}(\mu_3)} \right] \qquad \text{for } \xi \to \pm \infty , \tag{5.1.37}$$

which agrees with the asymptotic solution obtained by linearizing (5.1.5) around the high-conductivity steady state μ_3, cf., (4.1.11). The width of the domain $2\xi_0$ can be estimated from (5.1.36) with $\mu_d(\xi_0) = \mu_2$. It diverges logarithmically

$$\xi_0 \approx -\left[-\frac{\partial \tilde{\rho}}{\partial \mu}(\mu_1) \right]^{-1/2} \ln(\mu_{min} - \mu_1) \qquad \text{as } \mu_{min} \to \mu_1 . \tag{5.1.38}$$

In the generic case

$$\Delta\phi(\varepsilon_0) := \int_{\mu_1(\varepsilon_0)}^{\mu_3(\varepsilon_0)} \tilde{\rho}(\mu, \varepsilon_0) \, d\mu \sim \varepsilon_{co} - \varepsilon_0 \qquad \text{as } \varepsilon_0 \to \varepsilon_{co} , \tag{5.1.39}$$

the critical behavior of λ_0 and ξ_0 is given by

$$\lambda_0 \sim \varepsilon_{co} - \varepsilon_0 , \qquad \xi_0 \sim -\ln(\varepsilon_{co} - \varepsilon_0) , \tag{5.1.40}$$

where

$$\Delta\phi = \int_{\mu_1}^{\mu_{min}} \tilde{\rho}(\mu, \varepsilon_0) \, d\mu \approx \frac{1}{2} \frac{\partial \tilde{\rho}}{\partial \mu}(\mu_1)(\mu_{min} - \mu_1)^2$$

has been used.

Finally we note that a similar argument as in (5.1.24) can be applied to a sample with two wide depletion (or accumulation) layers. The bonding and antibonding linear combinations of the slowest two modes $\psi_0^{(i)}$, $\psi_G^{(i)}$ of the single layers $i = 1, 2$ lead to four modes, namely three unstable modes

$$\psi_0 := \psi_0^{(1)} + \psi_0^{(2)} , \qquad \psi_1 := \psi_0^{(1)} - \psi_0^{(2)} , \qquad \psi_2 := \psi_G^{(1)} - \psi_G^{(2)} \tag{5.1.41}$$

(a)

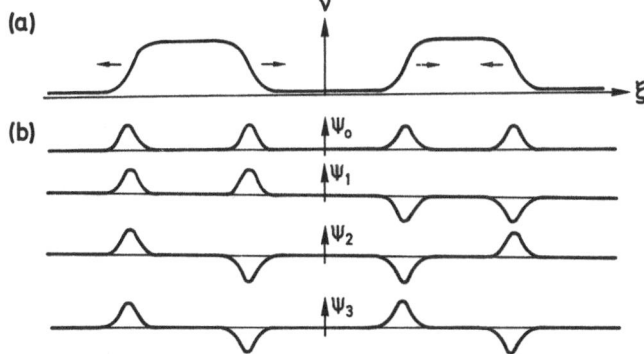

(b)

Fig. 5.6. Schematic representation of a double filament by a one-dimensional concentration profile $v(\xi)$ with two wide accumulation layers (a). The four slowest modes $\psi_0, \psi_1, \psi_2, \psi_3$ describing fluctuations $\delta v(\xi)$ from the stationary profile $v(\xi)$ are shown schematically in (b)

with $\lambda_0 > \lambda_1 > \lambda_2 > 0$, and the Goldstone mode

$$\psi_3 := \psi_G^{(1)} + \psi_G^{(2)}$$

with $\lambda_3 = 0$ (Fig. 5.6). Here ψ_0 is associated with either growth or shrinking of both layers, ψ_1 corresponds to growth of one layer at the expense of the other, and ψ_2 only increases or decreases the distance between the two layers. Under constant current conditions ψ_0 is forbidden, and ψ_1 is the slowest allowed mode. It describes how one of initially two layers disappears as a result of fluctuations (arrows in Fig. 5.6a). This configuration could serve as a simple one-dimensional approximation of a double filament.

(ii) Narrow Depletion Layer

For values ε_0 near the boundary of the bistability domain ε_h the homogeneous steady states μ_2, μ_3 are close to each other, and the charge density can be expanded near μ_3 into the universal quadratic form

$$\tilde{\rho}(\mu, \varepsilon_0) = -C \left[(\mu - \mu_c)^2 - \left(\frac{\delta}{2}\right)^2 \right], \qquad \text{where} \tag{5.1.42}$$

$$\delta(\varepsilon_0) := \mu_3 - \mu_2 \ll \mu_3$$

$$\mu_c := \tfrac{1}{2}(\mu_2 + \mu_3) = \mu_2|_{\varepsilon_0 = \varepsilon_h} = \mu_3|_{\varepsilon_0 = \varepsilon_h} \tag{5.1.43}$$

and C is a positive constant. From the generic S-shaped static current density-field characteristic it is evident that near the bistability boundary

$$\delta \sim (\varepsilon_0 - \varepsilon_h)^{1/2} . \tag{5.1.44}$$

This expansion in terms of the variable μ is equivalent to the expansion in terms of v used in Sect. 4.1.3(b), (4.1.29). The depletion layer is given explicitly by the exact solution of (5.1.27) with (5.1.42)

$$\mu(\xi) = \mu_3 - \tfrac{3}{2}\delta(\cosh q\xi)^{-2} , \qquad q := \tfrac{1}{2}\sqrt{C\delta} . \tag{5.1.45}$$

With (5.1.20) the eigenvalue equation (5.1.18) becomes

$$-\delta\mu'' + \left(\frac{\lambda}{D(\xi)} - \frac{\partial\tilde{\rho}}{\partial\mu}\right)\delta\mu = 0 . \tag{5.1.46}$$

Approximating $D(\xi)$ by an average value D and using (5.1.42, 45) one obtains from (5.1.46)

$$-\delta\mu'' + [V_0(\cosh q\xi)^{-2} - E]\delta\mu = 0 , \qquad \text{where} \tag{5.1.47}$$

$$V_0 := -12q^2 , \qquad E := -4q^2 - \lambda/D . \tag{5.1.48}$$

Equation (5.1.47) is isomorphic to a Schrödinger equation with a modified Pöschl-Teller potential [5.20] and can be solved exactly. There are three discrete eigenvalues

$$\lambda_0 = 5Dq^2 , \qquad \lambda_1 = 0 , \qquad \lambda_2 = -3Dq^2 \tag{5.1.49}$$

corresponding to the unstable mode

$$\delta\mu_0 = (\cosh q\xi)^{-3} , \tag{5.1.50}$$

the Goldstone mode

$$\delta\mu_1 = \sinh q\xi(\cosh q\xi)^{-3} . \tag{5.1.51}$$

and a stable mode

$$\delta\mu_2 = (1 - 4\sinh^2 q\xi)(\cosh q\xi)^{-3} , \tag{5.1.52}$$

respectively.

As ε_0 tends to ε_h, $q \to 0$ by (5.1.44, 45) and all modes (5.1.49) exhibit critical slowing down

$$\lambda \sim \sqrt{\varepsilon_0 - \varepsilon_h} \to 0 \tag{5.1.53}$$

as the boundary of the bistability domain is approached.

It should be stressed that all the results of Sect. 5.1.2 carry over to accumulation layers if the appropriate replacements are made.

5.1.3 Stable Modes of the Kink Profile

The discrete eigenvalues and the eigenfunctions of the kink-shaped stationary coexistence profile can be calculated analytically near the critical point. Assuming the coexistence condition (equal-areas rule), (4.1.21), the charge density can be expanded near the critical point into the universal cubic form

$$\tilde{\rho}(\mu, \varepsilon_0) = -C(\mu - \mu_c)[(\mu - \mu_c)^2 - \delta^2] \qquad \text{with}$$

$$\delta(\varepsilon_0) := \mu_3 - \mu_2$$

$$\mu_c := \tfrac{1}{2}(\mu_2 + \mu_3) \tag{5.1.54}$$

and a positive constant C. This expansion is equivalent to a fourth-order expansion of ϕ as given in (4.1.33). The kink-shaped coexistence solution is given explicitly by

$$\mu(\xi) = \mu_c \pm \delta \tanh q\xi , \qquad q := \delta \sqrt{C/2} , \tag{5.1.55}$$

The eigenvalue equation (5.1.18) for the transverse fluctuations then assumes the form (5.1.47), albeit with $V_0 = -6q^2$. Hence there are two discrete modes, namely the Goldstone mode with $\lambda_0 = 0$ and a stable mode with

$$\lambda_1 = -3Dq^2 . \tag{5.1.56}$$

An analogous result has been obtained for the stationary kinks of the Schlögl model [5.3, 4].

5.2 Cylindrical Current Filaments

In this section we investigate the stability of cylindrical stationary dissipative structures $\mu(R)$ against fluctuations $\delta\mu(R, \varphi)$ depending upon the radial and azimuthal coordinates R, φ. With the separation ansatz

$$\delta\mu(R, \varphi) = \delta\mu(R)e^{im\varphi} , \qquad m = 0, \pm 1, \pm 2, \ldots \tag{5.2.1}$$

the eigenvalue equation (5.1.4) reduces to

$$-\frac{1}{R} \frac{\partial}{\partial R}\left(R \frac{\partial}{\partial R} \delta\mu\right) + \left[\frac{m^2}{R^2} + V(R, \lambda)\right]\delta\mu = 0 , \tag{5.2.2}$$

where $V(R, \lambda)$ is defined as in (5.1.7).

5.2.1 General Results

One solution of (5.2.2) is the Goldstone mode of infinitesimal translation $-\Delta R$ in the direction $\varphi = 0$

$$\delta\mu_G(R, \varphi) = \Delta R \frac{d\mu(R)}{dR} e^{i\varphi} \tag{5.2.3}$$

corresponding to $\lambda = 0$, $m = 1$. It is shown schematically in Fig. 5.7a. By the substitution $\chi(R) := R^{1/2}\delta\mu(R)$, (5.2.2) can be transformed into a one-dimensional "radial Schrödinger equation"

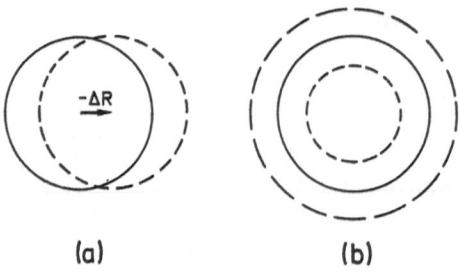

Fig. 5.7. Sketch of a cylindrical current filament (*full circle*) and of the effect of (a) the
Goldstone mode $\delta\mu_G$, and (b) the unstable
mode $\delta\mu_0$ (*dashed lines*). The Goldstone mode
shifts the current filament; the unstable mode
expands or compresses it isotropically

(a)   (b)

$$-\frac{d^2}{dR^2}\chi(R) + \left[\frac{m^2 - 1/4}{R^2} + V(R, \lambda)\right]\chi(R) = 0 \qquad (5.2.4)$$

with boundary conditions $\chi(0) = \chi(\infty) = 0$ and additional regularity conditions at
$R = 0$ [5.21].

For any fixed m, a similar Oscillation Theorem as in Sect. 5.1.1 holds. Hence for
monotonic concentration profiles $v(R)$ the eigenvalue $\lambda = 0$ of the Goldstone mode
(5.2.3) is the largest eigenvalue for $m = 1$, since $d\mu/dR$ does not vanish in the open
interval $(0, \infty)$. By the comparison Theorems of second-order linear differential
equations [5.22] any solution of (5.2.4) – as a one-point boundary value problem
– with $\lambda = 0, m = 0$ has a zero in the interval $(0, \infty)$. The argument used in the proof
of the Oscillation Theorem in Sect. 5.1.1 can then be applied to deduce the existence
of an eigensolution of (5.2.4) with $\lambda_0 > 0, m = 0$ and no zero in the interval $(0, \infty)$.
Thus there is an unstable mode $\delta\mu_0(R)$ with $\lambda_0 > 0, m = 0$ corresponding to uniform
expansion or shrinking of cylindrical current filaments as shown schematically in
Fig. 5.7b. Of all modes (5.2.1) only those with $m = 0$ change the total current through
the sample:

$$\delta I = \varepsilon_0 \int v(R)\delta\mu(R)e^{im\varphi}R \, dR \, d\varphi \neq 0 \ . \qquad (5.2.5)$$

Therefore the unstable mode $\delta\mu_0(R)$ is not allowed for fixed current: high- or
low-current filaments can be stabilized by imposing constant current conditions.

5.2.2 Unstable Mode

The unstable eigenvalue λ_0 can be calculated explicitly by a similar procedure as in
Sect. 5.1.2. Linearizing

$$V(R, \lambda) \approx [\lambda - \lambda_1(R)]/D(R) \qquad (5.2.6)$$

in (5.2.4), forming the Wronskian $\chi_0'\chi_G - \chi_0\chi_G'$ with the unstable mode χ_0 and the
Goldstone mode χ_G, one obtains by (5.2.4)

$$\lambda_0 = \int_0^\infty \frac{\chi_0\chi_G}{R^2} \, dR \Big/ \int_0^\infty \frac{\chi_0\chi_G}{D(R)} \, dR \ . \qquad (5.2.7)$$

Equation (5.2.7) can be further evaluated for wide (low- or high-) current filaments, i.e., for $\varepsilon_0 \lesssim \varepsilon_{co}$ or $\varepsilon_0 \gtrsim \varepsilon_{co}$, respectively. Then the stationary concentration profile $v_{fil}(R)$ varies considerably only in a thin cylindrical interface at large radius R_0 and assumes almost constant values v_1, v_3 in the interior ($R < R_0$) and the exterior ($R > R_0$) of the current filaments. The eigenfunctions $\chi_0 = R^{1/2}\delta\mu_0$ and $\chi_G = R^{1/2}\delta\mu_G$ are both sharply peaked at R_0 and approximately equal at sufficiently large R where the "centrifugal" term m^2/R^2 can be neglected. Therefore (5.2.7) becomes, by (5.2.3)

$$\lambda_0 \approx \int_0^\infty \left(\frac{d\mu}{dR}\right)^2 \frac{R\,dR}{R^2} \Bigg/ \int_0^\infty \left(\frac{d\mu}{dR}\right)^2 \frac{R\,dR}{D(R)}$$

$$\approx \frac{D(R_0)}{R_0^2} \tag{5.2.8}$$

where the slowly varying quantities D, R^2 have been extracted from the integrals. Note that the integral in the numerator of (5.2.8) is exactly the surface-tension term Σ which occurs in the equal-areas rule for cylindrical current filaments, (4.2.10, 11).

If a spontaneous initial fluctuation arises which shifts the filament walls radially by ΔR, then the evolution of the filamentary profile during the early stages of the instability is given approximately by

$$v(R,\tau) \approx v_{fil}(R) + \Delta R \frac{dv}{dR} e^{\lambda_0\tau} \approx v_{fil}(R + \Delta R e^{\lambda_0\tau}) , \tag{5.2.9}$$

provided that the total current is not held constant externally. Here we have used the approximation $\delta\mu_0(R) \approx \delta\mu_G(R, 0)$ as discussed above. Equation (5.2.9) describes a radially expanding or shrinking current filament wall.

Critical slowing down of the unstable mode is found when ε_0 tends to its coexistence value: For $\varepsilon_0 \to \varepsilon_{co}$ it follows $R_0 \to \infty$, and $\lambda_0 \to 0$ by (5.2.8). In particular, by the equal-areas rule (4.2.10), λ_0 scales like

$$\lambda_0 \sim \Sigma \sim |\varepsilon_{co} - \varepsilon_0| \to 0 , \tag{5.2.10}$$

where the generic case

$$\tilde{\phi}(\mu(\infty)) - \tilde{\phi}(\mu(0)) \sim \varepsilon_{co} - \varepsilon_0$$

is assumed near ε_{co}.

Finally, it should be recalled that the real spectrum of eigenvalues has been derived in the preceding two sections under the assumption that the dielectric-relaxation time is much smaller than the g-r times ("lifetime semiconductor"), and therefore the dielectric relaxation occurs practically instantaneously, and does not couple with the g-r instability. If, however, the dielectric-relaxation time is long, such as in high-purity materials at low temperatures ("relaxation semiconductors"), the dielectric-relaxation mode may couple with the g-r instability, and may well lead to complex, oscillatory unstable modes. This would manifest itself in a "breathing",

.e., periodic expansion and shrinking, of the filament. This possibility will be
liscussed in Sect. 6.1.1 in the context of possible mechanisms for self-sustained and
:haotic oscillations.

5.3 Finite Boundary Conditions

n this section we consider plane-transverse spatial profiles $\mu(\xi)$ of a finite width Λ
[5.23]. The eigenvalue equation (5.1.6) for the fluctuations $\delta\mu(\xi)$ has to be solved
n the finite interval $[0, \Lambda]$ subject to appropriate boundary conditions at $\xi = 0$ and
$\xi = \Lambda$.

The effect of the boundaries can substantially alter the eigenmodes, and change
he stability character of the profiles. This was already pointed out in the case of
Dirichlet boundary conditions for an electron overheating instability of SNDC-type
>y *Bass* et al. [5.24]. The influence of an external resistive circuit upon stability was
:tudied for Dirichlet boundary conditions in an electrothermal instability by *Jack-
:on* and *Shaw* [5.25]. Explicit solutions for both Neumann and Dirichlet boundary
:onditions and various circuit conditions were obtained for the hot-spot model of
ι superconducting microbridge [5.26], which also represents an electrothermal
nstability. The stability of finite reaction-diffusion systems was studied for Dirichlet
:onditions, using the linearization around the steady state [5.27–29], or WKB
nethods [5.30], or Lyapunov functions [5.31–33] (in the latter case for Neumann
:onditions). Surveys of results for both Dirichlet and Neumann conditions have
ιlso been given [5.34, 35]. Similar mathematical results were proved for a variety
>f population dynamical and epidemical problems [5.36, 37].

We shall now discuss the eigenmodes of the finite system in the case of our g-r
nstability. The Goldstone mode $\delta\mu_G(\xi) = \mu'(\xi) \cdot \Delta\xi$ of the infinite system is still a
'ormal solution of the differential equation (5.1.6), but it does not satisfy the correct
>oundary conditions. Therefore it is not an eigenfunction of (5.1.6), and the general
ιrgument of Sect. 5.1.1 cannot be applied. Still, we shall be able to use the Goldstone
node in a general stability argument [5.35], together with the observation that for
ιomogeneous Neumann or Dirichlet boundary conditions ($\delta\mu' = 0$ or $\delta\mu = 0$,
·espectively), (5.1.6) represents a regular Sturmian system. The Sturmian theory
[5.18] ensures the existence of a sequence of eigenfunctions $\delta\mu_i(\xi)$ $(i = 0, 1, 2, \ldots)$
with exactly i zeros in the interval $0 < \xi < \Lambda$, associated with a decreasing sequence
>f eigenvalues $\lambda_0 > \lambda_1 > \cdots > \lambda_i > \cdots$ In particular, there is a "ground state" $\delta\mu_0(\xi)$
without node, associated with the largest eigenvalue λ_0. As in Sect. 5.1.1 we assume
$\partial V(\xi, \lambda)/\partial\lambda > 0$, cf., (5.1.11), in the following.

5.3.1 Neumann Boundary Conditions

First we focus on homogeneous Neumann conditions. The stationary profiles satisfy

$$\mu'(0) = \mu'(\Lambda) = 0 \tag{5.3.1}$$

ιnd the fluctuations must also satisfy

$$\delta\mu'(0) = \delta\mu'(\Lambda) = 0 \ . \tag{5.3.2}$$

Without loss of generality we can assume $\delta\mu_0(\xi) > 0$ in the entire interval $0 \leq \xi \leq \Lambda$ for the eigenfunction of (5.1.6) with the largest eigenvalue λ_0. For stationary profiles satisfying (5.3.1), the Goldstone mode $\delta\mu_G(\xi) \sim \mu'(\xi)$ vanishes at $\xi = 0$ and $\xi = \Lambda$. Each extremum of $\mu(\xi)$ corresponds to a node of the Goldstone mode. Let the smallest node be denoted by ξ_0, and without restriction assume $\delta\mu_G \geq 0$ for $0 \leq \xi \leq \xi_0$. In case of a monotonic profile $\mu(\xi)$ set $\xi_0 = \Lambda$. Subtracting the values of the Wronskian of $\delta\mu_0$, $\delta\mu_G$ (i.e., $\delta\mu_0\delta\mu'_G - \delta\mu'_0\delta\mu_G$) at $\xi = \xi_0$ and at $\xi = 0$ yields, by (5.1.6),

$$-\delta\mu_0(\xi_0)\delta\mu'_G(\xi_0) + \delta\mu_0(0)\delta\mu'_G(0) = \int_0^{\xi_0} [V(\xi,\lambda_0) - V(\xi,0)]\delta\mu_0\delta\mu_G \, d\xi \ . \tag{5.3.3}$$

Since $\delta\mu_0(0)$, $\delta\mu_0(\xi_0)$, $\delta\mu'_G(0) > 0$, and $\delta\mu'_G(\xi_0) < 0$, the left-hand side of (5.3.3) is positive. Since the integrand on the right-hand side of (5.3.3) does not change sign, cf., (5.1.11), it follows that $\lambda_0 > 0$. Hence all transverse stationary profiles are unstable for homogeneous Neumann boundary conditions.

5.3.2 Dirichlet Boundary Conditions

Next we consider spatial profiles satisfying Dirichlet boundary conditions:

$$\mu(0) = \mu_{s1} \ , \qquad \mu(\Lambda) = \mu_{s2} \ . \tag{5.3.4}$$

If this holds for all times, the fluctuations must obey

$$\delta\mu(0) = \delta\mu(\Lambda) = 0 \ . \tag{5.3.5}$$

Assume without restriction $\delta\mu_0(\xi) > 0$ for $0 < \xi < \Lambda$ and $\delta\mu_0(0) = \delta\mu_0(\Lambda) = 0$.

For monotonic stationary profiles the Goldstone mode $\delta\mu_G$ has no zero in $0 \leq \xi \leq \Lambda$; assume without restriction $\delta\mu_G > 0$. Subtracting the Wronskian of $\delta\mu_0$, $\delta\mu_G$ at $\xi = \Lambda$ and at $\xi = 0$ yields

$$0 > \delta\mu'_0(\Lambda)\delta\mu_G(\Lambda) - \delta\mu'_0(0)\delta\mu_G(0) = \int_0^{\Lambda} [V(\xi,\lambda_0) - V(\xi,0)]\delta\mu_0\delta\mu_G \, d\xi \ . \tag{5.3.6}$$

Hence by (5.1.11) $\lambda_0 < 0$, which proves stability.

For oscillatory profiles with at least two extrema, assume that $\delta\mu_G > 0$ holds between two neighboring extrema $\xi_1 < \xi_2$. Use $\delta\mu_G(\xi_1) = \delta\mu_G(\xi_2) = 0$, and subtract the Wronskian at ξ_2 and at ξ_1:

$$0 < -\delta\mu_0(\xi_2)\delta\mu'_G(\xi_2) + \delta\mu_0(\xi_1)\delta\mu'_G(\xi_1) = \int_{\xi_1}^{\xi_2} [V(\xi,\lambda_0) - V(\xi,0)]\delta\mu_0\delta\mu_G \, d\xi \ . \tag{5.3.7}$$

Hence $\lambda_0 > 0$, which proves instability.

For profiles with one extremum, the above method of subtracting Wronskians fails. However, it has been shown for one-variable reaction-diffusion system in one spatial dimension that such profiles can be stable as well as unstable [5.28]: for $dE/d\Lambda > 0$ they are stable, and for $dE/d\Lambda < 0$ they are unstable, where $E := \mu'(0)^2/2$ corresponds to the "energy" in the associated "Hamiltonian" system. If follows from inspection of the diagrams of Λ versus $v'(0)$ in Sect. 4.3 (Fig. 4.29) that the middle solution branch (dashed), where $d[v'(0)]^2/d\Lambda < 0$, is unstable. Thus the middle one of three hump-shaped profiles is always unstable, while the upper and the lower profiles must be stable (Fig. 4.31). If the boundary condition is chosen at a value corresponding to one of the homogeneous steady states (Fig. 4.29a, b), one of the three spatial profiles is given by this homogeneous steady state. A careful inspection of Fig. 4.29a, b reveals that for $v_{s1} = v_{s2} \equiv v_R = v_2$ (center) there is a small range of values Λ where the homogeneous steady state v_2, which is always unstable in an *infinite* system, is stabilized by the finite boundary conditions, since it corresponds to the upper (a: $\varepsilon_0 < \varepsilon_{co}$) or the lower ($b$: $\varepsilon_0 > \varepsilon_{co}$) profile. The unstable states have been represented by dashed lines throughout Figs. 4.29–35.

Next we show, in extension of the results obtained in [5.28], that the profiles with one extremum and equal boundary values at both ends can never be unstable against more than one discrete mode. From the Oscillation Theorem of Sturmian Theory [5.18] it follows that the second eigenfunction, $\delta\mu_1$, has one node. Since the profile $\mu(\xi)$, and hence the potential $V(\xi, \lambda)$ in (5.1.6), is symmetric with respect to $\xi = \Lambda/2$, the eigenfunctions are symmetric or antisymmetric. Hence $\delta\mu_1$ has a zero at $\xi = \Lambda/2$, as does the Goldstone mode $\delta\mu_G$. Subtracting the Wronskian of $\delta\mu_1$, $\delta\mu_G$ at $\xi = \Lambda/2$ and $\xi = 0$ yields, by (5.1.6)

$$-\delta\mu_1'(0)\delta\mu_G(0) = \int_0^{\Lambda/2} [V(\xi, \lambda_1) - V(\xi, 0)]\delta\mu_1\delta\mu_G \, d\xi \ . \tag{5.3.8}$$

Since $\delta\mu_G(0)$, $\delta\mu_1'(0) > 0$, and $\delta\mu_1(\xi)$, $\delta\mu_G(\xi) > 0$ for $0 < \xi < \Lambda/2$, it follows that $\lambda_1 < 0$. Therefore $\delta\mu_1$ and all higher discrete modes $\delta\mu_2$, $\delta\mu_3$, ... are stable. The only mode which may become unstable, and in fact does, if $dE/d\Lambda < 0$, is the ground state $\delta\mu_0$, which either increases or decreases the concentration of the profile everywhere in space. Hence this mode changes the total current I and is excluded if the current is held constant. Thus unstable depletion or accumulation layer-like profiles can be stabilized also in finite systems by a heavily loaded external circuit.

5.4 Fluctuation-Induced Phase Transitions

In the preceding sections of this chapter we have found that hump-shaped transverse spatial profiles which lie between two stable profiles are unstable with respect to one discrete mode modulated perpendicular to the externally applied electric field. This applies to plane depletion or accumulation layers, and cylindrical high- or low-current filaments in infinite systems, and to the middle of three hump-shaped profiles in finite systems with symmetric Dirichlet boundary conditions. We shall now discuss the significance of these profiles as thresholds for noise-induced non-equilibrium phase transitions in a bistable system.

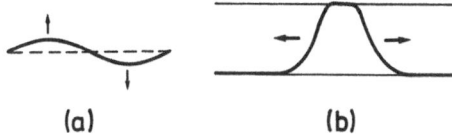

Fig. 5.8. (a) Homophase fluctuation of the unstable homogeneous steady state, leading to spinodal decomposition. (b) Heterophase fluctuation of a stable homogeneous steady state, leading to nucleation

(a)   (b)

5.4.1 Nucleation of Current Filaments

The dynamics of first-order phase transitions in general equilibrium and nonequilibrium systems has been recently reviewed [5.38]. An important feature is the development of phase separation, either by *spinodal decomposition* or by *nucleation*.

The first mechanism, spinodal decomposition, occurs if a system is rapidly quenched from above the critical point into an unstable homogeneous state below the critical point. In the unstable region of the phase diagram, which is bounded by the spinodal curve, arbitrarily small "homophase fluctuations" will grow (Fig. 5.8a), and the system eventually decomposes into two stable, spatially separated phase. In equilibrium systems spinodal decomposition can be described during the initial stages by a linear-mode analysis of the homogeneous unstable state, as in the theory of Cahn-Hilliard [5.39], or by more sophisticated nonlinear theories [5.40, 41]. Monte Carlo calculations are also available [5.42].

In our g-r instability the spinodal decomposition corresponds to a rapid quenching from a state of positive differential conductivity into a state of negative differential conductivity. In Sect. 3.4.4 we have found an unstable mode spectrum (Fig. 3.12a) quite similar to that of the linear Cahn-Hilliard theory, which can lead eventually – when the nonlinearities become effective – to a phase separation: current filamentation.

The second mechanism, nucleation, occurs if a spontaneous large-amplitude "heterophase fluctuation" is formed in the metastable regime of the phase diagram (Fig. 5.8b). A typical example is the formation of a critical liquid droplet in the supersaturated vapor phase of a Van der Waals' system. Droplets of smaller than critical radius shrink and die out, whereas droplets of larger radius grow. In other words, the heterophase fluctuation must have a sufficiently large amplitude in order not to be damped out; such fluctuations are the origin of the finite lifetime of metastable states in bistable systems. Droplet nucleation in equilibrium systems can be described by the Becker-Döring theory [5.43] and its improvements e.g., [5.44, 45]. As an example in nonequilibrium systems, the electron-hole droplet nucleation in highly excited semiconductors has been treated by generalizations of the classical nucleation theory [5.38].

In the following we will show that in the case of our g-r instability the hump-shaped stationary concentration profile, which was found unstable against exactly one discrete mode in Sect. 5.1–3, plays the role of a critical nucleus which is necessary for spontaneous current filamentation.

First we consider an infinite plane or cylindrical geometry. For $\varepsilon_{co} < \varepsilon_0 < \varepsilon_{th}$, the unstable linear mode either everywhere decreases or increases the electron concentration in the depletion layer (or current filament, in case of a cylindrical geometry), depending upon the initial fluctuation. In the first case the concentration

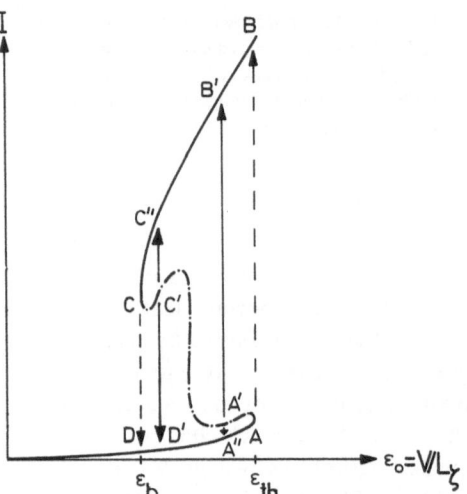

Fig. 5.9. Current-voltage characteristics, including stable (not necessarily absolutely stable!) homogeneous steady states (———) and inhomogeneous steady states of current-layer type (—·—·—). The arrows indicate non-equilibrium phase transitions under constant voltage conditions. In particular, A'-B' and C'-D' are fluctuation-induced transitions (schematic)

decreases until limited by the nonlinearities when the whole sample is filled homogeneously by the low-conductivity steady state v_1 ("OFF"). In the second case the concentration grows, and the walls of the accumulation layer (Fig. 5.3c) or filament expand until the whole sample is in the high-conductivity steady state v_3 ("ON"). Thus under constant voltage conditions a transition from the low- (OFF) to the high-conductivity (ON) state can be induced by a spontaneous localized fluctuation exceeding the concentration profile of the accumulation layer at the appropriate field ε_0. This is a nucleation process similar to the formation of droplets or bubbles in a gas-liquid phase transition. The larger ε_0, the narrower is the accumulation layer (filament), and the easier can a sufficiently large ("overcritical") localized fluctuation arise. Although the low-conductivity OFF phase v_1 is locally stable against small homogeneous fluctuations, it is not absolutely stable, and a switching transition can already occur before the maximum threshold field ε_{th} is reached (Fig. 5.9). For $\varepsilon_h < \varepsilon_0 < \varepsilon_{co}$ the role of v_3 (ON) and v_1 (OFF) is interchanged, and a depletion-layered or low-current filamentary spontaneous fluctuation can take the whole sample to the absolutely stable low-conductivity state, even above the holding field ε_h. Thus under constant voltage conditions the hysteresis cycle A-B-C-D-A in Fig. 5.9 can be shortened by spontaneous localized fluctuations (e.g., A" − A', C" − C') inducing switching transitions (A' − B', C' − D'). It should be noted that, although the spontaneous occurrence of a sufficiently large fluctuation is most favored near ε_h and ε_{th}, the initial growth increment λ_0 becomes maximum somewhere between ε_h and ε_{co}, or ε_{co} and ε_{th}, respectively.

Depletion and accumulation layers correspond to the "threshold solutions" found by a numerical analysis of the time evolution of finite localized fluctuations in reaction-diffusion systems [5.46]: for any continuous family of localized test fluctuations which include over- as well as undercritical fluctuations there is a critical fluctuation which for $t \to \infty$ evolves into the threshold solution.

For finite transverse dimensions and symmetric Dirichlet boundary conditions an analogous argument holds. Consider, e.g., the three profiles shown for a sequence of fields ε_0 and different boundary values in Fig. 4.31a–c. The middle of these three profiles is in each case unstable against exactly one mode. Hence the concentration uniformly grows or decreases until it reaches either the upper or the lower stable stationary concentration profile. Thus a transition from the lower to the upper state, or vice versa, can be induced by a spontaneous fluctuation exceeding the concentration profile of the middle stationary solution, which plays the role of a threshold or critical nucleus. Fluctuations which remain below the threshold solution will be damped out. Fig. 4.31a shows the case $v_s < v_1$. With increasing ε_0 the threshold profile comes closer to the lower solution, and hence the spontaneous occurrence of a fluctuation taking the system from the lower to the upper solution is more favored as ε_0 approaches the value 9.4. Similarly, as ε_0 decreases, "upper" and "lower" solutions exchange their roles, and a transition from the upper to the lower profile is more likely.

An essential difference in the nucleation behavior of the finite and the infinite system is that in finite systems *both* (locally) stable steady states are unstable against sufficiently large localized fluctuations, while in infinite systems (without fixed boundary values) only *one of* these two is unstable against such localized fluctuations: namely the upper state (v_3) for $\varepsilon_0 < \varepsilon_{co}$, and the lower state ($v_1$) for $\varepsilon_0 > \varepsilon_{co}$. The other state is absolutely stable against localized fluctuations. The origin of this lies in the topology of the phase portraits, cf., Fig. 4.2. The homoclinic trajectory, which corresponds to the critical nucleus in the infinite system, is connected to the singular point v_3 for $\varepsilon_0 < \varepsilon_{co}$, and to v_1 for $\varepsilon_0 > \varepsilon_{co}$. There is no homoclinic trajectory (i.e., critical nucleus) that is connected to the other singular point, respectively. For finite Dirichlet boundary conditions, in contrast, the trajectory corresponding to the critical nucleus lies more or less symmetrically between the two trajectories corresponding to the stable profiles. Therefore finite systems do not have *absolutely* stable steady states within their bistability regime.

The view of electronic switching through nucleation of current filaments as presented above has two different aspects: first, a stochastic spontaneous localized fluctuation of the carrier density exceeding the threshold profile must arise. Second, this supercritical fluctuation grows as determined by the deterministic semiconductor transport equations. This involves two time-scales that are relevant for the switching process. The statistical probability that such a fluctuation arises essentially determines the delay time of the switching transition, and is very important for the dynamics of threshold switching. There is a vast literature on the stochastic theory of noise-induced phase transitions in nonequilibrium systems, see e.g., [5.47, 48], but this is beyond the scope of the present book. The second aspect refers to the systematic evolution of the carrier-density profiles during the actual switching process; this will be touched upon at least qualitatively in the following subsection.

5.4.2 Transverse Solitary Waves

The preceding analysis has established the growth of supercritical fluctuations exceeding the threshold profile of the critical nucleus (current layer or filament). Initially – as long as the linearization around the threshold profile is a valid approximation – its temporal evolution is governed by the eigenvalue λ_0 of the unstable mode calculated in Sects. 5.1.2 and 5.2.2, respectively, see (5.1.26, 2.9). For longer times, the dynamics of the expanding filament wall is governed by the nonlinear transport equations. If the transverse dimension is wide enough, an asymptotic solitary wave with a stable profile and constant velocity can be expected to propagate laterally, until the whole sample is switched over to the final stable steady state. The wavefront corresponds to the filament wall.

Such solitary waves occur in a variety of physical [5.49, 50], chemical [5.2–4, 51–56], and population dynamical [5.57–60] systems. Their profiles and velocities have been investigated analytically and numerically for simple reaction-diffusion equations of Ginzburg-Landau type [5.49–55], and for other more general equations [5.56–61]. A classification of the possible solitary waves and their stability properties has also been given [5.58, 60]. We refer to these results also with respect to our g-r instability.

Finally, it should be mentioned that nonequilibrium phase transitions in a bistable system can not only be induced by localized fluctuations and subsequently developing diffusion-driven solitary waves, but also by spatially homogeneous statistical fluctuations. These two mechanisms give quite different regimes of absolute stability of the steady state (in an infinite system): For diffusion-driven transitions the coexistence condition given by an equal-areas rule divides the two regimes of absolute stability of either steady state. For homogeneous fluctuation-driven transitions the relative height of the two peaks of the probability distribution decides upon the absolute stability; this yields in general a different division. A resolution of this discrepancy has been attempted via a Monte Carlo simulation of the Schlögl model [5.62].

6. Self-Sustained Oscillations and Chaos

The subject of this chapter is time-dependent, undamped oscillatory behavior associated with g-r instabilities. While we have already dealt with transient temporal behavior during the initial stages of an instability in Chaps. 3, 5, we shall now study fully developed, nonlinear temporal dissipative structures. Stable periodic oscillations ("limit cycles") in NDC elements have been known for a long time, but the discovery of irregular, aperiodic ("chaotic") oscillations has only recently been reported. This has stimulated a rapidly growing number of experimental investigations of chaos in different semiconductor materials, and under a variety of conditions. A theoretical understanding of these phenomena is only just beginning to emerge.

6.1 Mechanisms for Oscillatory Behavior

6.1.1 Survey of Mechanisms

Spontaneous oscillations of the current or the voltage in semiconductors under dc bias can be induced by a variety of mechanisms, which may be classified into at least five main groups:

(i) *Circuit-induced oscillations* (due to the coupling of an NDC element with reactive circuit components, namely capacitors and inductors [6.1–6])
Here the NDC element simply acts as a nonlinear resistor. It has been well known since the age of the vacuum tube that such nonlinear circuits give rise to self-generated sustained oscillations, and can be modeled by a nonlinear oscillator equation, the Van der Pol equation [6.1–2]. The frequency is determined by the capacitance, inductance, and resistance of the circuit. We shall describe this in Sect. 6.2.1.

(ii) *Transit-time oscillations* (due to the motion of high-field domains in NDC elements from one electrode to the other [6.7–21])
Whenever a domain reaches the collecting electrode (the anode in case of negatively charged majority carriers), the field rises at the injecting electrode (the cathode, say) to nucleate another domain. This results in a periodic current oscillation, whose frequency is determined by the transit-time of the domain, i.e., the domain velocity and the transit length of the sample. The domain can be due to a drift instability, i.e., a nonmonotonic dependence of the mobility upon the field, as in the Gunn effect [6.7–15], or due to a g-r instability [6.16–21]. Transit-time oscillations might also be generated by quasi-neutral shock waves consisting of an electron-hole plasma [6.22]. In the well-known example of the Gunn oscillator the frequency is high,

typically in the microwave [GHz] range, while recombination domains are associated with low frequencies, from a few Hz to several kHz.

The common feature of all these oscillations is that they arise from a domain-type instability of the homogeneous steady state against longitudinal fluctuations of the form $\sim \exp[\lambda(k_\parallel)t + ik_\parallel x]$ with

$$Re\{\lambda\} > 0$$

$$Im\{\lambda\} = -vk_\parallel + O(k_\parallel^2) \tag{6.1.1}$$

resulting in a growing traveling wave $\sim \exp(Re\{\lambda\}t)\exp[ik_\parallel(x - vt)]$, an example of which we studied in Sect. 3.3. Here $k_\parallel$ is a longitudinal wave vector, x is the coordinate in the direction of the current flow, and v is the phase velocity of the traveling wave. By way of example, we shall discuss two mechanisms for such an instability.

The physical mechanism, which effects a domain instability of the type (6.1.1) in the Gunn effect can be understood as follows: the conduction-band structure of GaAs and other III–V compounds consists of a high-mobility central valley and a low-mobility satellite valley at higher energy. At low fields the drift velocity increases proportionally to the field. With increasing field more electrons are transferred from the lower high-mobility valley to the upper low-mobility valley, so that the average electron mobility, and the drift velocity decrease. Thus the velocity-field characteristic is nonmonotonic. Now consider a uniform field with a domain of increased field in the center of the element as shown in Fig. 6.1a, due to a spontaneous fluctuation. The carrier distribution that produces this field fluctuation is shown in Fig. 6.1b. There is an accumulation of negative charge on the right side of the domain, and a depletion layer to the left. Since we consider negatively charged carriers, the carriers and hence the domain will be moving to the left. Assuming that the higher field within the domain corresponds to a lower mobility than the lower field outside the domain, it is clear that the field fluctuation will initially grow with time, because the higher field in the center of the domain results in carriers moving more slowly than those at the edges. Mobile charge will therefore deplete on the left (leading) edge of the domain and accumulate at the right (trailing) edge. This charge will add to what is already there, increasing the field in the domain. This situation constitutes a runaway process. Given enough time, the domain field will

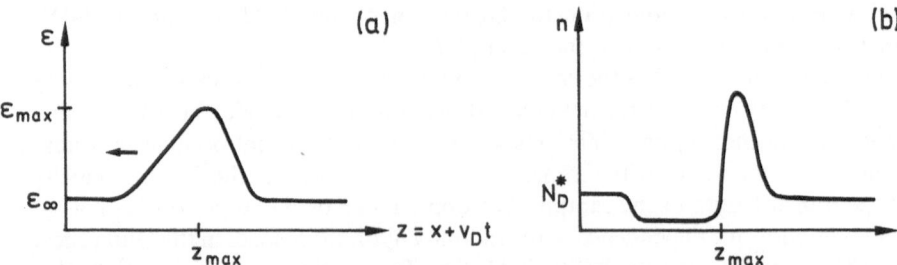

Fig. 6.1. (a) Electric-field profile, (b) electron-density profile of a moving Gunn domain (schematic). The domain is moving with velocity v_D in the negative x-direction

grow until the domain velocity $v \equiv -v_D$ is equal to the velocity of the carriers outside the domain.

Another mechanism that results in domain-transit oscillations is based upon contact extraction and trapping of carriers [6.19]: in high-resistivity GaAs at room temperature and with fields above 300 V/cm low-frequency transit-time oscillations have been observed which could be explained by the following mechanism: hole extraction at the negative electrode enables large space charge fields to build up due to ionized impurity centers in the depletion layer next to the contact. The voltage threshold for oscillations is reached when the width of the charge depletion layer at the negative electrode is equal to the hole diffusion length. Higher fields lead to a situation where a steady state is no longer possible, because the depletion region is too wide to be crossed by holes and the thermal generation from traps is too slow to provide the carriers needed to sustain the current. Hole trapping near the negative electrode instantaneously lowers the field and causes the region of maximum field to move away from the electrode. The velocity of this field domain, which moves progressively once it has been detached from the electrode, is determined by trapping in a deep acceptor level near the center of the energy gap.

Other models assume a field-enhanced capture of electrons associated with a repulsive potential barrier of deep traps [6.20] or with a configurational barrier [6.21] which is more easily penetrated by higher energy electrons.

(iii) *Bulk-dominated intrinsic oscillations* (due to nonlinear g-r processes)
A variety of physical mechanism can be comprised in this class. Their common feature is that these oscillations are not generated by the transit of moving domains but by a standing-wave like oscillation of the field or carrier profile, arising from a longitudinal oscillatory instability, e.g., of the type $\sim \exp[\lambda(k_\parallel)t + ik_\parallel x]$ with

$$\mathrm{Re}\{\lambda\} > 0$$

$$\mathrm{Im}\{\lambda\} = \omega + O(k_\parallel) \tag{6.1.2}$$

and hence $\sim \exp(\mathrm{Re}\{\lambda\}t)\exp(i\omega t)\exp(ik_\parallel x)$; an example was given in Sect. 3.5. The frequency of oscillation is independent of the sample length – provided that the sample is long enough so that the boundary conditions do not quench the instability.

The instability may either occur at a finite wavelength $k_\parallel^c > 0$, or in the long-wavelength limit $k_\parallel \to 0$, depending upon the $k_\parallel$-dependence of the real part of λ in (6.1.2). The first type is characterized by $\mathrm{Re}\{\lambda\} < 0$ for $k_\parallel < k_\parallel^c$, and $\mathrm{Re}\{\lambda\} > 0$ for $k_\parallel > k_\parallel^c$, and might be called *diffusion driven*, since homogeneous and long-wavelength fluctuations are damped out, while inhomogeneous ones with $k_\parallel \geq k_\parallel^c$ grow. The second type occurs in the two-level SNDC model of Sect. 3.5.1 and could be called *diffusion damped*, since homogeneous or long-wavelength fluctuations are undamped $[\mathrm{Re}\{\lambda\}(k_\parallel = 0) > 0]$, but short-wavelength fluctuations are damped out, cf., Fig. 3.13b. The essential features of the latter type of oscillation can be expected to be reasonably well approximated by a spatially homogeneous analysis, as it will be given in Sect. 6.2.2.

Among the physical mechanisms are field-enhanced trapping, recombination, or impact ionization in combination with dielectric relaxation [6.23–28, 100], photoexcitation [6.29], or heating of the electron gas [6.30–33]. The oscillations occur over a wide frequency range from 0.1 Hz [6.31] to 5 MHz [6.23]. In pin-diodes [6.23–27] double injection of electrons and holes from the two electrodes builds up space charges near both contacts. Relaxation of these injected space charges, coupled with trapping, diffusion, and drift of carriers drives these oscillations. In the case of single-carrier injection from one contact, the transient behavior can sometimes be modeled by dividing the sample into two regions in each of which spatial homogeneity is assumed [6.34]. If the influence of the contacts is not essential for the oscillatory mechanism, then models representing the bulk by a single spatially homogeneous region may be used [6.28, 32, 33, 35–37, 99, 100].

(iv) *Breathing of current filaments* (due to transverse oscillatory motion of the filament walls)

We have shown in Sect. 4.2 that SNDC gives rise to current filamentation, and have calculated the radius of stationary current filaments R_0 as a function of the applied field. If the field itself is considered as a dynamic variable, this might lead to a periodic shrinking and expanding ("breathing") of the current filaments, as the result of a transverse oscillatory instability.

Little is known experimentally [6.38] and theoretically about such an instability, but a possible mechanism could be provided by the coupling of impact ionization and dielectric relaxation with transverse diffusion in SNDC elements [6.39]. If dielectric relaxation occurs on a slower time-scale than g-r assisted transverse diffusion (for example, in high-purity relaxation semiconductors), then a quasi-stationary filamentary transverse carrier-density profile is formed for each instantaneous value of the electric field $\varepsilon(t)$. The quasi-stationary filament radius $R_0(\varepsilon)$, which depends upon the field according to (4.2.10, 19), is slaved by the slow dielectric-relaxation oscillations. This results in a breathing of the filament radius $R_0(t)$ and hence in oscillations of the current

$$I(t) \approx I_0 \varepsilon(t) \pi \{v_1 R_0(t)^2 + v_3 [a^2 - R_0(t)^2]\}$$

by (4.2.24), where v_1, v_3 are the carrier densities in the low- and the high-current density phase, respectively, and a is the radius of the (cylindrical) sample. Alternatively, oscillations of the carrier density in the filament wall could also lead to oscillations of the filament radius, and hence the current. This would be the result of an instability of the filamentary profile against transverse fluctuations of the form $\sim \exp(\lambda t)\psi(R)$ with complex eigenvalues λ,

$$\mathrm{Re}\{\lambda\} > 0 \ , \qquad \mathrm{Im}\{\lambda\} = \omega \neq 0 \ , \tag{6.1.3}$$

where $\psi(R)$ is a suitable radial eigenfunction. Note that in Sect. 5.2 we have obtained a *real* eigenvalue spectrum under the assumption of a *fast* dielectric-relaxation mode $v \gg |\lambda|$ only.

Large oscillations of the carrier density within a narrow filament wall lead to small oscillations of the filament radius and the current. Such small amplitude microampère-oscillations, superimposed on *dc* currents of several milliampères, have indeed been observed in the filamentary SNDC regime of p-Ge at low temperatures [6.38].

(v) *Helical waves* in an electron-hole plasma subjected to parallel electric and magnetic fields [6.40–45].

These are screw-shaped plasma density waves winding around the axis of the electric and magnetic field. The plasma is created by contact injection or light generation and exhibits a magneto-hydrodynamic instability. This so-called oscillistor effect [6.41] leads to current oscillations from a few kHz to about 10 MHz in different materials like n-Ge, p-Ge, p-Si, n-InSb, p-InSb at temperatures ranging from 4.2 to 300 K.

Mathematically, circuit-induced oscillations (i) can be treated by a system of nonlinear ordinary differential equations, while the mechanisms (ii–v) involve complicated partial differential equations including diffusion and drift terms. However, some characteristic features of these oscillations can often be discussed in terms of simpler ordinary differential equations. For example, in the moving coordinate system of a domain, the transport equations can be reduced to two ordinary differential equations. In this approximation, transit-time oscillations will be studied in Sect. 6.1.2. Bulk-dominated intrinsic oscillations can often be modeled by spatially homogeneous quasi-neutral oscillations of the carrier densities and the field in the bulk region. Within this approximation self-sustained periodic oscillations correspond to limit cycles of ordinary differential equations; these intrinsic limit cycles as well as circuit-induced limit cycles will be treated in Sect. 6.2. All of the above mechanisms can also give rise to aperiodic, chaotic oscillations. Those associated with g-r induced instabilities will be analysed in Sect. 6.3.

6.1.2 Transit-Time Oscillations

In Sect. 3.3 we have encountered one particular example of a domain-type instability, in which a propagating inhomogeneous longitudinal electric-field profile arose from the instability of a g-r induced anomalous tilted SNDC characteristic. Other examples of moving-field domains are recombination domains in g-r-driven NNDC sytems [6.16–22], and Gunn domains in mobility-driven NNDC systems [6.9–15]. Such solitary waves are observed as oscillations of the current in the external circuit.

By way of example, we shall analyse in the following the motion of Gunn domains in the simplest possible mode of operation, namely the "stable-domain mode", which is attainable in resistive-loaded circuits with dc bias.

The domains are special solutions of the standard semiconductor transport equations (1.1.5–11), subject to appropriate boundary conditions. These are complicated nonlinear partial differential equations and can not in general be solved analytically. Numerical solutions taking proper account of the boundary conditions, in particular the electric field at the cathode, have been given for Gunn domains elsewhere [6.14, 15]. Another tool of investigation is the topological phase-portrait analysis which gives qualitative insight into the nature of the solutions, and the effect of boundary conditions. It has been extensively used by *Böer* and others in the early investigations of recombination domains and Gunn domains [4.7–13]. Here it is our aim to discuss simple analytical relations which provide at least some essential partial information about the domains, like the peak field and the propagation velocity of the domains, even though they do not give the full spatial profiles of the field distributions. This is the purpose of equal-areas rules. Analogous rules for current-density filaments have been derived in Chap. 4. Such equal-areas rules allow one to extract this information from the constitutive equations by a simple geometrical construction without explicitly integrating the full transport equations. The characteristic constitutive equation in case of negative differential mobility, which leads to NNDC in the Gunn effect, is the drift velocity v of the majority carriers as a function of the local electric field ε.

Mathematically, equal-areas rules are first integrals of the transport equations, connecting external parameters like applied current or field with certain boundary or peak values of the profiles [6.46].

We discard minority carriers (holes, say) and trapped carriers, neglect the influence of the cathode and anode regions, and seek a solution representing a high-field domain that propagates with constant velocity v_D without change of shape, and is surrounded by an infinitely extended neutral medium with electron density $n = N_D^*$, where N_D^* is the effective doping concentration, see Fig. 6.1. The transport equations (1.1.5–11) can then be reduced to the one-dimensional form

$$J = env(\varepsilon) + eD\frac{d}{dz}n + \frac{\epsilon_s}{4\pi}v_D\frac{d}{dz}\varepsilon \tag{6.1.4}$$

$$\frac{d}{dz}\varepsilon = \frac{4\pi e}{\epsilon_s}(N_D^* - n) \ , \tag{6.1.5}$$

where J is the total longitudinal current density consisting of drift, diffusion, and displacement currents, D is the diffusion constant, and $z = x + v_D t$ is the relative coordinate in the moving frame of the domain.

We will now derive a first integral of (6.1.4, 5). Substituting (6.1.5) into (6.1.4) we obtain the differential equation

$$J - env(\varepsilon) - ev_D(N_D^* - n) + eD\frac{d}{dz}n = 0 \ . \tag{6.1.6}$$

In the neutral material outside the domain the current density J, by (6.1.4), consists of the drift current alone:

$$J = eN_D^* v_\infty \ , \tag{6.1.7}$$

where $v_\infty := v(\varepsilon_\infty)$ is the drift velocity corresponding to the field ε_∞ in the neutral material. Substituting (6.1.7) into (6.1.6) yields

$$N_D^*(v_\infty - v_D) - n(v - v_D) + D\frac{d}{dz}n = 0 \ . \tag{6.1.8}$$

Multiplying (6.1.8) by $n^{-1}\,d\varepsilon/dz$, and integrating over z from $z = -\infty$, or $z = +\infty$, to the position $z = z_{max}$ of the peak field $\varepsilon = \varepsilon_{max}$, we end up with the two relations

$$A_+ - B + C_+ = 0 \qquad \text{and} \tag{6.1.9a}$$

$$A_- - B + C_- = 0 \tag{6.1.9b}$$

representing first integrals of (6.1.4, 5), where

$$A_\pm := N_D^*(v_\infty - v_D) \int\limits_{\pm\infty}^{z_{\text{max}}} \frac{1}{n}\frac{d\varepsilon}{dz}\,dz \tag{6.1.10}$$

$$B := \int\limits_{\varepsilon_\infty}^{\varepsilon_{\text{max}}} [v(\varepsilon) - v_D]\,d\varepsilon \tag{6.1.11}$$

$$C_\pm := D\frac{4\pi e}{\epsilon_s} \int\limits_{\pm\infty}^{z_{\text{max}}} \frac{1}{n}(N_D^* - n)\frac{dn}{dz}\,dz \ . \tag{6.1.12}$$

Observing that dn/dz changes sign once in each of the intervals $-\infty < z < z_{\text{max}}$ and $+\infty > z > z_{\text{max}}$, and that

$$n(\pm\infty) = n(z_{\text{max}}) = N_D^* \ , \tag{6.1.13}$$

one can transform (6.1.12) into an integral over dn which vanishes:

$$C_+ = C_- = D\frac{4\pi e}{\epsilon_s} \oint \frac{1}{n}(N_D^* - n)\,dn = 0 \ . \tag{6.1.14}$$

With this we subtract (6.1.9a, 9b) and find

$$A_+ = A_- \ . \tag{6.1.15}$$

The integrals A_+, A_- can be transformed into integrals over $d\varepsilon$, where we have to distinguish the double-valued "function" $n(\varepsilon)$ by writing n_- in the interval $-\infty < z < z_{\text{max}}$ and n_+ in the interval $z_{\text{max}} < z < \infty$. From $n_+ > N_D^*$ and $n_- < N_D^*$ (Fig. 6.1) it follows that:

$$A_+ \sim \int\limits_{\varepsilon_\infty}^{\varepsilon_{\text{max}}} \frac{1}{n_+}\,d\varepsilon < \int\limits_{\varepsilon_\infty}^{\varepsilon_{\text{max}}} \frac{1}{n_-}\,d\varepsilon \sim A_- \ . \tag{6.1.16}$$

We can deduce from (6.1.15, 16) and the definition (6.1.10):

$$v_\infty = v_D \qquad \text{and} \tag{6.1.17}$$

$$A_+ = A_- = 0 \ . \tag{6.1.18}$$

Therefore (6.1.9a) or (6.1.9b) reduces to

$$B = \int\limits_{\varepsilon_\infty}^{\varepsilon_{\text{max}}} [v(\varepsilon) - v_D]\,d\varepsilon = 0 \ . \tag{6.1.19}$$

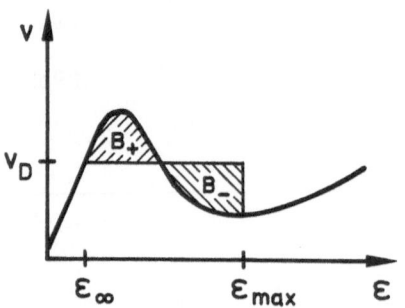

Fig. 6.2. Equal-areas rule for Gunn domains. The hatched areas B_+ and B_- must be equal. (v is the electron drift velocity, v_D the domain velocity.) Schematic

This is the *equal-areas rule* for Gunn domains [6.12], where the domain velocity is, by (6.1.17), equal to the low-field drift velocity of the electrons determined by the external current density (6.1.7). The equal-areas rule (6.1.19) states that the two hatched areas in Fig. 6.2 must be equal, and thereby determines the peak field ε_{max} of the domain. The equal-areas rule (6.1.19) has been generalized to the case where the diffusion constant D depends upon the field [6.13].

In the preceding discussion we have neglected the influence of the boundaries and the external circuit, and considered the steady motion of a stable domain in an infinitely extended sample. In practical cases, however, Gunn diodes are often operated in resonant circuits, such as resonant microwave cavities with a time-dependent applied voltage. The observed modes then depend upon the reactive circuit components yielding either domain-controlled or circuit-dominated (Sect. 6.2.1) modes of oscillation. Furthermore, for primarily resistive circuits, the cathode boundary condition of the electric field ε_c has been shown to be a principal determinant of the nature of the current instability, both numerically and experimentally [6.14, 15]. For low ε_c and *dc* bias a single high-field dipole domain is formed near the cathode boundary, or at other nucleation sites, and propagates to the anode. After its arrival, depending upon the applied bias either a new domain is nucleated near the cathode, or (for larger bias) the field distribution remains stationary. For intermediate ε_c, domains nucleate at the cathode, and the classic cathode-to-anode transit-time mode is obtained. For high ε_c, part of the dipole layer at the cathode detaches, moves a distance towards the anode and then disappears, usually before reaching the anode (for long enough samples). In reactive circuits at low ε_c the domains may be quenched as the circuit swings the voltage below the domain sustaining value, leading to the "delayed-domain" mode, the "quenched-domain" mode, or the "limited space-charge accumulation (LSA)" mode.

6.2 Limit-Cycle Oscillations

In this section we shall study homogeneous oscillations, represented by limit cycles in a low-dimensional phase space. Limit cycles are closed phase trajectories which are asymptotically stable in the sense that all phase trajectories starting in a neighborhood of these cycles asymptotically approach these, i.e., they are one-dimensional attractors. They manifest themselves as self-sustained periodic oscillations.

6.2.1 Circuit-Induced Oscillations

In practical cases an NDC semiconductor is always connected with controlled or spurious resistive and reactive circuit elements. These consist, for instance, of leads, contacts, and support components:

(i) The attachment of metallic leads to the NDC elements introduces a lead resistance R_1 and lead inductance L_1;

(ii) the contact regions themselves most often produce a nonlinear resistance R_c, apart from imposing specific boundary conditions at the interface of the NDC material;

(iii) supporting, mounting, or holding the NDC element introduces package capacitance C_p and package inductance L_p;

(iv) an external dc voltage source U_0 will contain its own internal resistance R_I, in addition to a load resistor R_L.

The NDC element itself can also be represented in terms of lumped circuit elements. Time-dependent electric and magnetic fields in the semiconductor lead to displacement currents and induced voltages, respectively. These can be modeled by defining an intrinsic capacitance

$$C_i := \frac{1}{V^2} \int \frac{\epsilon_s}{4\pi} \mathscr{E}^2 \, d^3x \tag{6.2.1}$$

and an intrinsic inductance

$$L_i := \frac{1}{I^2} \int \frac{1}{4\pi} H^2 \, d^3x \tag{6.2.2}$$

of the NDC element, where I is the current through, and V the voltage across the sample. It follows from the energy balance

$$IV = \int j\mathscr{E} \, d^3x + \frac{1}{2} \frac{d}{dt}(C_i V^2 + L_i I^2) \tag{6.2.3}$$

that the NDC element can then be approximated by a nonlinear resistor in series with an intrinsic inductance L_i, and parallel to an intrinsic capacitance C_i. This approximation assumes that the NDC element can be represented by a static $I_c - V_c$ characteristic of the conduction current I_c versus the conductive voltage V_c, which is only true at low frequencies when the mobility and the carrier densities instantaneously take on their steady state values. In Sect. 6.2.2 we shall consider the more general case that the dynamic degrees of freedom of the carrier densities are explicitly taken into account.

The lumped element approximation of the circuit containing the NDC element and its environment components (i)–(iv) is shown in Fig. 6.3. It is a complicated nonlinear circuit with five reactive components corresponding to five first-order differential equations. There are, however, a large number of common situations that allow further simplification of the circuit.

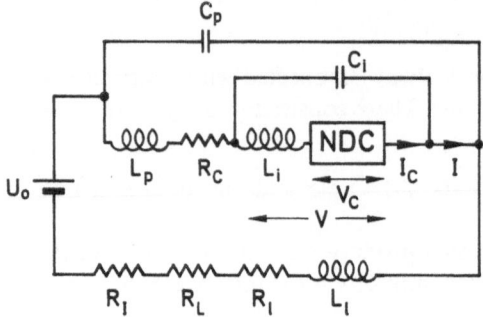

Fig. 6.3. Lumped element approximation of an NDC element in its circuit environment. U_0 is the applied static voltage, V is the total voltage across the NDC element, V_c its conductive part. I is the total current through the NDC element, I_c is the conduction current. C_i and L_i are the intrinsic capacitance and inductance of the NDC element, C_p and L_p are package capacitance and inductance, R_c is a contact resistance, R_l and L_l are lead resistance and inductance, R_L and R_l are load resistance and internal resistance of the voltage source, respectively

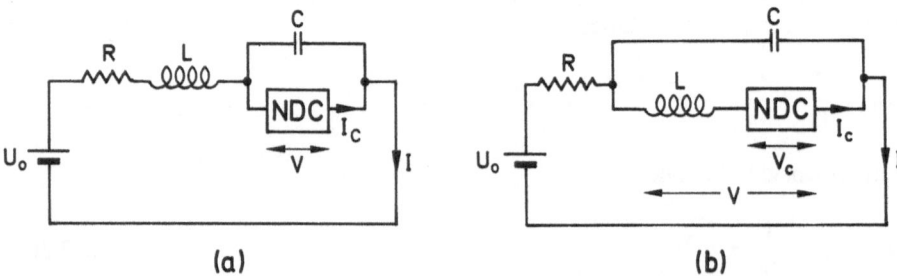

Fig. 6.4. Approximation of the circuit in Fig. 6.3 for (a) an NNDC element, (b) an SNDC element

First, consider an NNDC element. When such a system undergoes time-dependent NDC-driven processes, it is generally the case that the relative changes in voltage across it are large compared to the relative changes in current through it. Thus here the capacitive effect is much larger than the inductive effect. Because of this, a convenient approximation is to neglect L_i and L_p, and include R_c into the $I_c(V)$ characteristic. The circuit of Fig. 6.3 then reduces to that shown in Fig. 6.4a [6.4]. Here, $R := R_I + R_L + R_l$, $L := L_l$ and $C := C_p + C_i$. Kirchhoff's laws yield

$$U_0 = IR + L\dot{I} + V \qquad \text{and} \tag{6.2.4}$$

$$I = I_c(V) + C\dot{V} . \tag{6.2.5}$$

In the absence of high-field domains the conduction current I_c is a single-valued function of the voltage $V = \varepsilon L_z$, and $C_i = \epsilon_s A/(4\pi L_z)$ is a constant, where A is the cross section of the sample, and L_z its length. Scaling the time

$$T := t/(LC)^{1/2} \tag{6.2.6}$$

(6.2.4, 5) can be recast in the form of an autonomous dynamic system

$$\frac{d}{dT}I = \frac{1}{Z_0}(U_0 - V - IR) \tag{6.2.7}$$

$$\frac{d}{dT}V = Z_0[I - I_c(V)] \tag{6.2.8}$$

where $Z_0 := (L/C)^{1/2}$ is the ratio of voltage to current amplitude across either reactive circuit element in case of an oscillating circuit consisting of an inductor and a capacitor only. Thus, large values of Z_0 result in large-amplitude voltage oscillation. The steady states of (6.2.7, 8) are given by the intersection of the dc load line $V = U_0 - IR$ with the NNDC characteristic $I = I_c(V)$.

Next, we shall discuss an SNDC element. When such a system undergoes time-dependent processes, the relative changes in current through it are generally large compared to the relative changes in voltage across it. Thus here the inductive effect is much larger than the capacitive effect. Counterexamples are, however, furnished by chaotic oscillations in some relaxation semiconductors, see Sect. 6.3, where the displacement currents are essential in driving the instability. Usually values of R_L substantially greater than L_1/t_s, where t_s is the transition time from one conductive state to another, are employed so that the system is heavily loaded. Under these conditions a convenient approximation is to neglect L_1 and C_i, and include R_c into the $V_c(I_c)$ characteristics. The circuit of Fig. 6.3 then reduces to that shown in Fig. 6.4b [6.5, 6]. Here $R := R_1 + R_L + R_1$, $L := L_1 + L_p + M$, where M is the mutual inductance that may exist between L_p and L_1, and $C := C_p$. Kirchoff's laws yield:

$$U_0 = IR + V \tag{6.2.9}$$

$$I = I_c + C\dot{V} \quad \text{and} \tag{6.2.10}$$

$$V = V_c(I_c) + L\dot{I}_c . \tag{6.2.11}$$

In the absence of current filaments, the conductive voltage V_c is a single-valued function of the conduction current $I_c = jA$, and L_1 remains constant over an oscillatory cycle. Again, (6.2.9–11) can be cast in the form of a time-scaled dynamic system, using (6.2.6):

$$\frac{d}{dT}V = Z_0\left(\frac{U_0 - V}{R} - I_c\right) \tag{6.2.12}$$

$$\frac{d}{dT}I_c = \frac{1}{Z_0}[V - V_c(I_c)] . \tag{6.2.13}$$

Comparison of (6.2.12, 13) with (6.2.7, 8) shows that (6.2.12) is the dual of (6.2.7), and (6.2.13) is the dual of (6.2.8). In going from the NNDC to the SNDC system we simply have to replace V, I, I_c, Z_0, R, U_0 with $I_c, V, V_c, 1/Z_0, 1/R, U_0/R$, respectively. That is, in the context of the approximations noted, which are the common situa-

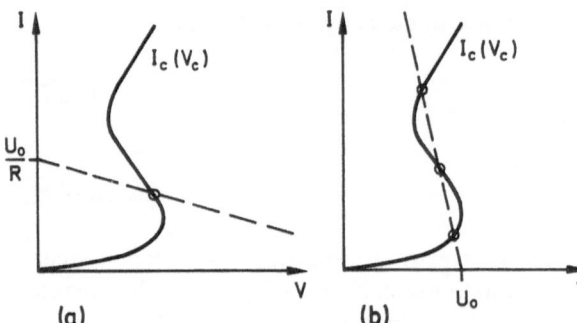

Fig. 6.5a, b. Steady states of an SNDC element, given by the intersection of the characteristic $I_c(V_c)$ with the load line (–––; schematic.) (a) Heavily loaded circuit (large R); (b) weakly loaded circuit (small R)

tions, the circuit theory transforms directly between the S and N cases when the important reactive components are identified for each NDC type. The important circuit parameters in each case produce a dual circuit system where the voltage across the NNDC element in its primary circuit behaves exactly as the current through the SNDC element in its primary circuit. By way of example, we shall perform the circuit analysis in the following for an SNDC system, but the same arguments also apply to NNDC systems when the appropriate replacements are made.

The steady states of (6.2.12, 13) are given by the intersection of the dc load line $V = U_0 - IR$ with the SNDC characteristic $V = V_c(I)$, where $I_c \equiv I$, as shown in Fig. 6.5. If the magnitude of the slope of the load line in the I–V plane is smaller than that of the $I_c(V_c)$ characteristic at any point of the NDC region,

$$R > R_n := -\text{Min}|dV_c/dI_c| \;, \tag{6.2.14}$$

there is a unique steady state (I_0, V_0) for any bias U_0. For smaller values of R there are three steady states for a range of bias U_0. The stability of the steady states is readily determined from linearization of (6.2.12, 13) around the steady state. We obtain with $\delta I_c = I_c - I_0$, $\delta V = V - V_0$:

$$\begin{pmatrix} \dfrac{d}{dT}\delta V \\[2ex] \dfrac{d}{dT}\delta I_c \end{pmatrix} = \begin{pmatrix} -\dfrac{Z_0}{R} & -Z_0 \\[2ex] \dfrac{1}{Z_0} & -\dfrac{1}{Z_0}\dfrac{dV_c}{dI_c} \end{pmatrix} \begin{pmatrix} \delta V \\[2ex] \delta I_c \end{pmatrix}, \tag{6.2.15}$$

where dV_c/dI_c is to be evaluated at the steady state. The trace of the matrix on the right-hand side is

$$T = -\left(\frac{Z_0}{R} + \frac{1}{Z_0}\frac{dV_c}{dI_c} \right) \tag{6.2.16}$$

and the determinant is

$$D = \frac{1}{R}\frac{dV_c}{dI_c} + 1 \;. \tag{6.2.17}$$

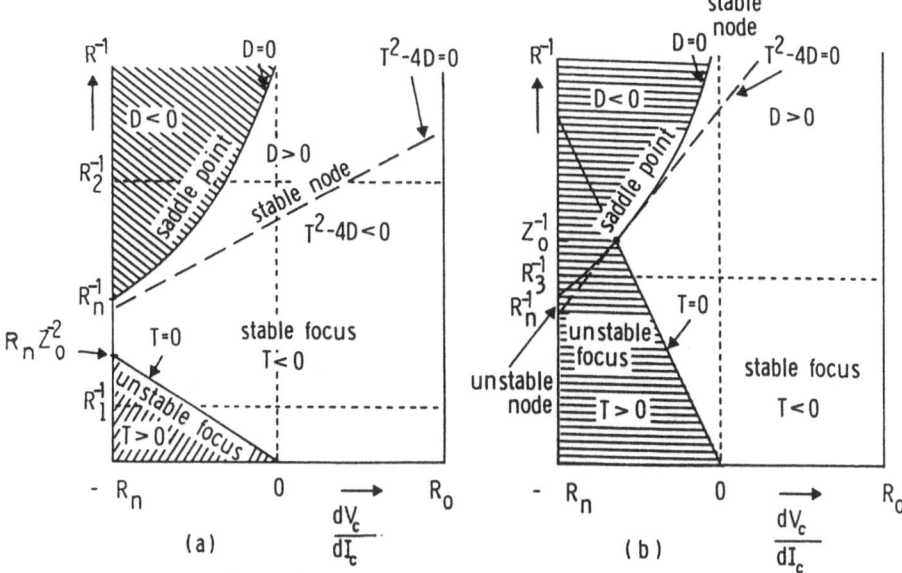

Fig. 6.6a, b. Regimes of stability of the steady state for an SNDC element in a reactive circuit. (a) $Z_0 > R_n$, (b) $Z_0 < R_n$. Hopf bifurcations of limit cycles occur on the line $T = 0$; saddle-node bifurcations on the curve $D = 0$. Regions of instability are hatched. (Schematic)

The steady state is stable if the eigenvalues λ of the matrix in (6.2.15), given by

$$\lambda^2 - T\lambda + D = 0 , \tag{6.2.18}$$

have negative real parts, i.e., if $T < 0$ and $D > 0$. The different regimes of stability can be conveniently discussed in the $(R^{-1}, dV_c/dI_c)$ plane as shown in Fig. 6.6. [6.47]. The positive differential conductivity states are always stable because this results in $D > 0$, $T < 0$ by (6.2.16, 17). It is remarkable that for $R > 0$ some of the NDC states also become stabilized by the circuit.

For $Z_0 > R_n$ (Fig. 6.6a) the two curves $D = 0$ and $T = 0$ do not intersect. When the $T = 0$ line is crossed, T changes sign, while D remains positive, and $T^2 - 4D < 0$. Hence the steady state, which is unique for $R^{-1} < R_n^{-1}$ (Fig. 6.5a), changes from stable focus to unstable focus. On the $T = 0$ line a Hopf bifurcation occurs in which a stable limit cycle bifurcates from the focus. This limit cycle exists in the whole regime where $T > 0$. Only for $R \to \infty$ (corresponding to a horizontal load line in Fig. 6.5a) does the limit cycle bifurcate right at the beginning of the NDC region, i.e., at the turning points $dV_c/dI_c = 0$. With decreasing load R, the bifurcation point is shifted further into the NDC regime. For $R_n^{-1} > R^{-1} > R_n/Z_0^2$, e.g., for large values of Z_0 and the situation of Fig. 6.5a, all NDC points of the $V_c(I_c)$ characteristic are stable, and no self-sustained oscillation is generated. When the bias U_0 is slowly increased at a constant value of the load $R = R_1 > Z_0^2/R_n$, corresponding to the lower dashed horizontal line in Fig. 6.6a, then dV_c/dI_c decreases from R_0 to its minimum value $-R_n$ and then increases again to zero, as the intersection point of

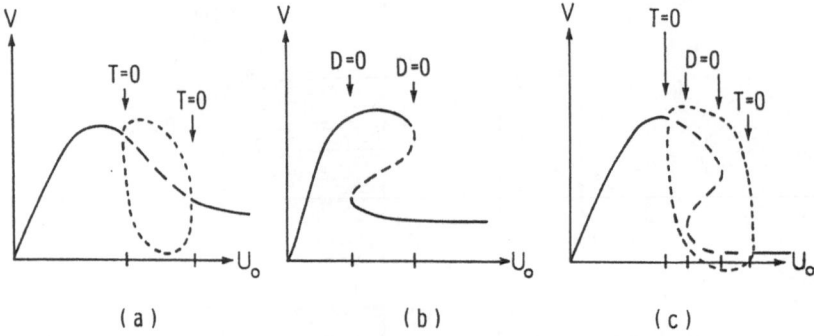

Fig. 6.7a–c. Schematic plot of voltage across the NDC-element V versus bias voltage U_0 for an SNDC element in a reactive circuit (**a**) at fixed load $R = R_1 > Z_0^2/R_n$ for $Z_0 > R_n$, (**b**) at fixed load $R = R_2 < R_n$ for $Z_0 > R_n$, (**c**) at fixed load $R = R_3$ with $R_n > R_3 > Z_0$ for $Z_0 < R_n$. Full lines denote stable steady states; dashed lines: unstable steady states; dotted lines: amplitude of the limit-cycle oscillations. The bifurcation points are marked by $T = 0$ and $D = 0$

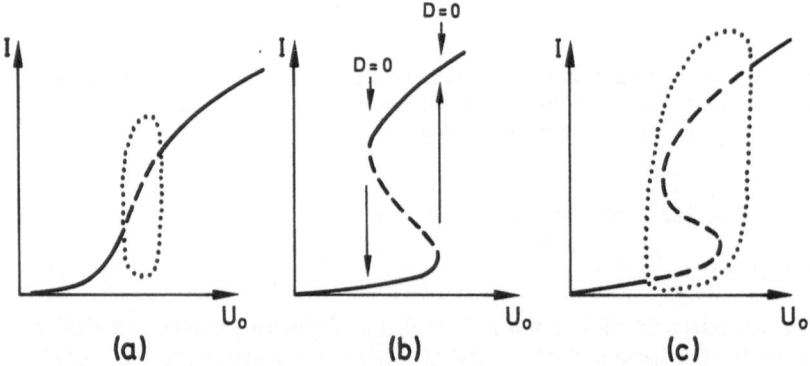

Fig. 6.8a–c. Schematic plot of current I versus bias U_0. Parameters as in Fig. 6.7

the load line and the SNDC characteristic travels along the $V_c(I_c)$ curve. Hence the line $T = 0$ is crossed twice, which marks the two boundaries of the regime of bias U_0 where a limit cycle exists. This is shown schematically in the bifurcation diagram of V versus U_0 in Fig. 6.7a. The associated plot of I versus U_0 is given in Fig. 6.8a. Note that the circuit quenches the hysteresis of the SNDC element, and introduces an oscillatory regime.

Next, let us consider a smaller value of $R \equiv R_2 < R_n$, such that the curve $D = 0$ can be crossed, while T is negative throughout (upper dashed horizontal line in Fig. 6.6a). For $D > 0$ and $T^2 - 4D > 0$ the steady state is a stable node; for $D < 0$ it is a saddle point, i.e., unstable. It follows from (6.2.17) that $D < 0$ is equivalent to $dV_c/dI_c < -R$. Hence (for $R < Z_0$) exactly those points are unstable for which the slope of the $V_c(I_c)$ characteristic is steeper than that of the load line. In this case there are two more intersections of the load line with the $V_c(I_c)$ curve satisfying $dV_c/dI_c > -R$ (as shown in Fig. 6.5b), hence these are stable steady states, no matter whether they have negative or positive differential conductivity. For values on the curve $D = 0$ the load line becomes tangent to the $V_c(I_c)$ characteristic. This corresponds to a saddle-node bifurcation in which two steady states (operating points on the dc load line) coalesce and disappear. This is illustrated by the corresponding $V(U_0)$ plot in Fig. 6.7b, and by the $I(U_0)$ plot in Fig. 6.8b. The current exhibits hysteresis as a function of bias U_0, but no Hopf bifurcation. The up- and down-transitions of the hysteresis loop, with increasing and decresing U_0, respectively, are indicated by the two arrows labeled $D = 0$.

Finally, we consider the case $Z_0 < R_n$ (Fig. 6.6b). Now a combination of limit-cycle and saddle-node bifurcations may occur if the load $R = R_3$ satisfies $R_n^{-1} < R_3^{-1} < Z_0^{-1}$ (dashed horizontal line in Fig. 6.6b). On the curve $D = 0$ a saddle point and an *unstable* node coalesce and disappear. Figures 6.7c, 6.8c show typical $V(U_0)$ and $I(U_0)$ plots, in which a limit cycle is surrounding up to three singular points (steady states).

In summary, the occurrence of limit-cycle oscillations is favored in SNDC systems with large R and small Z_0. Figure 6.6 can also be used to discuss the oscillatory circuit instability for NNDC systems if the labeling of the axes is replaced according to the duality theory. As a result, limit cycles are now favored by small R and large Z_0.

Next, we shall investigate the nonlinear limit-cycle oscillations in more detail. The governing equations (6.2.12, 13) can be combined into an oscillator equation with a nonlinear damping term and a nonlinear force:

$$\frac{U_0}{R} = \left(1 + \frac{V_c(I_c)}{I_c R}\right) I_c + \frac{1}{Z_0}\left(\frac{dV_c(I_c)}{dI_c} + \frac{LZ_0}{R}\right)\frac{dI_c}{dT} + \frac{d^2 I_c}{dT^2} . \qquad (6.2.19)$$

When biased in the region of NDC ($dV_c/dI_c < 0$), the damping can be negative, leading to growth rather than decay. In general the bracketed part of the damping term is of the order of the low-current resistance of the NNDC element $R_0 := (dV_c/dI_c)$ ($I_c = 0$), and the strength of the damping term is determined by the dimensionless parameter R_0/Z_0. Eq. (6.2.19) is a generalized Van der Pol equation [6.1]. For small R_0/Z_0 the damping term is a small perturbation and the limit-cycle solutions for $V(T)$ and $I_c(T)$ are nearly sinusoidal. This is shown in Fig. 6.9a for

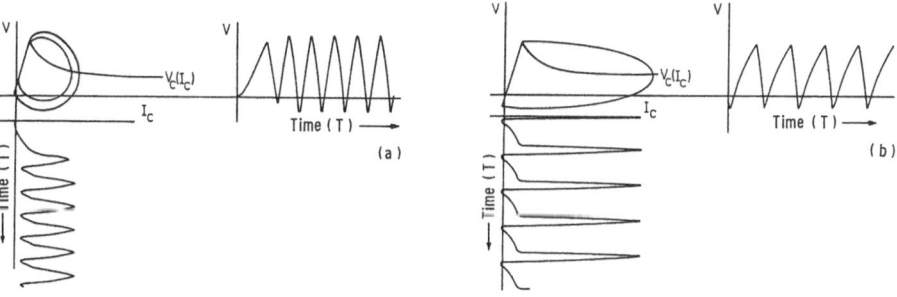

Fig. 6.9a, b. Circuit-induced oscillations in an SNDC element for (a) $R_0/Z_0 = 3$, (b) $R_0/Z_0 = 12$. The phase portraits V versus I_c of the limit-cycle oscillations are shown together with the NDC characteristic $V_c(I_c)$, and the corresponding time series $V(T)$ and $I_c(T)$. (Numerical solution [6.4])

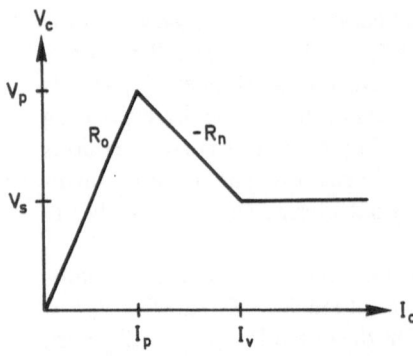

Fig. 6.10. Three-piece linear approximation of the NDC characteristic $V_c(I_c)$

large R and $R_0/Z_0 = 3$, where we plot $I_c(T)$, $V(T)$, and $V(I_c)$, obtained numerically. For large R_0/Z_0 the damping term is important and the solutions become well-defined relaxation oscillations[1]. Figure 6.9b illustrates this for $R_0/Z_0 = 12$. Here the current oscillations are almost sawtooth and the voltage oscillations exhibit sharp spikes. This is due to a time-scale separation between the variables V and I_c for small Z_0, as evident from (6.2.12, 13), such that I_c very rapidly attains a pseudo-steady state given by $V = V_c(I_c)$, and V changes slowly. Therefore, the relaxation-oscillation cycle consists of a slow rise close to the positive differential conductivity portion of the $V_c(I_c)$ characteristic, until the maximum voltage V_p is reached, and a fast, spiky second part follows.

In order to obtain more insight, we shall now employ a three-piece linear approximation for the $V_c(I_c)$ curve to find analytical solutions of the limit cycle. Such an approximation is shown in Fig. 6.10. The differential equation (6.2.19) can be solved in the three regions $I_c < I_p$, $I_p < I_c < I_v$, and $I_v < I_c$, which correspond to $dV_c/dI_c = R_0$, $-R_n$, and 0, respectively, and the solutions are joined continuously [6.6]. The composition of the solutions is evident in the relaxation oscillation shown in Fig. 6.9b: The current waveform begins with a slow exponential rise with time constant $(R_0/L + 1/RC)^{-1}$, followed by a sharp spike in current after $I_c = I_p$ is reached. The time needed to reach I_p is bias-dependent and if, for example, the minimum current is arbitrarily set equal to zero, the risetime to threshold is approximately given by

$$\Delta T = \left(\frac{R_0}{L} + \frac{1}{RC}\right)^{-1} \ln\left[1 - \frac{I_p R}{U_0}\left(1 + \frac{R_0}{R}\right)\right]^{-1} . \tag{6.2.20}$$

Increasing U_0, all other things being equal, decreases ΔT and increases the frequency of oscillation. Bias tuning is a characteristic of NDC relaxation oscillators.

The spike above threshold is composed of a fast exponential transit through the region of negative slope $(I_p < I_c < I_v)$, with time constant $(1/RC - R_n/L)^{-1}$ followed

---

[1] Note that the term *relaxation oscillation* is used here in the sense of a strongly nonlinear *self-sustained* oscillation as throughout the mathematical and electrical engineering literature, whereas it means a *damped* oscillation in laser physics.

by a damped sine wave for $I_c > I_v$, and another exponential transit for $I_p < I_c < I_v$. An exponential decay for $I_c < I_p$ completes the limit cycle.

As Fig. 6.9b indicates, the $V(I_c)$ limit-cycle trajectory, obtained by eliminating T from $I_c(T)$ and $V(T)$, has the appearance of a truncated ellipse, closed by the low-voltage part of the $V_c(I_c)$ characteristic. This Lissajous-type figure provides an indication of the magnitude of the current and voltage swings and their relation to the circuit parameters Z_0 and U_0, and the NDC element parameters V_p, I_p, V_s. It turns out that, besides Z_0/R_0, the saturation voltage V_s has a substantial influence on $V(I_c)$, higher V_s reduces both the current and voltage amplitudes considerably [6.5, 6].

The above analysis assumes a single-valued $V_c(I_c)$ curve and hence a uniform current density. This is a good approximation, as long as the circuit response is the dominant factor in controlling the formation and quenching of current filaments. Similarly, in case of NNDC, our analysis is a good approximation as long as the field is uniform, and the oscillatory behavior is dominated by the circuit and not by the transit of high-field domains. The above analysis can also be used to define competing limits of circuit and domain or filament domination.

6.2.2 Impact-Ionization Induced Oscillations

We shall now analyse a simple mechanism for intrinsic, i.e., not circuit-induced, limit-cycle oscillations of the field and the conduction current density in the bulk of the material [6.100, 101]. To this end we will neglect for the moment all reactive circuit components, but instead take into account explicitly the dynamic degrees of freedom of the carrier densities. General conditions for g-r induced oscillatory instabilities in single-carrier semiconductors with M trap levels have been derived in Chap. 3. In particular, the following necessary and sufficient condition for the Hopf bifurcation of a spatially homogeneous limit cycle in a model with two trap levels (ground and excited state of an impurity) was found:

$$v(\varepsilon_0)(K_1 \partial\varphi_{t_1}/\partial\varepsilon + K_2 \partial\varphi_{t_2}/\partial\varepsilon) + (\lambda_1 + \lambda_2)(\tilde{v} - \lambda_1)(\tilde{v} - \lambda_2) = 0 \; , \qquad (6.2.21)$$

where $v(\varepsilon_0)$ is the drift velocity, φ_{t_1} and φ_{t_2} are g-r rates (3.1.12), and $\lambda_1 > \lambda_2$ are the eigenvalues of the 2×2 g-r matrix $\tilde{A}$, defined in (3.1.29), and

$$\tilde{v} := v \, dv/d\varepsilon \qquad (6.2.22)$$

$$K_1 := \tilde{v} - \tilde{A}_{11} - \tilde{A}_{21} \qquad (6.2.23)$$

$$K_2 := \tilde{v} - \tilde{A}_{22} - \tilde{A}_{12} \; , \qquad (6.2.24)$$

all evaluated at the homogeneous steady state. The quantities $\tilde{v}$, K_1, K_2 are positive, and $\partial\varphi_{t_1}/\partial\varepsilon$, $\partial\varphi_{t_2}/\partial\varepsilon$ are negative, if a positive differential mobility $dv/d\varepsilon$ and standard g-r kinetics, with impact-ionization coefficients monotonically increasing with field, and trapping coefficients nonincreasing with field, are assumed. As we have shown in Chap. 3, condition (6.2.21) singles out high-purity relaxation semiconductors of SNDC type, with low electron concentration v on the NDC branch of the static current-field characteristic, and with small differential mobility. Thus the time-scale given by the dielectric relaxation time is much slower than in the conventional case of a lifetime semiconductor, and low-frequency oscillations are expected.

In the following the limit cycles will be investigated in the nonlinear regime beyond the Hopf bifurcation point ε_0, given by (6.2.21), for the two-level model. This is in a sense the simplest model which exhibits spatially homogeneous impact-ionization induced periodic oscillations. We have shown in Sect. 3.5 by a linear analysis that homogeneous (i.e., with wavevector $k_\parallel = 0$) self-sustained oscillations are not possible in a one-carrier g-r model with a single impurity level and positive differential mobility. The simplest mechanism which allows for such oscillations, involves (at least) two impurity levels. The nonlinear transport equations (3.1.8–10) can in this case be reduced to a system of three nonlinear ordinary differential equations. The assumption of a spatially homogeneous electric field reduces Poisson's equation (3.1.10) to the condition of local neutrality

$$0 = 1 - v - v_{t_1} - v_{t_2} \ . \tag{6.2.25}$$

This, together with the assumption of homogeneous carrier densities, reduces the equation for the total longitudinal current density J (3.4.1a) and the continuity equations (3.1.8, 9) to

$$\dot{v} = \psi_0(v, v_{t_1}, \varepsilon) \tag{6.2.26}$$

$$\dot{v}_{t_1} = \psi_1(v, v_{t_1}, \varepsilon) \tag{6.2.27}$$

$$\dot{\varepsilon} = J - vv(\varepsilon) \ , \tag{6.2.28}$$

where ε is the longitudinal time-dependent electric field and the g-r rates are

$$\psi_0(v, v_{t_1}, \varepsilon) := \varphi_0(v, v_{t_1}, v_{t_2} = 1 - v - v_{t_1}, \varepsilon) \tag{6.2.29a}$$

$$\psi_1(v, v_{t_1}, \varepsilon) := \varphi_{t_1}(v, v_{t_1}, v_{t_2} = 1 - v - v_{t_1}, \varepsilon) \tag{6.2.29b}$$

using local neutrality (6.2.25). Note that v_{t_2} is now a dependent variable. The left-hand side of (6.2.28) is the displacement current density which relaxes very fast to zero for standard lifetime semiconductors. However, for relaxation semiconductors – and the Hopf bifurcation condition actually requires this less conventional type of semiconductor, as we have shown in Sect. 3.5 – the displacement current is comparable to the other current contributions, and must not be neglected.

For a time-independent external current density J (6.2.26–28) represent an autonomous dynamic system, whose steady state $(\varepsilon_0, v, v_{t_1})$ is determined by

$$\psi_0(v, v_{t_1}, \varepsilon_0) = \psi_1(v, v_{t_1}, \varepsilon_0) = 0 \tag{6.2.30}$$

and

$$J = vv(\varepsilon_0) \ . \tag{6.2.31}$$

The latter equation represents the static current density-field characteristic, and indicates that either the current density J or the static field, or a combination of both in case of a loaded circuit, may be chosen as the external control parameter. In Sect. 2.1.2 the steady states (6.2.30) of the two-level model were analyzed in detail.

It is instructive to investigate the stability of this steady state against charge-neutral homogeneous fluctuations directly from the spatially homogeneous equations (6.2.26–28), and thereby re-derive the condition (6.2.21) for a Hopf bifurcation. This gives additional information about the topology and the eigenvectors of the steady state in the three-dimensional phase space of $(v(t), v_{t_1}(t), \varepsilon(t))$. The linearization of (6.2.26–28) around $(v, v_{t_1}, \varepsilon_0)$ for small fluctuations $(\delta v, \delta v_{t_1}, \delta \varepsilon)$ yields:

$$
\begin{pmatrix} \delta \dot{v} \\ \delta \dot{v}_{t_1} \\ \delta \dot{\varepsilon} \end{pmatrix} = \begin{pmatrix} A_{11} & A_{12} & \dfrac{\partial \varphi_0}{\partial \varepsilon} \\ A_{21} & A_{22} & \dfrac{\partial \varphi_{t_1}}{\partial \varepsilon} \\ -v(\varepsilon_0) & 0 & -\tilde{v} \end{pmatrix} \begin{pmatrix} \delta v \\ \delta v_{t_1} \\ \delta \varepsilon \end{pmatrix}, \tag{6.2.32}
$$

where

$$
(A_{ij}) := \begin{pmatrix} \dfrac{\partial \psi_0}{\partial v} & \dfrac{\partial \psi_0}{\partial v_{t_1}} \\ \dfrac{\partial \psi_1}{\partial v} & \dfrac{\partial \psi_1}{\partial v_{t_1}} \end{pmatrix} \tag{6.2.33}
$$

is the g-r stability matrix introduced in Sect. 2.1 for charge-neutral fluctuations of δv and δv_{t_1} alone. It is connected with the g-r matrix $\tilde{A}_{ij} := B_{ij} - d_i$ defined in (3.1.29), which describes charge-neutral fluctuations of δv_{t_1} and $\delta v_{t_2} = -(\delta v + \delta v_{t_1})$, in the following way:

$$
A_{11} = \tilde{A}_{12} + \tilde{A}_{22}
$$

$$
A_{12} = -\tilde{A}_{11} + \tilde{A}_{12} - \tilde{A}_{21} + \tilde{A}_{22}
$$

$$
A_{21} = -\tilde{A}_{12}
$$

$$
A_{22} = \tilde{A}_{11} - \tilde{A}_{12} \ . \tag{6.2.34}
$$

Note that $\det\{A\} = \det\{\tilde{A}\} = \lambda_1 \lambda_2$ and $\operatorname{tr}\{A\} = \operatorname{tr}\{\tilde{A}\} = \lambda_1 + \lambda_2$ holds. The eigenvalue equation resulting from (6.2.32) with $(\delta v, \delta v_{t_1}, \delta \varepsilon) \sim \exp(\lambda t)$ is

$$
\lambda^3 + \lambda^2 g_2 + \lambda g_1 + g_0 = 0 \ . \tag{6.2.35}
$$

Here

$$
g_0 = \alpha_2 v(\varepsilon_0) + \tilde{v} \det\{A\} \tag{6.2.36a}
$$

$$
g_1 = \alpha_1 v(\varepsilon_0) + \det\{A\} - \tilde{v} \operatorname{tr}\{A\} \tag{6.2.36b}
$$

$$g_2 = \tilde{v} - \text{tr}\{A\} \qquad \text{with} \tag{6.2.36c}$$

$$\alpha_1 := \frac{\partial \varphi_0}{\partial \varepsilon} = -\left(\frac{\partial \varphi_{t_1}}{\partial \varepsilon} + \frac{\partial \varphi_{t_2}}{\partial \varepsilon}\right) \qquad \text{and} \tag{6.2.37}$$

$$\alpha_2 := \frac{\partial \varphi_{t_1}}{\partial \varepsilon} A_{12} - \frac{\partial \varphi_0}{\partial \varepsilon} A_{22} = \frac{\partial \varphi_{t_1}}{\partial \varepsilon}(\tilde{A}_{22} - \tilde{A}_{21}) + \frac{\partial \varphi_{t_2}}{\partial \varepsilon}(\tilde{A}_{11} - \tilde{A}_{12}) \tag{6.2.38}$$

which agrees with (3.5.2–5). The eigenvectors of (6.2.32) are given by

$$\begin{pmatrix} \delta v \\ \delta v_{t_1} \\ \delta \varepsilon \end{pmatrix} \sim \begin{pmatrix} -(\lambda + \tilde{v}) \\ \left[\frac{\partial \varphi_{t_1}}{\partial \varepsilon} v(\varepsilon_0) - A_{21}(\lambda + \tilde{v})\right] \Big/ (\lambda - A_{22}) \\ v(\varepsilon_0) \end{pmatrix}. \tag{6.2.39}$$

In the following we shall assume

(i)   standard mass-action g-r kinetics, with impact-ionization coefficients mono-tonically increasing with field, and trapping coefficients decreasing or constant with increasing field,
(ii) a positive differential mobility $dv/d\varepsilon$
(iii) a standard (i.e., not tilted, as in Sect. 3.3) SNDC characteristic.

With this $\text{tr}\{A\} < 0$, $\alpha_1 > 0$, $g_0 > 0$, $g_2 > 0$ holds, as was shown in Sect. 3.5. Additionally, $\det\{A\} < 0$ holds on the NDC branch of the $j_0 - \varepsilon_0$ characteristic, and $\det\{A\} > 0$ on the positive differential conductivity branch. The character of the singular point $(v, v_{t_1}, \varepsilon_0)$ in phase space can then be classified as follows. If

$$a := g_2^2 - 3g_1 = \tilde{v}^2 + (\lambda_1 - \lambda_2)^2 + \lambda_1 \lambda_2 + \tilde{v}(\lambda_1 + \lambda_2) - 3\alpha_1 v(\varepsilon)$$

$$= (\tilde{v} + \lambda_1)(\tilde{v} + \lambda_2) + (\lambda_1 - \lambda_2)^2 - 3\alpha_1 v(\varepsilon) \tag{6.2.40}$$

is positive, then the eigenvalue equation (6.2.35) may have three real, or one real and two complex roots. In the first case it follows from the sign of the coefficients in (6.2.35) that the singular point is a stable node (all eigenvalues negative) if it belongs to positive differential conductivity ($g_0, g_1, g_2 > 0$), but might be a saddle point (with one negative and two positive eigenvalues) if it has NDC.

If $a \leq 0$, then two of the eigenvalues are always complex, and one is real. We set these eigenvalues

$$\lambda^{(1)} = \lambda_\parallel$$

$$\lambda^{(2,3)} = \lambda_0 \pm i\omega \tag{6.2.41}$$

with real $\lambda_\parallel$, λ_0, ω. The eigenvalue equation can be written as

$$(\lambda - \lambda^{(1)})(\lambda - \lambda^{(2)})(\lambda - \lambda^{(3)}) = 0 . \tag{6.2.42}$$

Comparison with (6.2.35) yields

$$g_0 = -\lambda_\parallel(\lambda_0^2 + \omega^2) \tag{6.2.43a}$$

$$g_1 = \lambda_0^2 + \omega^2 + 2\lambda_0\lambda_\parallel \tag{6.2.43b}$$

$$g_2 = -\lambda_\parallel - 2\lambda_0 \ . \tag{6.2.43c}$$

From (6.2.43a) it follows with $g_0 > 0$ that always

$$\lambda_\parallel < 0 \ . \tag{6.2.44}$$

The singular point is therefore associated with a flow which combines attracting motion along the eigenvector direction corresponding to $\lambda^{(1)}$ with a spiral-type motion in the plane corresponding to the eigenvalues $\lambda^{(2)}$, $\lambda^{(3)}$. The spirals wind inward if $\lambda_0 < 0$, and outward if $\lambda_0 > 0$, corresponding to a stable or unstable focus, respectively. This alternative can be decided from inspection of the expression

$$g_1 g_2 - g_0 = -2\lambda_0[(\lambda_0 + \lambda_\parallel)^2 + \omega^2] \ , \tag{6.2.45}$$

by (6.2.43). For $g_1 g_2 > g_0$ it follows that $\lambda_0 < 0$, whence the saddle-focus is stable. For $g_1 g_2 < g_0$ the saddle-focus is unstable. The focus changes from stable to unstable when λ_0 changes sign from negative to positive, i.e., when the pair of complex-conjugate eigenvalues $\lambda^{(1)}$, $\lambda^{(2)}$ crosses the imaginary axis. This gives the condition for a Hopf bifurcation of a limit cycle:

$$g_1 g_2 - g_0 = 0 \ . \tag{6.2.46}$$

The eigenvalues at the Hopf bifurcation point are

$$\lambda^{(1)} = -g_2 \tag{6.2.47a}$$

$$\lambda^{(2,3)} = \pm i\sqrt{g_1} \tag{6.2.47b}$$

Inserting (6.2.36) into (6.2.46), using (6.2.37, 38), yields

$$0 = g_1 g_2 - g_0$$
$$= (\tilde{v} - \text{tr}\{A\})[\alpha_1 v(\varepsilon_0) - \tilde{v}\,\text{tr}\{A\}] - \alpha_2 v(\varepsilon_0) - \text{tr}\{A\} \cdot \det\{A\}$$
$$= v(\varepsilon_0)\left[-\frac{\partial\varphi_{t_1}}{\partial\varepsilon}(\tilde{v} - A_{11} + A_{12}) - \frac{\partial\varphi_{t_2}}{\partial\varepsilon}(\tilde{v} - A_{11})\right]$$
$$- (\lambda_1 + \lambda_2)(\tilde{v} - \lambda_1)(\tilde{v} - \lambda_2) \ . \tag{6.2.48}$$

Observing $K_1 \equiv \tilde{v} - A_{11} + A_{12}$ and $K_2 \equiv \tilde{v} - A_{11}$ by (6.2.34), we can see that (6.2.48) is identical with the condition (6.2.21). We may consider the steady state field ε_0, or equivalently, the external static current density $J = vv(\varepsilon_0)$ as a control parameter. Equation (6.2.48) determines the Hopf bifurcation point ε_0^H. For values of ε_0 such that $g_1 g_2 - g_0 < 0$, a stable limit cycle is expected to exist around the unstable saddle-focus. Near the bifurcation point (where $\lambda_0 = 0$) its angular frequency of oscillation is approximately given by

$$\omega = \sqrt{g_1} = \left[v(\varepsilon_0)\frac{\partial \varphi_0}{\partial \varepsilon} + \det\{A\} - \tilde{v}\,\mathrm{tr}\{A\} \right]^{1/2} \tag{6.2.49}$$

and its amplitude in phase space grows proportionally to $(g_0 - g_1 g_2)^{1/2}$. The approach of the trajectories to the surface in which the limit cycle is embedded may be very roughly approximated by an exponential decay

$$\sim \exp(-g_2 t) = \exp(-\tilde{v} + \lambda_1 + \lambda_2)t \tag{6.2.50}$$

along the eigenvector direction associated with $\lambda^{(1)}$. This decay is dominated in relaxation semiconductors by the large negative g-r eigenvalue λ_2, i.e., $\sim \exp(-|\lambda_2|t)$.

For the simple two-step impact-ionization mechanism studied in Sect. 2.1.2, and a choice of typical numerical parameters guided by the Hopf bifurcation condition (6.2.21), we shall now compute the phase space trajectories of the nonlinear system (6.2.26–28). The g-r rates (6.2.29) are given by

$$\psi_0/\tau_M = X_1 N_D^* v v_{t_1} + (X_1^S + X^* N_D^* v) v_{t_2} - T_1^S N_D^* p_D v \tag{6.2.51a}$$

$$\psi_1/\tau_M = -(X^* + X_1 N_D^* v) v_{t_1} + T^* v_{t_2} \tag{6.2.51b}$$

with $v_{t_2} = 1 - v - v_{t_1}$, and $p_D = N_A/N_D^* + v$. For a definition of the g-r coefficients see Fig. 2.4. The impact-ionization coefficients of ground and excited state ionization are approximated by the Shockley formula

$$X_1 = X_1^0 \exp(-6E_t/\varepsilon)$$

$$X_1^* = X_1^{*0} \exp(-1.5E_t/\varepsilon) \tag{6.2.52}$$

with a normalized impurity ground state energy E_t. The other g-r coefficients are in the simplest approximation assumed to be independent of field. The drift velocity is modeled by the empirical saturable from [6.34]

$$v(\varepsilon) = \frac{(\arctan r_2 \varepsilon)}{r_2} \tag{6.2.53}$$

which increases linearly with field for small ε and saturates at $v(\varepsilon) = \pi/(2r_2)$ for large ε. Here r_2 is a dimensionless saturation parameter $r_2 := (\pi \mu_0 kT)/(2v_s eL_D)$, where μ_0 is the low-field mobility, and v_s is the saturation velocity in physical units. With these definitions the differential mobility is

$$dv/d\varepsilon = [1 + (r_2 \varepsilon)^2]^{-1} . \tag{6.2.54}$$

The g-r matrix $\tilde{A}$ and the expressions for α_1, α_2 have been given in Table 3.1. In particular, if the field dependence of the g-r coefficients occurs merely through impact ionization,

$$\frac{\partial \varphi_{t_1}}{\partial \varepsilon} = -\tau_M \frac{\partial X_1}{\partial \varepsilon} N_D^* v v_{t_1} < 0 \tag{6.2.55a}$$

$$\frac{\partial \varphi_{t_2}}{\partial \varepsilon} = -\tau_M \frac{\partial X_1^*}{\partial \varepsilon} N_D^* v v_{t_2} < 0 . \tag{6.2.55b}$$

From (6.2.51a) it follows in the steady state that

$$-X_1 N_D^* v_{t_1} - X_1^* N_D^* v_{t_2} + T_1^S N_D^* p_D = X_1^S \frac{v_{t_2}}{v} > 0 \ .$$

(6.2.56)

This can be used to simplify the quantities (6.2.23, 24) obtained from Table 3.1:

$$K_1 = \tilde{v} - \tilde{A}_{11} - \tilde{A}_{21}$$

$$= \tilde{v} + [X_1 N_D^*(v - v_{t_1}) - X_1^* N_D^* v_{t_2} + T_1^S N_D^*(v + p_D)]\tau$$

$$= \tilde{v} + \left(X_1 N_D^* v + T_1^S N_D^* v + X_1^S \frac{v_{t_2}}{v} \right) \tau_M$$

$$K_2 = \tilde{v} - \tilde{A}_{22} - \tilde{A}_{12}$$

$$= \tilde{v} + [X_1^S + T_1^S N_D^*(v + p_D) - X_1 N_D^* v_{t_1} - X_1^* N_D^*(v_{t_2} - v)]\tau$$

$$= \tilde{v} + \left(X_1^S + T_1^S N_D^* v + X_1^* N_D^* v + X_1^S \frac{v_{t_2}}{v} \right) \tau_M \ .$$

(6.2.57)

This clearly shows that $K_1, K_2 > 0$. The condition $g_1 \cdot g_2 - g_0 < 0$ for the existence of a limit cycle requires, by (6.2.48), that the positive, square-bracketed term is overcompensated by a sufficiently large $(\lambda_1 + \lambda_2)(\tilde{v} - \lambda_1)(\tilde{v} - \lambda_2) > 0$. Since $\lambda_1 + \lambda_2 = \text{tr}\{A\} < 0$, and $\lambda_2 < 0$ always hold, this requires a small v and small $dv/d\varepsilon$ such that

$$\tilde{v} - \lambda_1 < 0$$

(6.2.58)

holds. Thus we have explicitly proved for this mechanism the necessary condition for oscillatory instabilities that was mentioned in Chap. 3, and which singles out relaxation semiconductors of high purity as candidates for spontaneous oscillations. Furthermore, small $\partial X_1/\partial \varepsilon$ and $\partial X_1^*/\partial \varepsilon$ are desirable, by (6.2.48). All these restrictions guide the choice of the numerical parameters for our model: a sufficiently large field ε_0 and small generation coefficients X^*, X_1^S should be chosen.

For an appropriate set of numerical parameters, the real part of the complex eigenvalues of (6.2.35) is plotted as a function of the static field ε_0 in Fig. 6.11a. The solution of the full longitudinal dispersion relation, including diffusion and drift, is shown in Fig. 6.11b.

As ε_0 is increased, $\text{Re}\{\lambda\} = \lambda_0$ changes from negative to positive values at a critical field $\varepsilon_c = 173.675$, as given by the Hopf bifurcation condition (6.2.48). At ε_c a limit cycle bifurcates from the NDC state. This limit cycle is shown as a numerical solution of (6.2.26–28) in Figs. 6.12, 13. Close to the Hopf bifurcation point (Fig. 6.12) the amplitude is small and the oscillation is almost harmonic (sinusoidal), while further away the amplitude increases, and the oscillation becomes strongly non-linear (Fig. 6.13). The transient phase trajectories starting near the unstable NDC steady state $(v, v_{t_1}, \varepsilon_0)$ are shown for a series of increasing fields in Fig. 6.14. The projections of the three-dimensional trajectories onto the $v - \varepsilon$ (left column), $v - v_{t_1}$

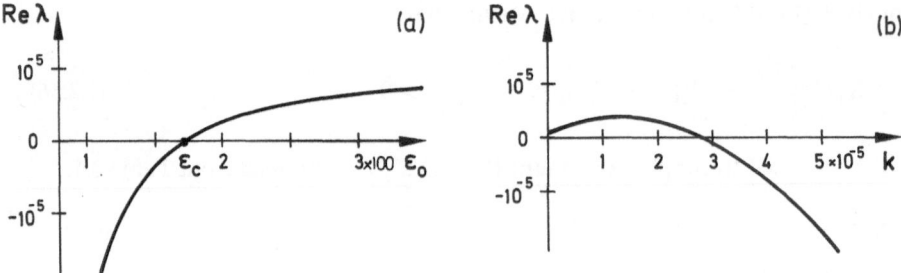

Fig. 6.11. (a) Real part of the eigenvalue λ describing small longitudinal fluctuations from the NDC state as a function of the control parameter ε_0, for the two-level mechanism of Sect. 2.1.2. Re$\{\lambda\}$ is in units of $1/\tau_M$, and ε_0 is in units of kTE_t/eL_D. The numerical parameters are $T_1^S N_D^* = 10^{-2}/\tau_M$, $T^* = 10^{-5}/\tau_M$, $X_1^S = X^* = 5 \times 10^{-6}/\tau_M$, $X_1^0 N_D^* = 5 \times 10^{-4}/\tau_M$, $X_1^{*0} N_D^* = 10^{-2}/\tau_M$, $E_t = 1$, $N_A/N_D^* = 0.5$, $r_2 = 0.3$. (b) Re$\{\lambda\}$ in units of $1/\tau_M$ versus k in units of $1/L_D$ for $\varepsilon_0 = 180$ $[kTE_t/eL_D]$ and the same parameters as in (a)

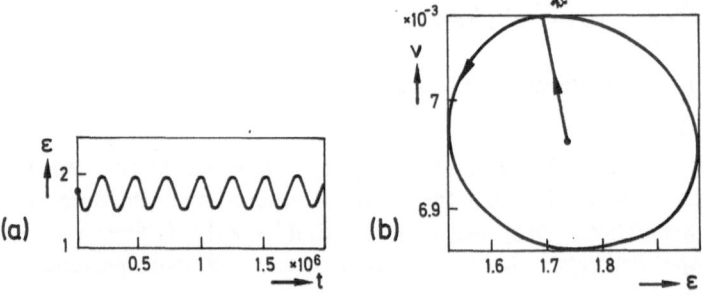

Fig. 6.12a, b. Limit cycle for $\varepsilon_0 = 174$ (corresponding to $J = 3.60 \times 10^{-4}$) and the same parameters as in Fig. 6.11. (a) Electric field ε in units $100 \times kTE_t/(eL_D)$ versus time in units τ_M. (b) Phase portrait of electron concentration v in units N_D^* versus ε in units $100 \times kTE_t/(eL_D)$

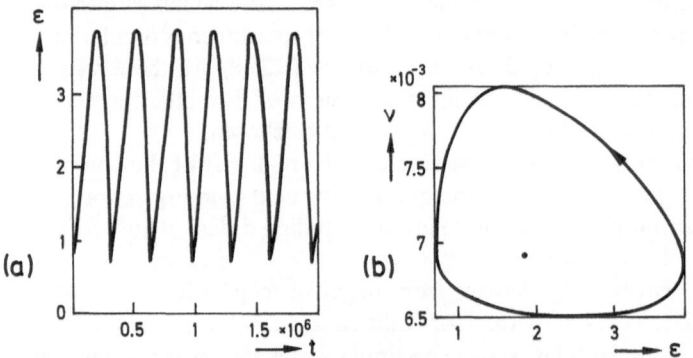

Fig. 6.13a, b. Same as Fig. 6.12 for $\varepsilon_0 = 190$ (corresponding to $J = 3.56 \times 10^{-4}$)

Fig. 6.14a–f. Projections of the transient phase trajectories onto the phase planes of free electron density v versus field ε (*left column*), free electron density v versus trapped electron density in the ground state v_{t_1} (*center column*), and v_{t_1} versus ε (*right column*) for the numerical parameters of Fig. 6.11; v and v_{t_1} are in units N_D^*, ε is in units $kTE_t/(eL_D)$. The control parameter ε_0 is (a) 150, (b) 174, (c) 175, (d) 180, (e) 190, (f) 200

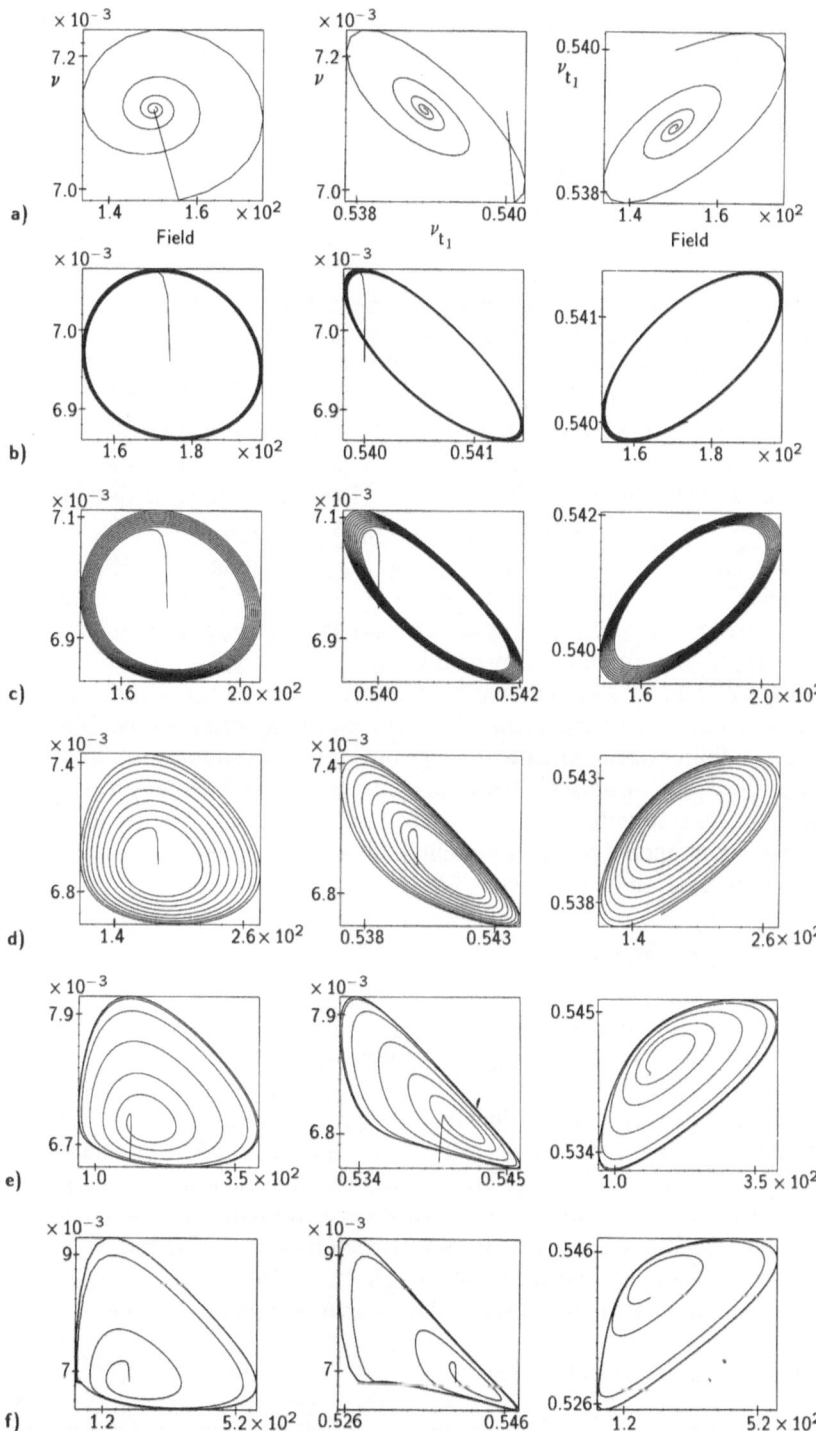

Fig. 6.14a–f. Caption see opposite page

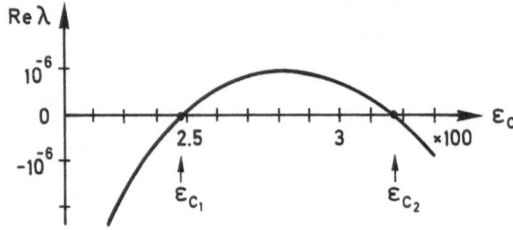

Fig. 6.15. Same as Fig. 6.11a for $X_1^0 N_D^* = 6 \times 10^{-4}/\tau_M$. All other numerical parameters are the same as in Fig. 6.11a

(center column), and $v_{t_1} - \varepsilon$ (right column) planes are plotted. Fig. 6.14a corresponds to $\varepsilon_0 < \varepsilon_c$, where the steady state is a stable focus, and the trajectories spiral inward, while Figs. 6.14b–f depict limit cycles around unstable focuses. It is characteristic that the electron density v initially changes much more rapidly than v_{t_1} and ε, until the surface of slow oscillatory motion is reached. This reflects the slaving of the electrons by the slow dielectric-relaxation oscillations of ε and v_{t_1}. There is a clear separation of time-scales between the fast decay of v toward the "slow manifold", determined essentially by the g-r eigenvalue $\lambda_{\parallel} \approx -|\lambda_2|$ according to (6.2.50), and the low-frequency oscillations, characterized by $\omega \approx \sqrt{g_1}$ according to (6.2.49). The approach of trajectories to the limit cycle depends sensitively upon the initial condition, and small deviations from the cycle can result in large excursions in phase space due to the rapid motion off the slow manifold.

For a slightly different set of numerical parameters, Re$\{\lambda\}$ versus ε_0 and some associated phase portraits are shown in Figs. 6.15 and 6.16, respectively. With increasing field ε_0, a limit cycle bifurcates at ε_{c_1}, grows up to a maximum amplitude, and then shrinks again, and disappears in a second (inverted) Hopf bifurcation at ε_{c_2}.

When the dynamic system (6.2.26–28) is biased far enough into the nonlinear oscillatory regime, then aperiodic chaotic oscillations may spontaneously appear. This phenomenon will be studied in Sect. 6.3.4.

6.2.3 Exciton-Induced Oscillations

In the preceding section the main autocatalytic process which drives the oscillatory instability was impact ionization of electrons. Another autocatalytic process is the stimulated production of bosons. The Bose distribution law implies that the generation rate of bosons of density x due to incident energy has the form $A + Bx$ with spontaneous and stimulated generation coefficients A and B, which is familiar in the case of photons. Excitons are also of boson character at not too high densities. Hence stimulated exciton production should also exist, though it has not been observed so far, and its importance is uncertain. Nonequilibrium phase transitions based upon this mechanism have been mentioned in Sect. 2.3. Here we shall consider a mechanism for stimulated exciton-induced limit-cycle oscillations proposed by *Pimpale* et al. [6.35].

The model consists of the following processes:

(i) Photogeneration of electron-hole (e-h) pairs $\gamma \rightarrow e + h$ at a rate Y^S, which can be controlled by the intensity of laser light (γ) at appropriate frequency.

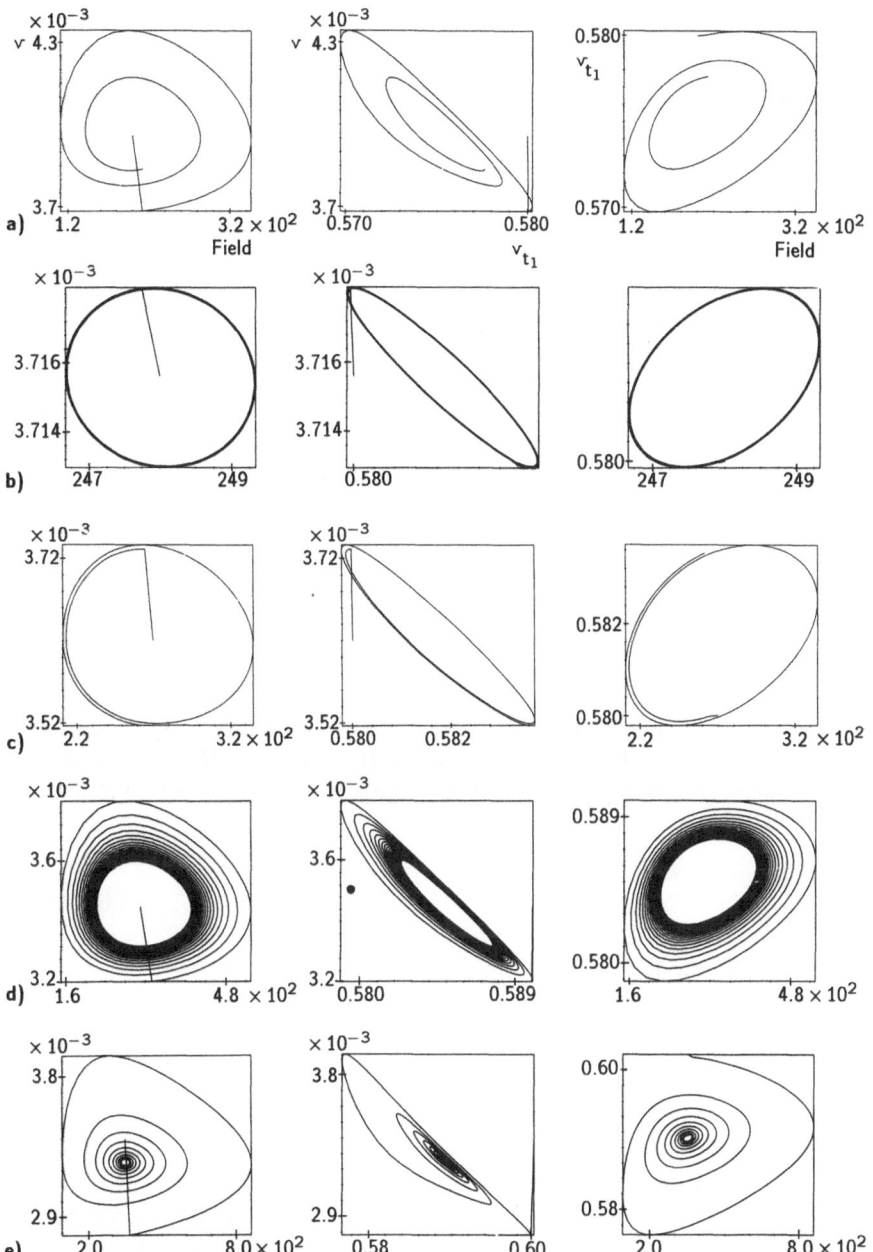

Fig. 6.16a–e. Same as Fig. 6.14 for the numerical parameters of Fig. 6.15. The control parameter ε_0 is (a) 200, (b) 248, (c) 270, (d) 310, (e) 350

(ii) Stimulated production of excitons (X) $e + h + X \rightarrow 2X$ of rate En^2x.

(iii) Radiative decay of excitons at a recombination center $X \rightarrow \gamma$ of rate $kx/(1 + qx)$.

Here the semiconductor has been assumed nondegenerate and under high-excitation conditions so that electron and hole concentrations may be taken to be equal ($= n$). The excitonic decay process (iii) has a rate $\sim kx$ for small exciton densities x, but saturates at a rate k/q for large x due to complete filling of the recombination centers. This decay law, which is similar to the Michaelis-Menten law in chemical kinetics, can be derived explicitly from the kinetics of bound excitons [6.36]: Assume the presence of N_t impurity centers per unit volume, n_x of which have captured an exciton. The concentration of bound excitons is then governed by the rate equation

$$\dot{n}_x = k'n_t x - k''n_x \; , \tag{6.2.59}$$

where k' and k'' are the rate coefficients for capture of a free exciton by a center, and for radiative decay of bound excitons, respectively, and

$$n_t = N_t - n_x \tag{6.2.60}$$

is the concentration of impurity centers which have not captured an exciton. The recombination rate of free excitons is given by

$$\dot{x} = -k'n_t x \; . \tag{6.2.61}$$

Assuming that the bound excitons relax much faster to a pseudosteady state than the free excitons and the electrons (this is another variant of *Haken*'s "slaving principle" for adiabatic elimination of fast variables), we can set $\dot{n}_x = 0$ in (6.2.59). This gives, with (6.2.60)

$$n_t = \frac{N_t}{1 + (k'/k'')x} \; . \tag{6.2.62}$$

Substituting (6.2.62) into (6.2.61), we obtain

$$\dot{x} = -\frac{k'N_t x}{1 + (k'/k'')x} \; , \tag{6.2.63}$$

which represents the Michaelis-Menten decay law (iii) with $k \equiv k'N_t$, $q \equiv k'/k''$.

For simplicity all recombination processes, except (i)–(iii), are ignored. These include various electron-hole and electron (or hole) – impurity transitions as well as excitonic processes such as the formation of excitonic molecules and electron-hole droplets. The reverse processes of (i)–(iii) are ignored since the semiconductor is assumed to be far from equilibrium.

The basic rate equations for electrons and free excitons subjected to processes (i)–(iii) are

$$\dot{n} = G - Cn^2x \tag{6.2.64}$$

$$\dot{x} = Cn^2x - x/(1 + x) \; , \tag{6.2.65}$$

where n and x have been scaled in units of q^{-1}, the time has been scaled in units of k^{-1}, and

$$G := Y^S q/k \tag{6.2.66}$$

$$C := E/(q^2 k) \ . \tag{6.2.67}$$

The steady state values of n and x are solutions of

$$G = Cn^2 x = x/(1 + x) \tag{6.2.68}$$

which yields

$$x_0 = G/(1 - G) \tag{6.2.69}$$

$$n_0 = [(1 - G)/C]^{1/2} \ , \tag{6.2.70}$$

where G must be confined to the range

$$0 < G < 1 \ . \tag{6.2.71}$$

To analyse the stability of the steady state (x_0, n_0), we put

$$x(t) = x_0 + \delta x \ , \qquad n(t) = n_0 + \delta n \tag{6.2.72}$$

and linearize (6.2.64, 65) around (x_0, n_0):

$$\begin{pmatrix} \delta \dot{x} \\ \delta \dot{n} \end{pmatrix} = A \begin{pmatrix} \delta x \\ \delta n \end{pmatrix} \tag{6.2.73}$$

with

$$A = \begin{pmatrix} G(1 - G) & 2G[C/(1 - G)]^{1/2} \\ -(1 - G) & -2G[C/(1 - G)]^{1/2} \end{pmatrix} \ . \tag{6.2.74}$$

From (6.2.74) it follows

$$\det\{A\} = 2G(1 - G)^{3/2} C^{1/2} > 0 \tag{6.2.75}$$

$$\text{tr}\{A\} = G\{(1 - G) - 2[C/(1 - G)]^{1/2}\} \ . \tag{6.2.76}$$

Since $\det\{A\}$ is always positive for physically meaningful solutions, the steady state changes from a stable focus to an unstable focus surrounded by a stable limit cycle when $\text{tr}\{A\}$ changes from negative to positive values. The Hopf bifurcation of a limit cycle occurs at

$$\text{tr}\{A\} = 0 \ , \tag{6.2.77}$$

hence

$$(1 - G)^3 = 4C \ . \tag{6.2.78}$$

For

$$G < 1 - (4C)^{1/3} \tag{6.2.79}$$

a stable limit cycle exists. An approximate analytical form of this cycle was obtained by *Pimpale* et al. [6.35].

The limit cycle would manifest itself experimentally in the form of oscillating exciton and electron-hole recombination light; and, if a small voltage is applied to the semiconductor, in the form of an oscillating current. Although such an oscillating current induced by bound exciton recombination has not yet been observed, *Gross* et al. [6.37] have reported oscillating exciton recombination light in CdS at $4.2 - 77$ K. Typical numerical parameters for CdS are a photon absorption rate of 10^{18} cm^{-3} s^{-1}, an exciton lifetime of the order of 10^{-6} s, and $N_t \approx 10^{15}$ cm^{-3}, yielding $G \approx 10^{-3}$. The angular frequency of the limit cycle can then be expected to be of the order of 1 μs^{-1}, and the amplitudes of oscillation of n and x lie within several percent of the steady state values.

6.3 Chaos

Examples of chaotic time-dependent behavior occur in a great variety of disciplines [6.48–56] and there is also considerable literature on various mathematical aspects [6.57–70]. Among these examples are turbulence in hydrodynamic systems, chemical reaction systems, lasers at very high pumping, optically bistable elements, Josephson junctions, nonlinear electronic circuits, semiconductors. Chaos in semiconductors can arise in different ways: either due to a reactive external circuit, or due to an intrinsic instability of the NDC element. In principle, all five classes of mechanisms surveyed in Sect. 6.1.1 can give rise to chaotic oscillations.

To the first class belongs chaos induced by the coupling of a negative differential conductivity element with inductors and capacitors in an external circuit, as we have considered in Sect. 6.2.1. If a periodic driving voltage is applied to the circuit, it can be modeled by a forced Van der Pol oscillator, which is well known to exhibit chaos if driven sufficiently hard [6.71].

The other classes include chaos associated with domain propagation in NNDC devices, namely Gunn diodes [6.72], as well as intrinsic chaos associated with helicon instabilities in semiconductor plasmas [6.73–76] or with generation-recombination processes [6.77–105]. Chaos in semiconductors induced by irradiation, electric, or magnetic fields, or a combination of these, represents one of the most recent examples of chaotic behavior in physical systems, and is not yet fully understood at the present time. In the following we shall focus on these recent experimental findings and the attempts at their theoretical explanation.

6.3.1 Routes to Chaos

First, we shall clarify some of the basic notions and concepts of chaos. In two-variable autonomous dynamic systems, such as we have mainly encountered so far, the field of directions of the phase portrait is uniquely defined, and therefore trajectories cannot cross. All trajectories in a bounded system must tend asymptotically to a singular point, a limit cycle, or similar limit continuum. Now, we shall consider nonautonomous (that is, explicitly time-dependent) two-variable systems, or autonomous systems involving more than two variables. The latter are more general since nonautonomous systems

$$\dot{q} = F(q, t) \tag{6.3.1}$$

can always be represented as autonomous systems with the time as an additional dynamic variable:

$$\dot{q} = F(q, \theta)$$

$$\dot{\theta} = 1 . \tag{6.3.2}$$

These systems may have completely irregular trajectories in phase space which do not tend asymptotically to a singular point or a limit cycle, i.e., an attractor of low, integer dimension. Such irregular nonperiodic motion of a deterministic system with few degrees of freedom, which depends very sensitively upon the initial conditions, is called *chaotic*. It is distinct from the irregularities caused by stochastic fluctuations in systems with many microscopic degrees of freedom.

The third variable adds a whole wealth of qualitatively new phenomena of the dynamic system, in addition to the singular points and limit cycles familiar from two-variable autonomous systems, where the trajectories in the phase plane can never cross. In three dimensions, however, the trajectories can be wrapped or twisted around each other in quite intricate ways. Consider, for example, the combination of a two-dimensional unstable focus with a "bistable slow manifold" folded up in the third dimension as shown in Fig. 6.17. The "slow manifold" is a surface in which the phase point moves slowly, whereas in the rest of the phase space the motion is fast. Thus, when a trajectory reaches the edge of the upper sheet of the slow manifold, it instantly jumps down to the lower sheet, and when it reaches the edge of the lower sheet, it jumps up to the upper sheet. The re-injection into the upper sheet depends sensitively on its position before the jump which explains intuitively the origin of the seemingly random jumps. This produces the complicated entangling of trajectories which is characteristic of chaotic motion. Such *spiral-type chaos* represents the

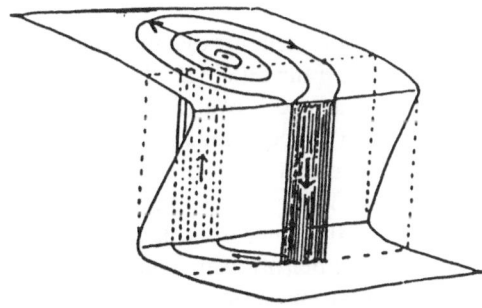

Fig. 6.17. Chaotic motion produced by the combination of a slow bistable manifold with an unstable focus (Spiral-type chaos). [6.58]

simplest form of chaotic behavior in continuous dynamic systems. Its prototype is the Rössler attractor [6.57] which is described by a three-variable system with a single nonlinear term (zx):

$$\dot{x} = -(y + z)$$

$$\dot{y} = x + 0.2y$$

$$\dot{z} = 0.2 + z(x - 5.7) \ . \tag{6.3.3}$$

This system describes a so-called *strange attractor*.

A strange attractor can be characterized as follows: it consists of a set of phase points embedded in a finite region of phase space. Trajectories outside that region but close enough are attracted into that region. Trajectories within that region will remain in it. The "strangeness" of the attractor consists in the fact that it is neither a singular point (dimension zero) nor a limit cycle (dimension 1) nor a torus (dimension 2) but has in general a noninteger "fractal" dimension. Trajectories that are initially very close to each other on the attractor, are separated exponentially as time goes on. Intuitively, this implies an unstable motion *within* the attractor.

The onset of chaos is usually preceded by other bifurcations, like those discussed earlier in this book. For example, with increasing value of the control parameter, the system may first exhibit a Hopf bifurcation of a limit cycle, and then, at still higher values of the control parameter, a transition to chaos. Different routes to chaos have been found in various physical systems. In the following we discuss some of the most important scenarios [6.49, 54, 63–66]. They represent sequences of global bifurcations leading to chaos.

First, we consider the transition to chaos by *period doubling*, or *subharmonic generation* (the Feigenbaum scenario [6.63]). When a limit cycle in a three-dimensional phase space becomes unstable, it may bifurcate into another limit cycle on which the system takes twice the time to return to its original state, that is the period has doubled, or the original frequency has been cut by half. The projection of the phase trajectory onto a plane is a twofold loop (as shown, for example, in Fig. 6.29b below). Upon further increase of the control parameter another period-doubling bifurcation occurs, so that the original period T is multiplied by four, and so on. Thus a whole sequence of period-doubling bifurcations occurs at successive values of the control parameter k_n, $n = 1, 2, 3, \ldots$, with period $2^n T$. For $n \to \infty$ the k_n converge to a critical value k_∞, at which the motion ceases to be periodic, and becomes chaotic.

A convenient description of period doubling and other chaotic dynamic effects is in terms of discrete, iterated maps [6.53]. These arise if the set of variables $q(t)$ is not taken as a function of the continuous time t, but instead is considered only at a certain discrete sequence of times, $t_0, t_1, t_2, t_3, \ldots$. The sequence $q_n := q(t_n)$ can be constructed, for example, by taking the transverse intersections of the trajectories in N-dimensional phase space with an $N - 1$ dimensional hypersurface. This is called a *Poincaré section*. The phase space flow through the Poincaré surface induces an $N - 1$ dimensional invertible map $q_n \to q_{n+1}$, the *Poincaré map*, whereby the differential equations are replaced by difference equations. For example, as the

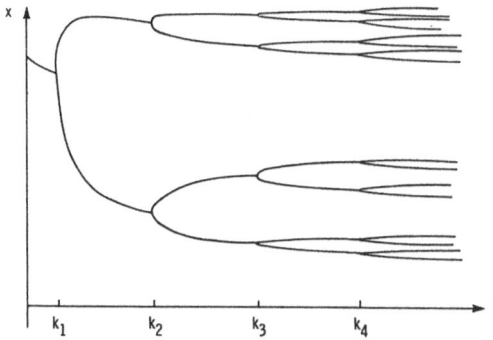

Fig. 6.18. Bifurcation diagram of a one-dimensional map (Poincaré section of limit cycles). The dynamic variable x is plotted versus the control parameter k (logarithmic scale). There is a sequence of period-doubling bifurcations at k_1, k_2, k_3, k_4. [6.53]

Poincaré section of a two-dimensional system oscillating around the origin one may choose the intersections with one of the coordinate half-axes, x_n. Each intersection point x_{n+1} is determined, once the previous point x_n is fixed. Therefore a relation $x_{n+1} = f(x_n)$ exists: a one-dimensional "map" (*Poincaré map*, or first-return map). A limit cycle is then represented by a single point of intersection x_∞, to which the intersection points of all neighboring phase trajectories converge for $n \to \infty$.

The period-doubling bifurcation sequence corresponds to a cascade of pitchfork bifurcations of the Poincaré map. This is illustrated in Fig. 6.18 for the (noninvertible) one-dimensional map $x_{n+1} = f(x_n) = 1 - kx_n^2$. Such noninvertible one-dimensional maps can generally be obtained from higher-dimensional invertible Poincaré maps g if one component of g is decoupled from the others. The cascade of bifurcation values k_n has been shown to follow a universal law

$$\lim_{n \to \infty} \frac{k_{n+1} - k_n}{k_{n+2} - k_{n+1}} = \delta \equiv 4.6692016\ldots \tag{6.3.4}$$

for a large class of one-dimensional maps f, independent of the detailed form of the function f, as long as it has a quadratic maximum and a negative Schwarzian derivative $d^2[f'(x)]^{-1/2}/dx^2 < 0$. [6.63]. Beyond k_∞, the chaotic regime is interrupted by narrow bands of k-values in which the motion is again periodic. In such "periodic windows" further period-doubling cascades occur. Generally these periodic sequences first appear with some period NT ($N = 3, 4, 5, \ldots$) and then go through a sequence of periods $2^n NT$, with an accumulation point at $n \to \infty$ ending that particular periodic window. A variety of universal features and scaling laws have been established for these one-dimensional maps [6.53].

In semiconductor systems the dynamic variable q is often the electric current, or the field, or the carrier concentration. A convenient method of reconstructing the Poincaré map from an experimental signal is to monitor the set of successive maxima or minima in the chaotic oscillations of this dynamic variable. This should not be confused with a stroboscopic study, where one looks at the system at integer multiples of a fixed time interval. Sometimes there is a natural frequency (for instance, in periodically driven chaotic systems) which will stabilize the stroboscopic image, but in general this will not be the case.

A second route to chaos is the *quasi-periodic* breakdown. A limit cycle with a single frequency ω may undergo a second Hopf bifurcation leading to a doubly

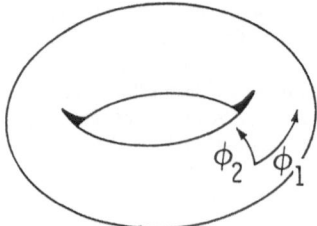

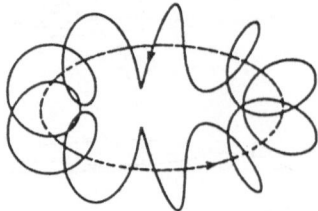

Fig. 6.19. Two-dimensional torus parametrized by two angular coordinates ϕ_1, ϕ_2

Fig. 6.20. Three-dimensional limit cycle, representing motion on a torus. [6.49]

periodic motion which is characterized by two fundamental frequencies ω_1, ω_2. It is convenient to parametrize the phase point by the two angular coordinates $\phi_1 = \omega_1 t$ and $\phi_2 = \omega_2 t$. The doubly periodic motion can thus be visualized as the motion on a two-dimensional torus (Fig. 6.19). If the ratio ω_1/ω_2 is rational, the trajectory is closed, as shown in Fig. 6.20. If, on the other hand, ω_1 and ω_2 are incommensurate (i.e., of irrational ratio), the motion is called *quasi-periodic* and the trajectory fills up the whole torus. A subsequent bifurcation may either take the system back to a limit cycle with a single frequency (which is called *frequency-locking* or phase-locking), or add a third fundamental frequency ω_3 (which transforms the two-dimensional torus into a "three-dimensional torus"). A condition for the latter is that the frequencies are sufficiently irrational with respect to each other. A more rigorous mathematical definition is given by the so-called KAM-condition [6.49]. Subsequent bifurcations may produce tori of higher and higher dimensions. At each step a new frequency ω_j is added to the set of fundamental frequencies. Chaos may finally be described by motion on a torus of infinite dimension. This is the Landau-Hopf scenario. Another, more likely possibility of transition to chaos (the Ruelle-Takens-Newhouse scenario [6.64]) is that the bifurcation from the two-dimensional torus leads directly to a strange attractor.

Finally, a transition to chaos may occur via *intermittency* (the Pomeau-Manneville scenario [6.65]). This means that the phase trajectory appears approximately as a limit cycle during long periods of time; but these periods are interrupted by sudden outbursts of chaotic motion. The associated Poincaré sections show saddle-node bifurcations, in which a stable node and a saddle point collide and annihilate.

Next, we shall outline some criteria that allow us to characterize chaotic motion, and distinguish it from complicated multiply periodic motion as well as from irregular behavior caused by stochastic fluctuations, for instance thermal noise. The latter are true random processes, and are thus in principle unpredictable, while chaotic motion is in principle fully determined by a few nonlinear differential equations, and is hence deterministic, although the actual time dependence of the trajectories looks extremely irregular and most of the analysis has to be carried out numerically.

One characteristic feature of chaos is its *low-dimensional phase space*. Consider the dynamic variable q at discrete times t_n, $n = 1, 2, 3, \ldots$: this yields a sequence of values $q_1, q_2, q_3, \ldots$ Each value q_{n+1} is then uniquely determined by its predecessor

q_n and we may represent the pairs (q_n, q_{n+1}) for all possible values of n as points in a (q_n, q_{n+1}) plane. This is a graph of the one-dimensional Poincaré map. Similarly, we might plot the triplets (q_n, q_{n+1}, q_{n+2}) as points in a three-dimensional space, or more generally, conceive the p-tuples $(q_n, q_{n+1}, \ldots, q_{n+p-1})$ as points in a p-dimensional embedding space. If the irregular motion were stochastic, we would expect all these p-dimensional spaces to be filled uniformly and randomly by points, since the sequences $q_n, \ldots, q_{n+p-1}$ would be random. Chaotic motion resulting from deterministic equations, however, will fill only certain low-dimensional regions of the p-dimensional spaces, even if p is chosen to be very large.

A quantitative measure of the low dimensionality of chaotic motion is provided by the concept of *fractal dimension* of a chaotic attractor [6.67–70]. There are two general types of definitions of dimension, those that depend only on metric properties, and those that depend on the frequency with which a typical trajectory visits different regions of the attractor. For typical strange attractors, all of the metric dimensions (i.e., the "capacity" and the slightly more general "Hausdorff dimension" [6.67]) take on a common value, the fractal dimension d. For any bounded set of points in $\mathbb{R}^p$, representing the strange attractor, it is defined as

$$d := \lim_{r \to 0} \frac{\log N(r)}{\log(1/r)} \, , \tag{6.3.5}$$

where $N(r)$ is the minimum number of p-dimensional cubes or balls of radius r needed to cover the set. Thus this number increases for $r \to 0$ like

$$N(r) \sim r^{-d} \, , \tag{6.3.6}$$

which gives $d = 0, 1$, and 2 for a point, a line, and a surface, as expected. However, for a strange attractor, d is in general noninteger. For sufficiently large $p > 2d + 1$, d is independent of the embedding dimension p. In contrast, for stochastic motion $d = p$ would be expected.

In practical experiments or numerical simulations, the computation of d is often faced with difficulties because a very large number of points on the attractor are needed to insure the asymptotic scaling law (6.3.6). A more rapidly converging algorithm, which gives a generally very good lower bound $v \le d$ of the fractal dimension, has been proposed by *Grassberger* and *Procaccia* [6.69]. Here the *correlation dimension* v is defined via the scaling law for $r \to 0$

$$C(r) \sim r^v \, , \quad \text{or} \quad v := \lim_{r \to 0} \frac{\log C(r)}{\log r} \tag{6.3.7}$$

where the correlation integral

$$C(r) :- \lim_{N \to \infty} \frac{1}{N^2} \sum_{i,j=1}^{N} \theta(r - |q_i - q_j|) \tag{6.3.8}$$

measures the number of pairs of points $q_n := \{q_n, \ldots, q_{n+p-1}\}$ whose distance is less than r in a p-dimensional embedding space with sufficiently large p [θ is the Heaviside function]. The exponent v gives also a lower bound for the *information dimension* d_I [6.67], which is defined by

$$d_I := \lim_{r \to 0} \frac{I(r)}{\log(1/r)} \quad, \tag{6.3.9}$$

where

$$I(r) := -\sum_{i=1}^{N(r)} P_i \log P_i \tag{6.3.10}$$

is a Kolmogorov-entropy-like information measure, and P_i is the fraction of time that the trajectories spend in the ith cube of the partition used in (6.3.5). Thus d_I is a "probabilistic dimension" in which the different parts of the attractor are weighted with the probability that they are visited.

A second characteristic feature of chaotic motion shows up in the correlations of the variables at different times: The *auto-correlation function*

$$\langle q(t)q(t+\tau) \rangle = \lim_{T \to \infty} \frac{1}{2T} \int_{-T}^{T} q(t)q(t+\tau)\,dt \tag{6.3.11}$$

tends to zero for $\tau \to \infty$.

Such behavior is different from simply and even multiply periodic motion. For periodic motion we obtain always an undamped oscillatory correlation function, e.g.,

$$\langle q(t)q(t+\tau) \rangle = \tfrac{1}{2}\cos \omega\tau \qquad \text{for} \tag{6.3.12}$$

$$q(t) = \sin \omega t \ . \tag{6.3.13}$$

In case of thermal fluctuations the auto-correlation function is often zero or negligible for τ greater than a finite auto-correlation time τ_c.

A third feature of chaos is related to the Fourier spectrum of the temporal evolution. The Fourier transform of $q(t)$ in a finite, but sufficiently long observation interval of time T is given by

$$\hat{q}(\omega; T) := \frac{1}{2\pi} \int_{-T}^{T} q(t)e^{i\omega t}\,dt \ . \tag{6.3.14}$$

We define the spectral power density (often referred to as "*power spectrum*") by

$$S(\omega) := \lim_{T \to \infty} \frac{\pi}{T}|\hat{q}(\omega; T)|^2 \tag{6.3.15}$$

such that the integral power density is

$$\int_{-\infty}^{\infty} S(\omega)\,d\omega = \lim_{T \to \infty} \frac{1}{2T} \int_{-T}^{T} q(t)^2\,dt \ . \tag{6.3.16}$$

It then follows by inserting (6.3.14) into (6.3.15) and using the definition of the auto-correlation function (6.3.11) that

$$S(\omega) = \frac{1}{2\pi} \int\limits_{-\infty}^{\infty} d\tau e^{i\omega\tau} \langle q(t)q(t+\tau)\rangle \ . \tag{6.3.17}$$

Here we have used the assumption that $q(t)q(t+\tau)$ decays fast with increasing τ so that we may extend the integral over τ from $-\infty$ to $+\infty$.

Equation (6.3.17) states that the spectral power density is the Fourier transform of the auto-correlation function (Wiener-Khinchin-Theorem). For a simply or multiply periodic motion the power spectrum has discrete lines at the fundamental frequencies. Chaotic motion, however, is characterized by a continuous broad band of frequencies.

A fourth indication of chaos is given by the so-called *Lyapunov exponents* (or characteristic exponents). These are generalizations of the eigenvalues of the linear stability matrix of a singular point,

$$\lambda := \lim_{t\to\infty} \sup \left\{ \frac{1}{t} \ln |q(t) - q^*(t)| \right\} \ , \tag{6.3.18}$$

where $q(t)$ are trajectories in the neighborhood of a reference trajectory $q^*(t)$. The sign of λ indicates whether or not $q^*(t)$ is asymptotically approached by other trajectories $q(t)$. If, in case of a three-variable dynamic system, one of the Lyapunov exponents is positive, one is zero and one is negative, this reflects a very sensitive dependence upon initial conditions, and the motion is likely to be chaotic (a strange attractor), although some more detailed analysis may be necessary [6.49].

Finally, as a fifth distinguishing feature of chaos, the onset of chaos is very often preceded by one of the characteristic scenarios discussed earlier on: a period-doubling cascade, quasi-periodic breakdown, or intermittency.

6.3.2 Single-Carrier Effects

We shall now review some recent experiments which discovered intrinsic chaos in semiconductors. Chaotic oscillations in high-purity Ge [6.73–76, 84–86, 88–94, 96], GaAs [6.77–83, 95, 97], and InSb [6.87] have been observed under a wide variety of experimental conditions, ranging from low temperatures [6.73–94, 96, 97] to room temperature [6.95], and including weak infrared [6.84, 87] or visible [6.77–82] irradiation as well as complete shielding against external irradiation [6.88–94], and in some cases parallel [6.73–76] or transverse [6.87, 88] magnetic fields.

Upon variation of the applied bias, taken as a control parameter, different routes to chaos were observed: the period-doubling (Feigenbaum) scenario, quasi-periodic (Ruelle-Takens-Newhouse) breakdown, and intermittent switching (Pomeau-Manneville) between two oscillatory states. The oscillation frequencies were typically quite low, between a few Hz and several kHz.

The physical mechanism of these chaotic oscillations is – with exception of the helical instability [6.73–76], which will be discussed in Sect. 6.3.5 – not completely understood, but there is strong evidence that impact ionization from impurity levels is involved in a majority of these experiments. The onset of chaos occurred either just below [6.77, 84, 95] or above [6.88] the threshold field for impurity breakdown. The similarity of the mechanism in various materials is indicated by the observation of an empirical scaling law [6.95] between the impurity level energy and the breakdown field which ranged from a few V/cm for shallow donor or acceptor states at helium temperatures to several kV/cm for deep levels in semi-insulating GaAs at room temperature.

The experiments can be divided into two classes: (i) driven chaos [6.77–83, 85, 87, 96b] which is induced by periodically chopped external radiation or ac-modulated external currents or pulsed voltage, (ii) self-generated chaos [6.84, 87–97] which is observed under static applied electric fields and time-independent, if any, irradiation, and is widely independent of external circuit conditions. A theoretical model for the first class of phenomena was recently proposed by *Teitsworth* and *Westervelt* [6.98]; it will be reviewed in Sect. 6.3.3. A model for impact-ionization induced self-generated chaos, as observed in the experiments of class (ii) was developed by *Schöll* [6.100], and will be presented in Sect. 6.3.4.

Quasi-periodic oscillations and turbulence in high-purity n-GaAs at 4.2 K was first reported by *Aoki* et al. [6.77, 78]. They used samples of linear dimensions 5.5 mm × 4 mm × 12 μm, doped with $N_D = 2 \times 10^{14}$ cm^{-3} donors, and planar ohmic contacts. They applied static electric fields ε in the SNDC range of 4–6 V/cm (which corresponds to the field strength necessary for shallow donor impact ionization) and weak monochromatic illumination around the bandgap ($\lambda = 815.5$ nm, corresponding to the exciton resonance, intensity $J_p \sim 2$ μW/cm^2).

With increasing field the power spectrum showed an increasing number of discrete lines, which could be expressed as the harmonics of two incommensurate fundamental frequencies $f_1 = 3.5 \times 10^4$ s^{-1}, $f_2 = 2.2 \times 10^5$ s^{-1} (at $\varepsilon = 3.56$ V/cm). Illumination with a He-Ne laser ($\lambda = 633$ nm, $J_p = 3.2$ mW/cm^2) also produced a discrete power spectrum, which changed to continuous above 3.6 V/cm. The authors initially tried to explain the continuous spectrum by random fluctuating forces. *Aoki* et al. [6.78, 79] later reported a sequence of period-doubling bifurcations of the current oscillations under similar conditions, but with a pulsed voltage of 38 kHz, and illumination at $\lambda = 812$ nm, when the voltage was decreased. Although the authors presented the Poincaré map $I_{n+1}(I_n)$ of the current oscillations $I(t)$ as evidence for deterministic chaos, as opposed to stochastic noise, the nature of the observed irregular current oscillations was somewhat obscured by the fact that the Poincaré map was rather smeared out, and the authors themselves [6.80] invoked random fluctuating forces in an attempt to explain the oscillations by a Langevin-type equation for the formation of current filaments ("stochastic firing of filamentary channels"). The chaotic oscillations were attributed to an impact-ionization induced firing-wave instability of current filaments, analogous to threshold firing of nerve pulses in neurophysics, and were modeled by nonlinear interactions of two fila-mentary channels [6.81]. Further experimental work revealed quasi-periodic break-down, intermittency, and chaos-chaos transitions as a function of the driving frequency f_{dr} of the chopped photoexcitation in the range 0.2 Hz $< f_{dr} <$ 3 Hz, under simultaneous application of a pulsed voltage of 38 kHz. The fractal dimension of the strange attractor was determined using the Grassberger-Procaccia algorithm [6.69], and values in the range $v = 3.6$ to 4.2 were found [6.82]. The routes to chaos were studied numerically [6.83] with a model for impact-ionization induced dielectric relaxation oscillations similiar to the one proposed in [6.100].

Self-generated nonlinear oscillations and chaos in ultrapure Ge at 1.5 ... 4.2 K were reported by *Teitsworth* et al. [6.84]. The crystals were p-doped with acceptor concentrations $N_A \approx 10^{10}$ to 10^{13} cm^{-3} and compensated with artio $N_D/N_A \approx 0.1$ to 1. The sizes ranged from $0.5 \times 0.5 \times 8$ mm^3 to $20 \times 20 \times 8$ mm^3, and electrical contacts were made with injecting p$^+$ layers. With a static field ε_0 and time-independent weak far-infrared illumination, the current abruptly increased without hysteresis at a breakdown field $\varepsilon_b \sim 3$ to 40 V/cm as a result of impact ionization of shallow acceptor levels. Above a well-defined threshold dc field $\varepsilon_t < \varepsilon_b$ current oscillations at low frequencies $f < 10^4$ s^{-1} were observed. The character of these oscillations was essentially independent of the type of external circuit and type of current or voltage bias. With increasing field, sequences of progressively complex oscillatory instabilities were observed, such as the period-doubling cascade shown in Fig. 6.21. In Fig. 6.21a, the oscillation is simply periodic; in Fig. 6.21b the period has doubled; Fig. 6.21c shows a period-four oscillation, and at a larger field still well below the breakdown field ε_b the oscillation appears chaotic (Fig. 6.21d). The experimental bifurcation diagram for this sequence of instabilities is shown in Fig. 6.22. This diagram was produced by plotting the observed minima of $I(t)$ along the vertical axis while slowly increasing the applied voltage, plotted along the horizontal axis. The transitions from simply periodic to period-two and period-four regimes clearly show up, as well as a broad region of chaotic oscillations. The diagram resembles that for the one-dimensional map shown in Fig. 6.18, although the scaling parameter δ, defined in (6.3.4), and obtained experimentally from the first few period doublings, is different from the (asymptotic) universal Feigenbaum number. In the corresponding power spectrum the onset of chaos appears as a rising level of broad-band noise which eventually covers the sharp peaks due to periodic oscillations. In other samples Teitsworth et al. observed quasi-periodic oscillations

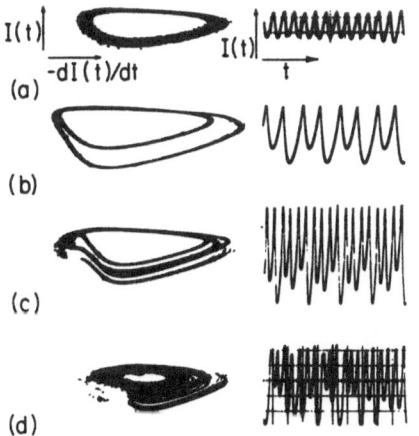

(a)

(b)

(c)

(d)

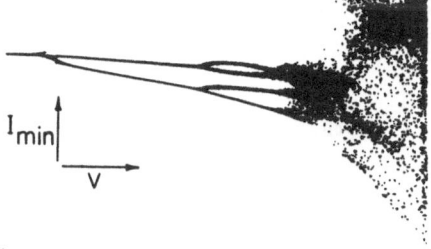

▲
Fig. 6.22. Bifurcation diagram of measured current minima I_{min} vs. applied voltage V, for the range $V = 11.1$ to 18.4 V; breakdown occurred at $V_b = 20$ V. [6.84]

Fig. 6.21a–d. Measured current oscillations $I(t)$ and phase portraits $I(t)$ vs. $-dI/dt$, for a p-Ge ($N_A = 2.5 \times 10^{10}$ cm^{-3}) photoconductor at $T = 4.2$ K with increasing voltage from (**a**) to (**d**) as in Fig. 6.22. [6.84]

at two incommensurate frequencies, frequency locking, and intermittent switching between different modes of oscillation [6.85, 86].

The frequency and shape of the simple current oscillations was found to depend upon the far-infrared (FIR) illumination intensity [6.84]. At low intensities, such that the generated free carrier density was $p < 10^6$ cm^{-3}, sharp spikes of relaxation-oscillation type were observed. Their frequency increased with FIR illumination from $f \sim 10^{-1}$ s^{-1} to $f \sim 10^2$ s^{-1}. For moderate FIR illumination intensity ($p \sim 10^6$ cm^{-3}) smooth periodic oscillations with frequency $f \sim 10^2$ to 10^4 s^{-1} developed, as shown in Fig. 6.21a. As a qualitative explanation Teitsworth et al. suggested oscillatory dielectric relaxation of trapped space charge near the injecting contact in combination with impact ionization of shallow acceptor levels.

A period-doubling route to chaos arising from impact ionization of shallow donor impurities and space charge injection in high-purity n-InSb around 1.9 K was found by *Seiler* et al. [6.87]. The effective donor concentration was $N_D - N_A \sim 10^{14}$ cm^{-3}. Nonlinear voltage oscillations occurred when a transverse magnetic field $B \perp \varepsilon$ of the order of 10 kG, or an *ac* sinusoidal driving current of frequency f_0 around 3×10^3 s^{-1} was applied. At low currents the oscillations could be quenched by CO$_2$ laser irradiation. Period doublings were observed either with increasing driving current, or increasing magnetic field, or decreasing lattice temperature.

Chaotic oscillations in the postbreakdown regime of p-Ge were observed at temperatures between 1.7 and 4.2 K by *Peinke* et al. [6.88]. The samples were p-doped with an acceptor concentration of about 10^{14} cm^{-3} and of size $0.5 \times 2 \times 4$ mm^3, with Ohmic contacts. Under complete shielding against external radiation an SNDC characteristic induced by impurity impact ionization was observed (Fig. 6.23). Three different routes to chaos were found, depending upon the applied electric (ε) and transverse magnetic field (B): Intermittent switching between different oscillatory states was observed for $B = 0$ at point A of the characteristic (Fig. 6.23), as shown in Fig. 6.24. The amplitude of the current oscillation in Fig. 6.24a is only a few microamperes (whereas the *dc* current is some milliamperes). Figures 6.24b, c are marked by sudden outbursts of large amplitude oscillations. A period-doubling sequence was also found. In Fig. 6.25 a sequence somewhat similar to quasi-periodic

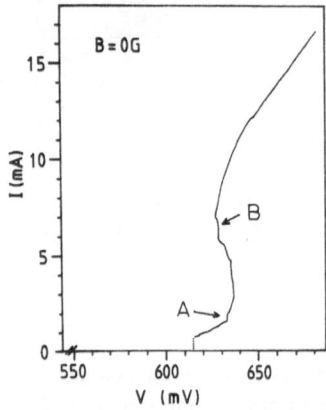

Fig. 6.23. Measured current versus voltage for p-Ge at 4.2 K with zero magnetic field B. Points A and B denote regions where chaos was observed. [6.88]

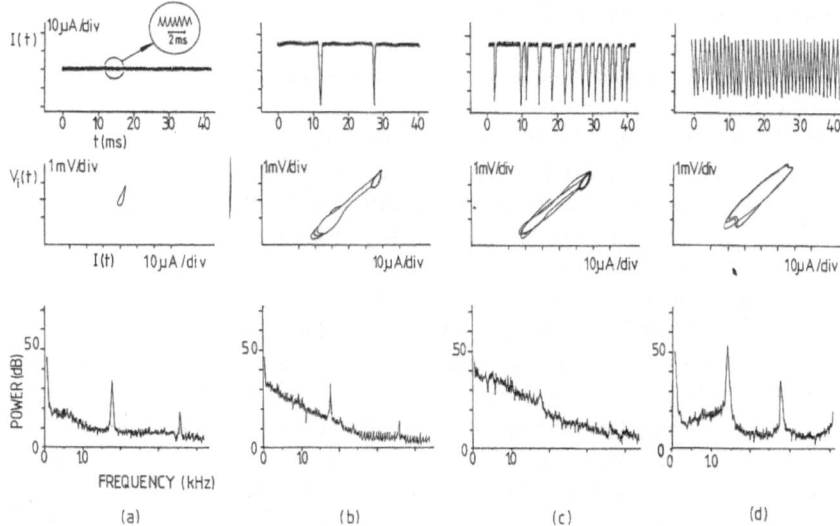

Fig. 6.24a–d. Intermittency observed at $B = 0$, $T = 4.2$ K (point A of Fig. 6.23). The time series of the current $I(t)$ (*top*), phase portraits of the voltage across the sample versus current (*middle*), and power spectra (*bottom*) are shown for a sequence of increasing electric fields [by a total of 4 mV/cm from (**a**) to (**d**)]. [6.88]

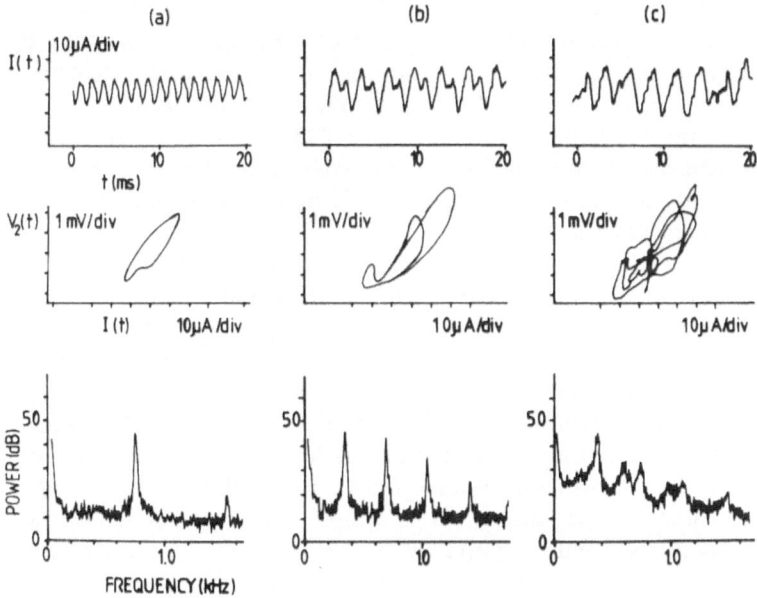

Fig. 6.25a–c. Quasi-periodic breakdown observed at $B = 0$, $T = 4.2$ K (point B of Fig. 6.23). The sequence (**a**)–(**c**) was obtained by increasing the electric field by a total of 35 mV/cm. [6.88]

breakdown is shown. Increasing the field first resulted in a period doubling (Fig. 6.25b), followed by the appearance of one additional incommensurate frequency immediately before a strong increase in spectral noise (Fig. 6.25c). The frequencies were typically between some 100 Hz and some kHz. Upon variation of the magnetic field from $B = 31.56$ to 46.5 G, the chaotic attractor was found to increase in dimensionality, reflecting a transition from ordinary chaos to hyperchaos as shown in Fig. 6.26 [6.89]. As the origin of this transition the coupling between different parts of the system was suggested, where each part of the system resides in an ordinary chaotic state [6.90, 91].

In order to investigate the physical nature of the coupling, the authors [6.90–93] used an experimental configuration where two parts of a single crystal were electrically separated in order to exclude coupling via charge carriers. This left diffusive phonons as the probable coupling mechanism. By applying different voltages to the two subsystems, spontaneous periodic and chaotic oscillations could be induced. Spatial correlations ranging from phaselocking via phase lagging to complete phase reversal of the oscillations in the two subsystems were observed. This behavior showed a striking similarity to the dynamics of a two-cellular chemical reaction system, the Rashevsky-Turing oscillator [6.94]. A semiconductor model which can explain the experiments in terms of energy relaxation oscillations of two hot carrier subsystems, driven by impact ionization, and coupled by phonons, has been proposed elsewhere [6.102–104].

Low-frequency chaotic oscillations in unintentionally doped semi-insulating GaAs at room temperature were reported by *Maracas* et al. [6.95]. When a dc voltage was applied between two ohmic contacts of a 60 μm long sample, a transition occurred from simple periodic to chaotic oscillations, as the field increased from 1.175 to 2.4 kV/cm. The threshold field for impurity impact ionization lies within this range. From the temperature dependence of the oscillation frequency it was inferred that two deep-trap levels at 0.70 and 0.69 eV are involved. Impact ionization of these traps in a slowly moving high-field domain was suggested as a possible qualitative explanation.

Bumelienė et al. [6.96a] observed chaotic oscillations of a hot electron plasma in nickel-compensated n-Ge with 1.2×10^{15} cm^{-3} donors, and a nickel impurity concentration of 8×10^{14} cm^{-3} at 77 K. The samples were dumbbell shaped with the dimension of the central part $4 \times 1 \times 1$ mm^3. An excess electron density of 10^{11} cm^{-3} was generated by optical illumination with a 200 W lamp. When dc electric fields of up to 500 V/cm were applied, the onset of periodic oscillations of the voltage across a load resistor occurred at 60 V/cm, and was followed by a transition to chaos characterized by a broad-band continuous spectrum in the frequency range of 4–12 kHz. An explanation was attempted by a model based on field-enhanced trapping at three types of repulsive impurity centres. This model for spatially

---

Fig. 6.26. Transition from chaos (**a**) to hyperchaos (**d**), observed in p-Ge at increasing magnetic fields (**a**) $B = 31.5$ G, (**b**) $B = 34$ G, (**c**) $B = 40$ G, (**d**) $B = 46.5$ G. The sample contained two outer contacts C_1, C_2, and two inner contacts C_3, C_4 between these. The bias voltage between C_1 and C_2 was kept at $V_0 = 2.145$ V ($T = 4.2$ K). The voltages V_1 and V_2 were measured between C_1 and C_3, and between C_3 and C_4, respectively. Top: phase portraits of $V_2(t)$ versus $V_1(t)$; bottom: power spectra of $V_2(t)$. [6.88]

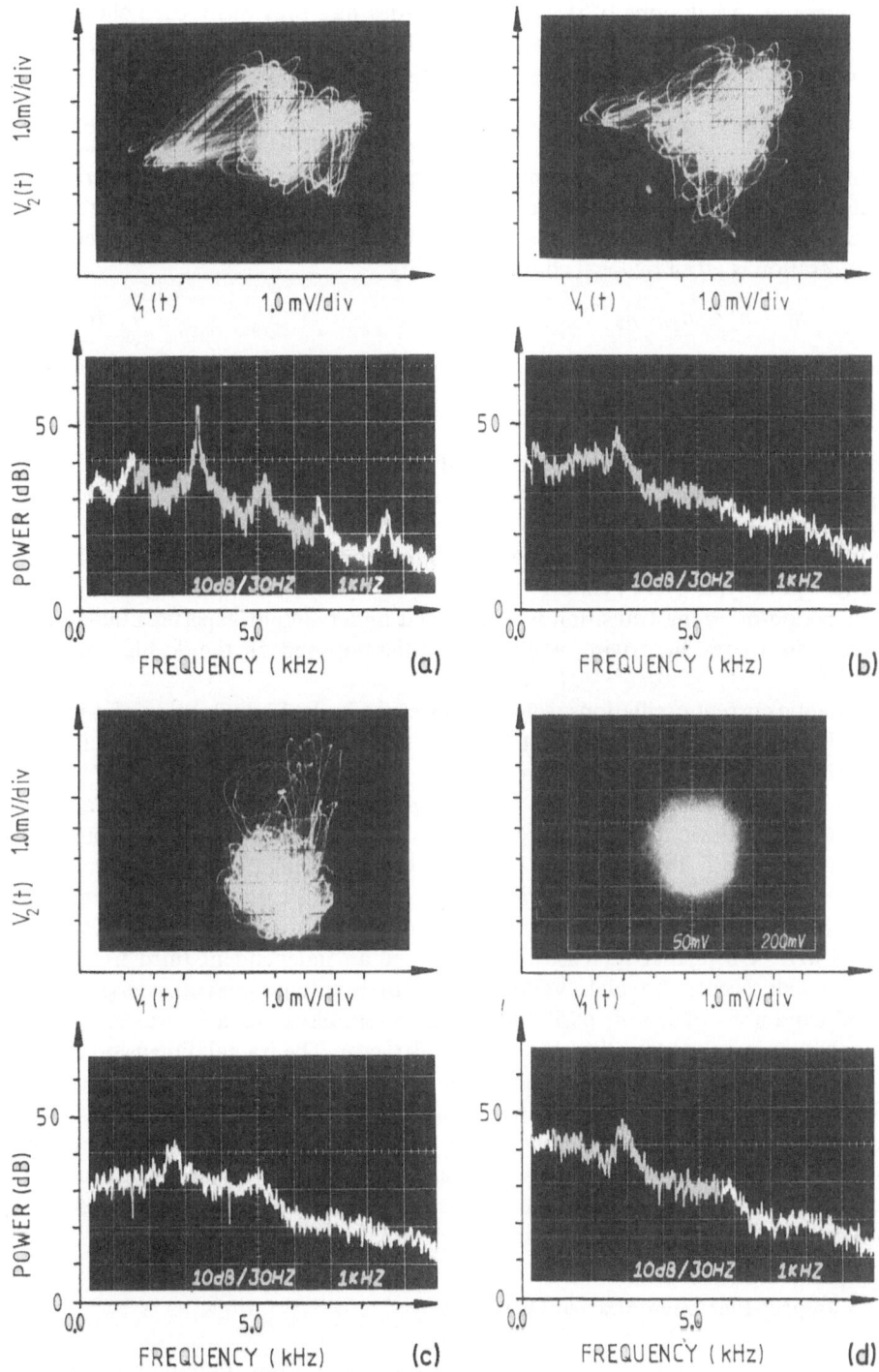

Fig. 6.26. Caption see opposite page

homogeneous oscillations of the carrier densities had previously been shown to exhibit a strange attractor by *Pyragas* [6.99]. The concentrations of occupied traps n_i are governed by the following set of three rate equations:

$$\dot{n}_i = T_i^S(\varepsilon)(N_i - n_i)n - X_i^S n_i , \qquad i = 1, 2, 3 \tag{6.3.19}$$

where N_i is the total density of impurity centers of type i, X_i^S is the generation coefficient, and $T_i^S(\varepsilon)$ is the trapping coefficient, which is assumed to be a strongly increasing function of the field ε, e.g., of the form $\sim \exp(\varepsilon/\varepsilon_0)^2$. The free electron concentration is given by local charge neutrality

$$n = N_D^* - n_1 - n_2 - n_3 \tag{6.3.20}$$

and the carrier densities and the field are coupled via the equation for the load line

$$U_0 = R_L n e \mu_n \varepsilon A + \varepsilon L , \tag{6.3.21}$$

where U_0 is the *dc* bias voltage, R_L the load resistance, μ_n the mobility, A the cross section, and L the length of the sample. The essential nonlinearity which drives the instability is contained in the ε-dependence of the trapping coefficient. It leads to an unstable saddle-focus on the slow two-dimensional manifold defined by $\dot{n} = 0$, and to "spiral-type" [6.57] chaos.

A period-doubling transition was observed under similar experimental conditions when a periodic driving voltage was superimposed on the dc bias voltage [6.96b].

Chaotic current oscillations were also observed in the SNDC region of n-GaAs at 4.2 K under static external conditions, with shielding against visible and infrared radiation, and perpendicular magnetic fields in the range 0.1 ... 1 T [6.97]. The donor concentration was 2.7×10^{14} cm^{-3} at 70% compensation. A Ruelle-Takens-Newhouse transition to chaos was observed upon increasing the voltage from 3 to 7.8 V at $B = 0.83$ T. Two successive Hopf bifurcations giving rise to periodic motion on a limit cycle, and to quasiperiodic motion on a 2-torus showed up in the corresponding power spectra at the incommensurate frequencies $f_1 = 91$ kHz and $f_2 = 5$ kHz. On further increase of the voltage an independent third frequency $f_3 = 41$ kHz occurred, which, within a very small voltage interval, merged with broad band noise of chaotic oscillations. This is reminiscent of a 3-torus becoming unstable against the formation of a strange attractor. The fractal dimension of the reconstructed attractor, as estimated from the Grassberger-Procaccia algorithm [6.69], was found to increase stepwise from $d = 0$ to 1.1 to 2.0 and to 2.7. As a possible physical explanation *Brandl* et al. suggested *Schöll's* dielectric relaxation oscillation mechanism [6.100]. They argued that the role of the magnetic field is crucial in slowing down the dielectric relaxation time such that the Hopf bifurcation condition (6.2.21, 58) can be satisfied.

6.3.3 Impact-Ionization Assisted Driven Chaos

A simple rate equation model for driven chaos, based upon trapping and impact ionization from a single acceptor level at low temperature in combination with a periodic driving current, and dielectric relaxation, was advanced by *Teitsworth* and *Westervelt* [6.98], and later compared to experiments on extrinsic p-Ge FIR photoconductors at 4.2 K driven with an ac voltage of suitable frequency. They neglected space-charge effects, and considered a time-dependent driving current density

$$J_{ext}(t) = J_0 + \Delta J \sin(\omega_{dr} t) \ . \tag{6.3.22}$$

This ac current might also serve as a very crude simulation of the effect of a periodically modulated external radiation. The transport equations for a spatially homogeneous p-type semiconductor are

$$\frac{dp}{dt} = X^S(N_A - n_A) + X(\varepsilon)p(N_A - n_A) - T^S(\varepsilon)pn_A \ . \tag{6.3.23}$$

$$\frac{\epsilon_s}{4\pi} \cdot \frac{d\varepsilon}{dt} = J_{ext}(t) - epv(\varepsilon) \tag{6.3.24}$$

where p, n_A and N_A denote the hole concentration, the ionized acceptor concentration, and the total acceptor density, respectively; ε is the electric field, ϵ_s the permittivity, $v(\varepsilon) \equiv \mu_p(\varepsilon)\varepsilon$ the drift velocity; X^S is the (thermal or optical) acceptor ionization coefficient, X and T^S are the field-dependent rate coefficients for impact ionization of acceptors, and capture of holes, respectively.

The ionized acceptor concentration can be eliminated from (6.3.23) via the charge-neutrality condition

$$n_A = N_D + p \ , \tag{6.3.25}$$

where N_D is the compensating donor density. Note that as a result of the low free-carrier concentration in high-purity materials at low temperatures (typically $p_0 \lesssim 10^5$ cm^{-3}), the dielectric-relaxation time

$$\tau_M = \epsilon_s/(4\pi e\mu_p p) \tag{6.3.26}$$

is large, and comparable with, or larger than, the recombination lifetime τ_r. Therefore, the displacement current density $\epsilon_s(d\varepsilon/dt)/(4\pi)$ varies on the same slow time-scale as the generation-recombination rate dp/dt, and cannot be neglected. This is the case of a relaxation semiconductor. It is the opposite of the conventional case of a lifetime semiconductor, where dielectric relaxation occurs on a much faster time-scale than recombination, and restores space-charge neutrality before the effective onset of recombination. Under this aspect the assumption of charge neutrality, (6.3.25), is inconsistent.

Equations (6.3.23, 24) constitute a two-variable nonautonomous nonlinear dynamic system. It is similar to a damped driven nonlinear oscillator, where $p(t)$ is analogous to the position, and $\varepsilon(t)$ is analogous to the momentum, X and T^S play the role of restoring forces, and J_{ext} is the driving force. The system (6.3.23, 24) represents a generalized, periodically driven Van der Pol oscillator. We have already discussed an example of a free Van der Pol oscillator in Sect. 6.2.1.

The Van der Pol oscillator can produce limit-cycle oscillations. Furthermore, the periodically forced Van der Pol oscillator is known to exhibit chaotic behavior under certain conditions [6.54, 71]. In the present example, the RC time constant of the electric oscillator is given by the dielectric-relaxation time τ_M, and the nonlinear feedback is the impact-ionization process. Typically, the oscillations correspond to periodic charging and discharging of the acceptors.

The time-dependent solutions of (6.3.23, 24) are qualitatively similar for a variety of functional forms for $X(\varepsilon)$, $T^S(\varepsilon)$, $v(\varepsilon)$, as long as the impact-ionization coefficient $X(\varepsilon)$ is negligible for ε much smaller than the breakdown field, and then abruptly increases, up to a local maximum, after which it decreases. The capture coefficient $T^S(\varepsilon)$ is roughly constant for small fields and then drops as ε^{-3}, reflecting a decrease in capture cross section with increasing carrier velocity. The drift velocity $v(\varepsilon)$ increases linearly in the low-field ohmic regime, and later on saturates at about 10^7 cm/s.

The numerical solution of (6.3.23, 24) is shown in Fig. 6.27 for parameter values appropriate to ultrapure Ge samples [6.84]: $N_A = 10^{11}$ cm^{-3}, $N_D = 2 \times 10^{10}$ cm^{-3}, $J_0 = 1.3 \times 10^{-7}$ A/cm^2, $\epsilon_s = 16$, $T = 4.2$ K. For a time-independent driving force $J_{ext} = J_0$, $\Delta J = 0$, and positive differential mobility $dv/d\varepsilon$, there is a single steady state, a stable focus in the (p, ε) phase plane, around which the phase trajectories spiral with angular frequency

$$\omega_0 = (h_0 - \tfrac{1}{4}h_1^2)^{1/2} \tag{6.3.27}$$

where

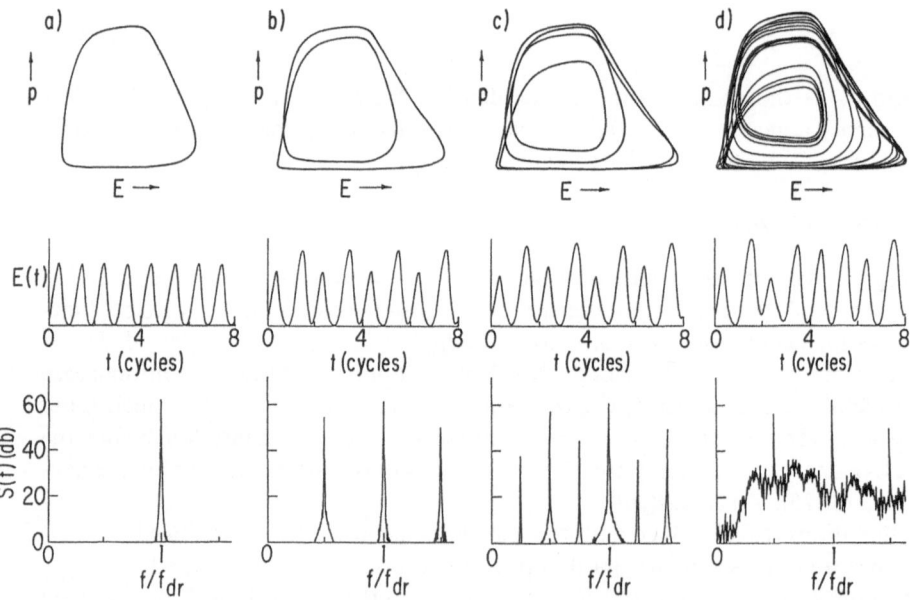

Fig. 6.27a–d. Calculated phase portraits of hole density p vs. field, time record of field vs. t, and power spectrum S vs. frequency for driving frequency $\omega_{dr} = 1.45\omega_0$ and four increasing drive amplitudes [6.98]: (a) $\Delta J/J_0 = 1.10$; (b) $\Delta J/J_0 = 1.25$; (c) $\Delta J/J_0 = 1.35$; (d) $\Delta J/J_0 = 1.45$

$$h_0 = \frac{4\pi e}{\epsilon_s} p \left\{ \frac{dv}{d\varepsilon} [X^s (N_A - N_D)/p + (X + T^s)p] \right.$$

$$\left. + v \left[(N_A - N_D - p) \frac{dX}{d\varepsilon} - (N_D + p) \frac{dT^s}{d\varepsilon} \right] \right\} > 0 \qquad \text{and}$$

$$-h_1 = - \left[\frac{4\pi e}{\epsilon_s} p + X^s (N_A - N_D)/p + Xp + T^s p \right] < 0$$

are the determinant and the trace, respectively, of the linear stability matrix obtained by linearizing (6.3.23–25); they correspond to the coefficients of the eigenvalue equation (3.5.18) derived for the case of an n-type semiconductor in Sect. 3.5.2. Thus only damped oscillations may occur. If, however, an ac driving current $\Delta J \cdot \sin(\omega_{dr} t)$ is superimposed on the dc current J_0, sustained oscillations are possible. Figure 6.27 shows the associated phase portraits for increasing values of $\Delta J/J_0$, and a driving frequency $\omega_{dr} = 1.45 \omega_0$ where $\omega_0 = 6422$ s^{-1}. Figure 6.27a shows a simply periodic limit cycle for $\Delta J/J_0 = 1.1$. Initially $\varepsilon(t)$ increases, which causes $p(t)$ to rise due to increased impact ionization. Then $\varepsilon(t)$ drops due to the enhanced dielectric relaxation that results from larger $p(t)$. Period doubling is observed in Fig. 6.27b where the drive strength has been increased to $\Delta J/J_0 = 1.25$. Period four is shown in Fig. 6.27c for drive strength $\Delta J/J_0 = 1.35$. Higher-order period doublings including period eight are observed as the drive is further increased; they accumulate at a critical value $\Delta J/J_0 = 1.39$, at which point the flow becomes chaotic.

Chaos is shown in Fig. 6.27d for $\Delta J/J_0 = 1.45$. The oscillatory character is not lost, but the motion is no longer periodic. In the associated power spectrum only the fundamental frequency, the first subharmonic, and their harmonics are still visible. The other discrete lines have disappeared with a concomitant increase in broad-band noise characteristic of the presence of deterministic chaos.

A Poincaré section in p and ε, and a first-return map of $\varepsilon(t)$ are shown in Fig. 6.28. The Poincaré section was obtained by plotting the phase points $(p(t), \varepsilon(t))$

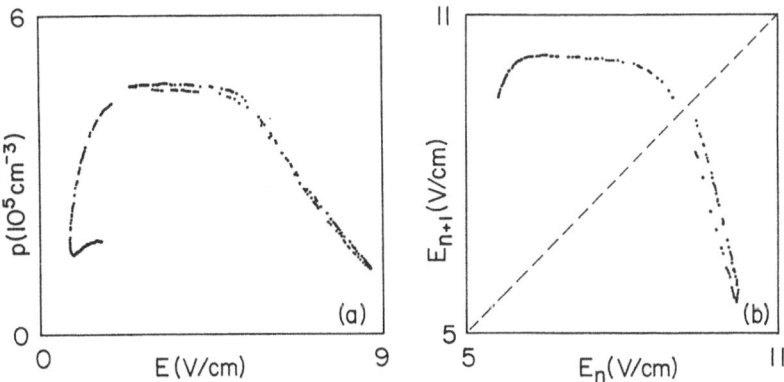

Fig. 6.28. (a) Poincaré section in p and field, constructed by strobing the phase portrait at constant drive phase. (b) Return map constructed from successive field maxima E_n. The numerical parameters are the same as in Fig. 6.27d. [6.98]

at fixed times $t_n = (2\pi/\omega_{dr})n$, $n = 1, 2, 3, \ldots$, where ω_{dr} is the drive frequency. Thus the sequence of obtained phase points $(p(t_n), \varepsilon(t_n))$ corresponds to the momentary positions of the phase flow at constant drive phase ϕ. The Poincaré section shows both the beginning of a fold in the lower left corner of Fig. 6.28a and the same fold one cycle later in the right side. This folding and stretching provides the mixing of trajectories which is a feature of chaos. The return map (Fig. 6.28b) was constructed from successive maxima ε_n of the field $\varepsilon(t)$. The obtained points $(\varepsilon_{n+1}, \varepsilon_n)$ are concentrated on a low-dimensional submanifold of the whole $(\varepsilon_{n+1}, \varepsilon_n)$ plane, which is characteristic for deterministic chaos, as opposed to stochastic motion. The obtained graph is reminiscent of one-dimensional maps that have been extensively studied [6.53, 63].

6.3.4 Impact-Ionization Induced Self-Generated Chaos

In this section we shall demonstrate that the two-level impact-ionization mechanism which was shown to exhibit limit-cycle oscillations in Sect. 6.2.2 can be used to explain self-generated chaos, as observed in the experiments described in Sect. 6.3.2 [6.100, 101]. The phase space of this model, defined by (6.2.26–28), consists of the electric field ε, the free-carrier concentration v, and the bound-carrier concentration v_{t_1} in the impurity ground state, and thus has the minimum dimension which is necessary to allow for chaos in autonomous, i.e., not externally driven, systems. Physically, the proposed mechanism invokes slow dielectric-relaxation oscillations driven by impact ionization from two impurity levels, and therefore applies to the typical experimental situations described in Sect. 6.3.2 above. For convenience, the formulas are given for an n-type semiconductor, although the model can readily be applied to p-type semiconductors by making the appropriate replacements. The temperature is assumed to be sufficiently low that the impurities are not thermally ionized. This implies helium temperatures and breakdown fields of only a few volts per centimeter in the case of shallow donors or acceptors, e.g., in InSb [6.87], n-GaAs [6.77–83, 97], and p-Ge [6.84–86, 88–94], or room temperatures and breakdown fields $\gtrsim 1$ kV/cm in case of deep traps, e.g., in semi-insulating GaAs [6.95].

The understanding of the mechanism for self-generated chaos is of great importance for the design of far-infrared photodetectors and integrated circuits, where oscillatory instabilities and broad-band noise are detrimental. Thus the analysis of a model for chaos in semiconductors might be useful in guiding the tailoring of low-noise electronic devices.

In the numerical solutions of the rate equations (6.2.26–28) with (6.2.51, 52, 53) presented in Figs. 6.29–39 below, the dimensionless g-r parameters were chosen as given in Table 6.1. The Hopf bifurcation of a limit cycle occurs at $\varepsilon_0 \approx 98$ for these parameters, compare the dispersion relation given in Fig. 3.13a.

Table 6.1. Dimensionless g-r parameters used in Figs. 6.29–39

$T_1^S N_D^* \tau_M$	$T^* \tau_M$	$X_1^S \tau_M$	$X^* \tau_M$	$X_1^0 N_D^* \tau_M$	$X_1^{*0} N_D^* \tau_M$	E_t	N_A/N_D^*	r_2
10^{-2}	10^{-5}	10^{-7}	10^{-7}	5×10^{-4}	10^{-2}	1	0.3	0.3

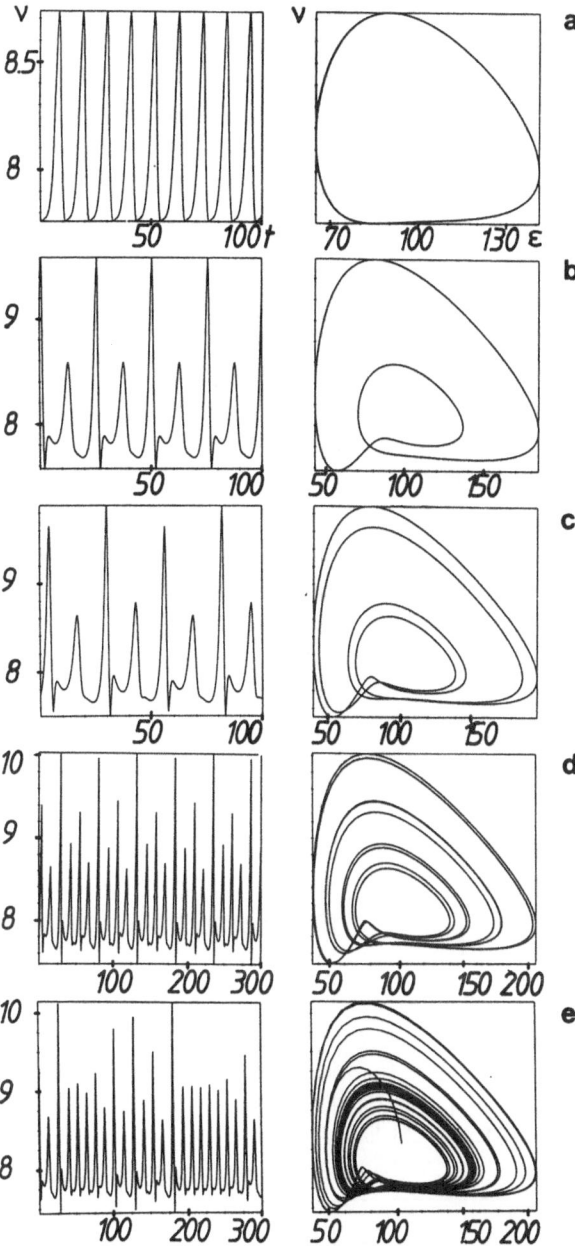

Fig. 6.29a–e. Numerical solutions of the two-level model, exhibiting a period-doubling route to chaos. The carrier concentration v in units $10^{-3} N_D^*$ versus time in units $10^4 \tau_M$ (*left column*), and phase portraits of v versus the field ε in units $(kTE_t)/(eL_D)$ (*right column*) are shown for the following static fields ε_0: (**a**) 102, (**b**) 105, (**c**) 105.3, (**d**) 105.42, (**e**) 105.5. Note the different time-scale in (**d**) and (**e**). Numerical parameters are given in Table 6.1

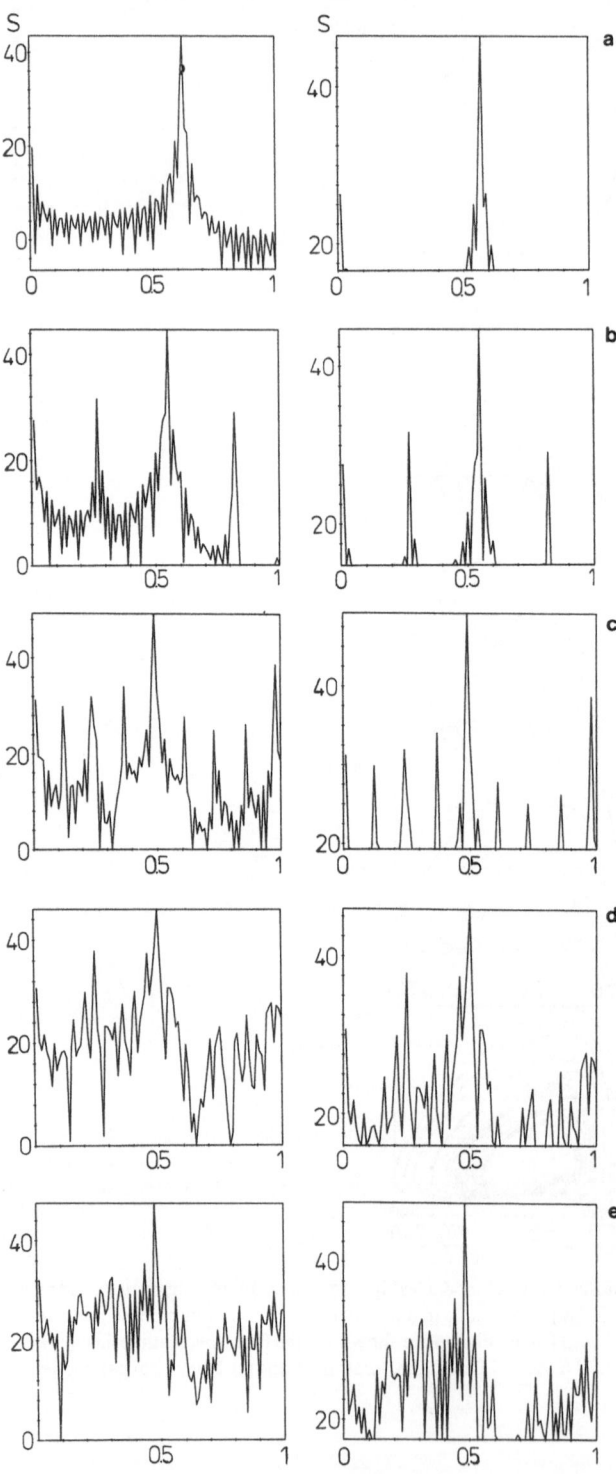

Fig. **6.30a–e.** Power spectra $S(\omega)$ of $\varepsilon(t)$ for
(**a**) $\varepsilon_0 = 103$ (limit cycle of period one),
(**b**) $\varepsilon_0 = 104$ (limit cycle of period two),
(**c**) $\varepsilon_0 = 105.4$ (limit cycle of period four),
(**d**) $\varepsilon_0 = 105.5$ (chaos),
(**e**) $\varepsilon_0 = 105.7$ (chaos).
The power S is in dB, and the angular frequency ω is in units $10^{-4}/\tau_M$. Numerical parameters as in Fig. 6.29. The Fourier transform was performed over a time interval $T = 10^7\tau_M$ of $\varepsilon(t)$. The vertical scale is different on the left and the right column

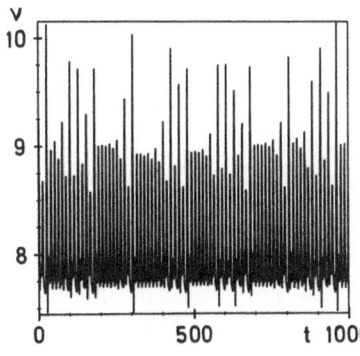

Fig. 6.31. Chaotic oscillations of the electron concentration versus time for a time interval $T = 10^7 \tau_M$ at $\varepsilon_0 = 105.5$. Numerical parameters as in Fig. 6.29

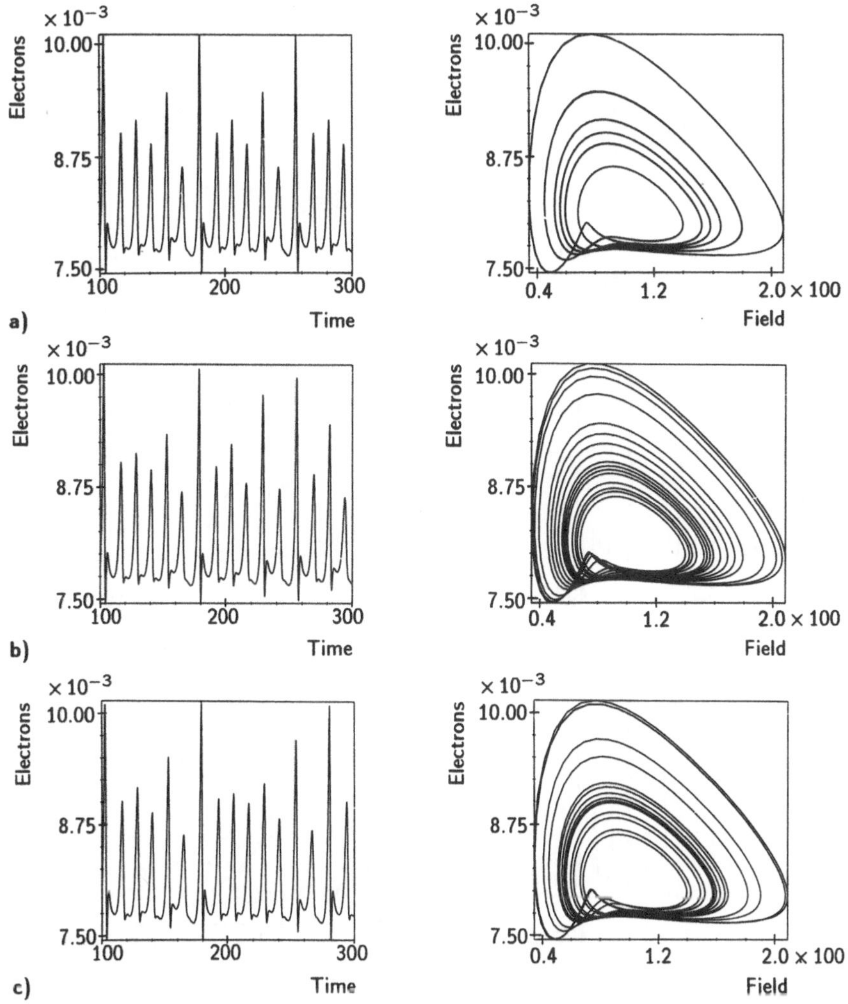

Fig. 6.32a–c. Time series and phase portraits for ε_0 in the chaotic regime: (a) 105.47 (period six window), (b) 105.48, (c) 105.49. Plot as in Fig. 6.29. The time interval is $2 \times 10^6 \tau_M$

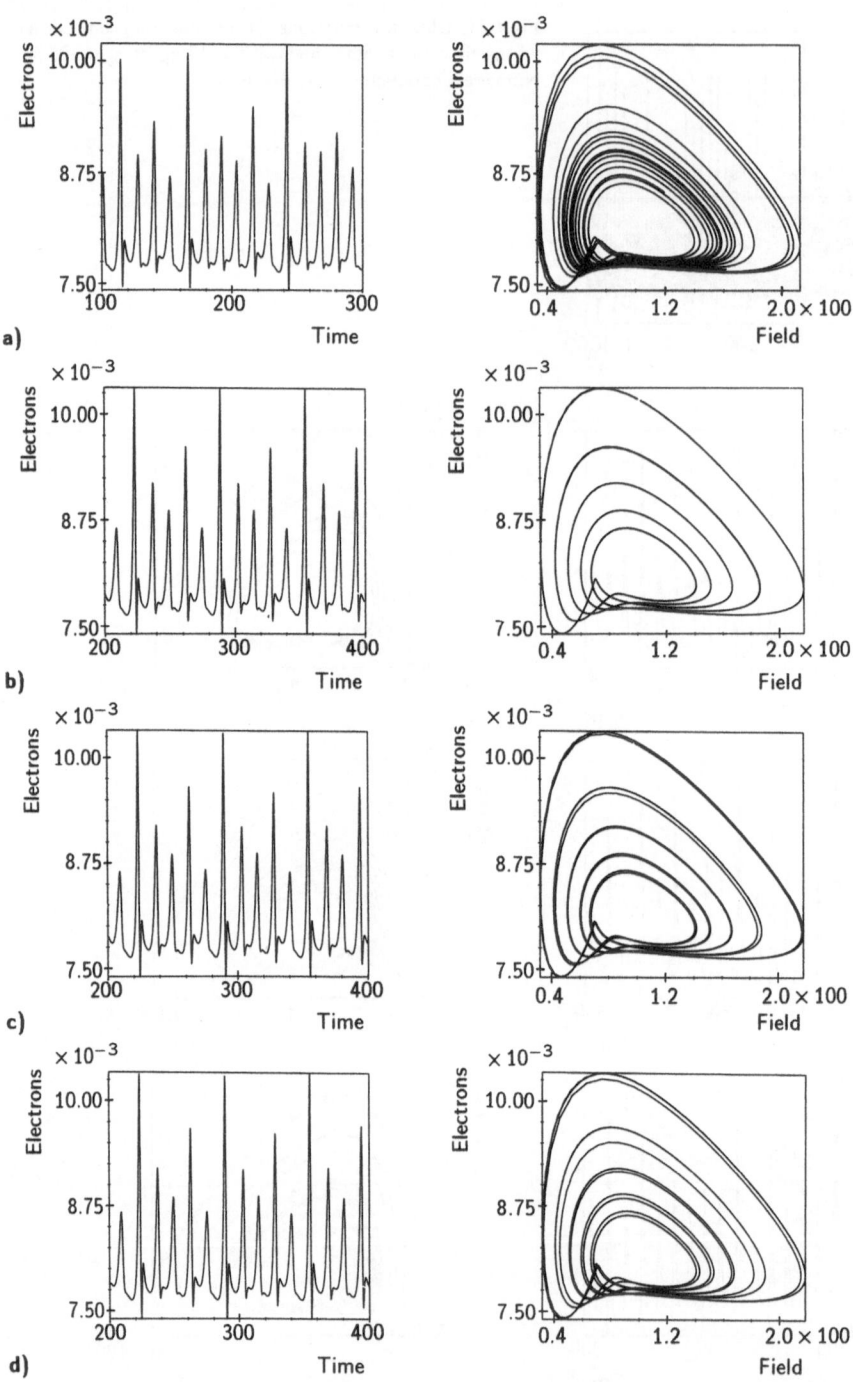

Fig. 6.33. As Fig. 6.32, for the following values of ε_0: (a) 105.525, (b) 105.585, (c) 105.590, (d) 105.595

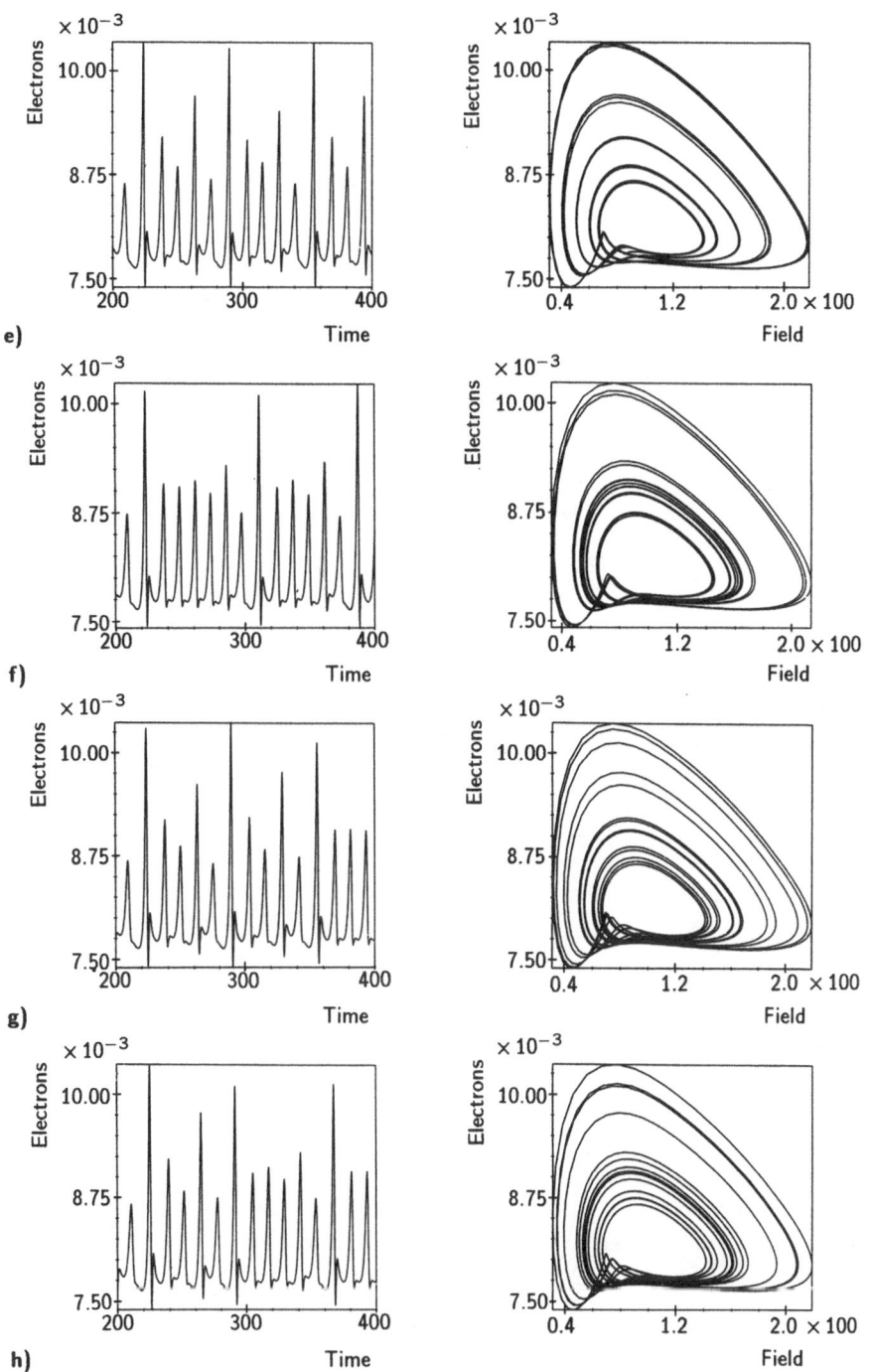

Fig. 6.33 (cont.). (e) 105.600 (f) 105.605 (g) 105.610 (h) 105.615. The time interval is $2 \times 10^6 \tau_M$

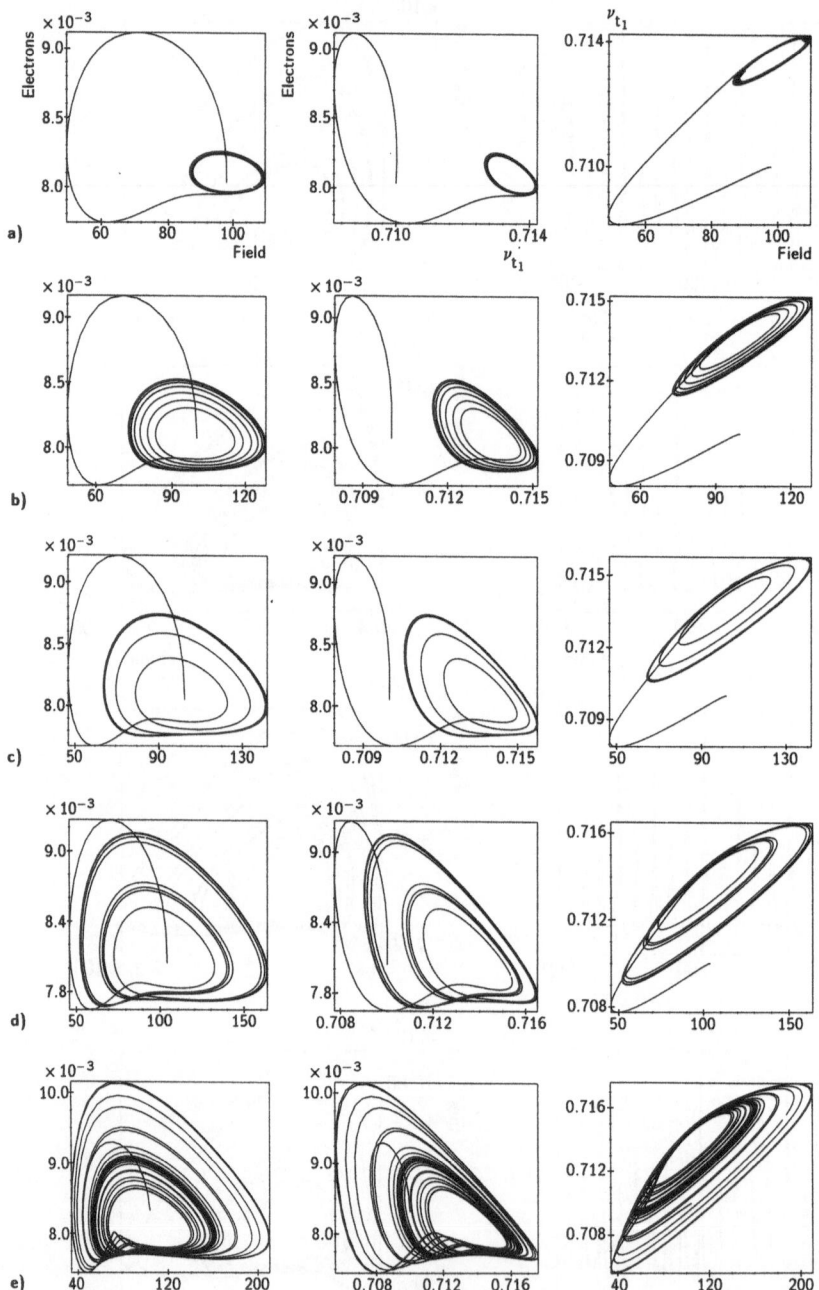

Fig. 6.34a–e. Phase portraits of transient oscillations, projected onto the $v - \varepsilon$ (*left column*), $v - v_{t_1}$ (*center column*), and $v_{t_1} - \varepsilon$ (*right column*) phase planes, for the following values of ε_0: (**a**) 98 (period one), (**b**) 100 (period one), (**c**) 102 (period one), (**d**) 104 (period two), (**e**) 105.5 (chaos). Numerical parameters as in Fig. 6.29. The starting point of the trajectories is $\varepsilon(0) = \varepsilon_0$, $v(0) \approx 8 \times 10^{-3}$, $v_{t_1}(0) = 0.71$; in (**a**)–(**d**) the time interval is $10^6 \tau_M$, in (**e**) $3 \times 10^6 \tau_M$

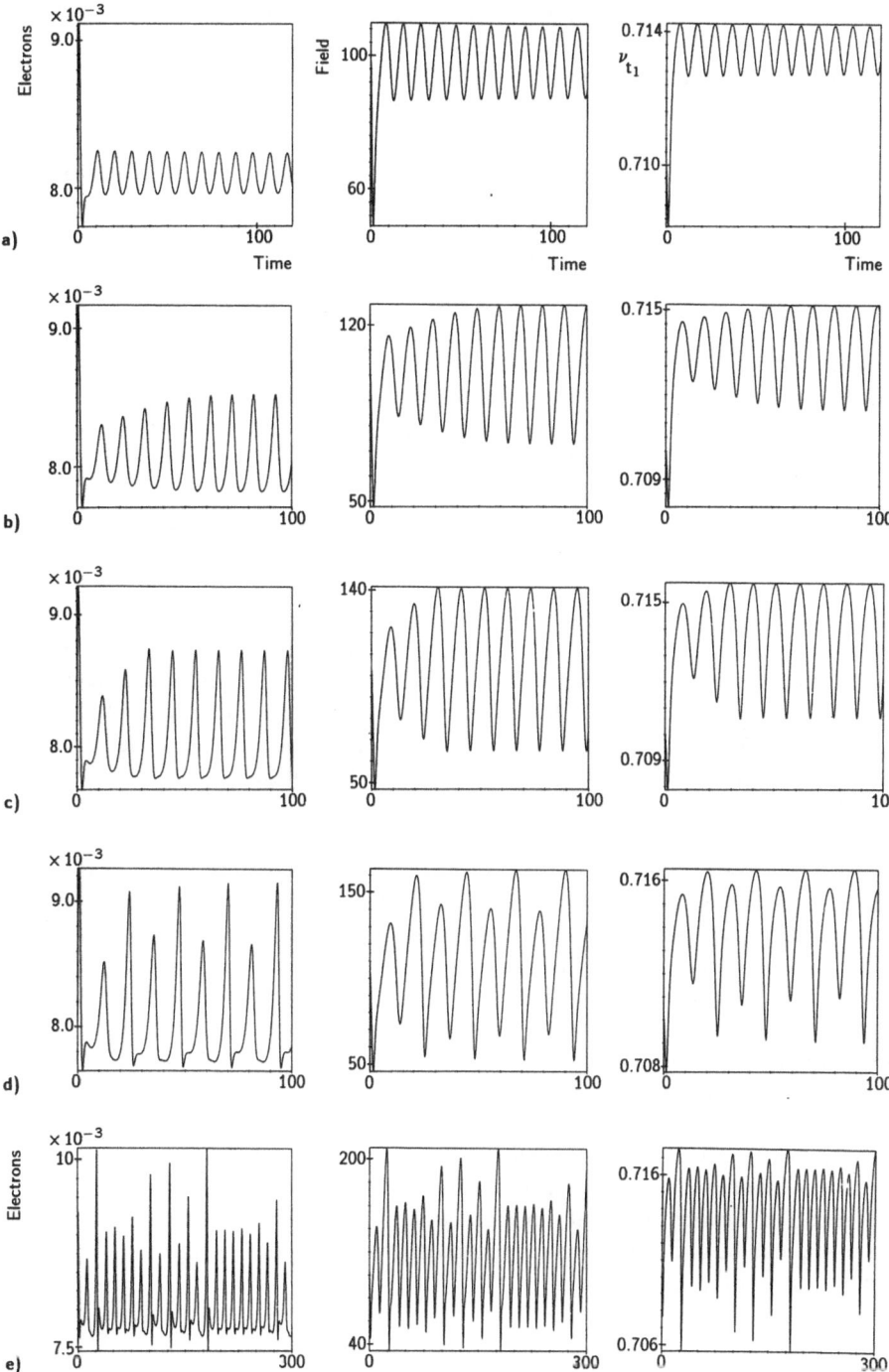

Fig. 6.35a–e. Time series $v(t)$ (*left column*), $\varepsilon(t)$ (*center column*) and $v_{t_1}(t)$ (*right column*) for the same parameters as in Fig. 6.34

Figure 6.29 shows a period-doubling route to chaos. The time series $v(t)$ and the phase portraits of v versus ε are depicted for a sequence of steady state fields ε_0 increasing from $\varepsilon_0 = 102$ (a) to $\varepsilon_0 = 105.5$ (e). The field ε_0 may be considered as a control parameter; it corresponds to an externally applied current density $J = v_0 v(\varepsilon_0)$, where v_0 denotes the steady state value of v at ε_0 on the NDC branch of the static current density-field characteristic. As ε_0 is increased, the amplitude of the limit cycle grows, and oscillations of period two (b), four (c), eight (d), and chaotic oscillations (e) are successively displayed. Corresponding power spectra of $\varepsilon(t)$ are shown in Fig. 6.30. For a limit cycle of period one the spectrum is sharply peaked at the intrinsic angular oscillation frequency $\omega_0 = 0.57 \times 10^{-4}/\tau_M$ (a). A second, lower peak at $\omega_0/2$ appears for a limit cycle of period two (b). Two further, still lower, peaks at $\omega_0/4$ and $3\omega_0/4$ emerge at period four (c), while in the chaotic regime a high level of broad-band noise is present (d, e). The right column shows the same sequence in a different power scale, which supresses the background noise which is due to the finite number of cycles (≈ 100) used in the Fourier transform.

The chaotic regime is analysed in more detail in Figs. 6.31–35 [6.101]. The chaotic oscillations of the electron concentration are shown in Fig. 6.31 for a longer time interval than in Fig. 6.29e, exhibiting characteristic erratic outbursts of large ampli-tude oscillations. The time series $v(t)$ and the phase portraits $v(\varepsilon)$ are plotted in Figs. 6.32, 33 for a sequence of fields ε_0 within the chaotic regime. Figure 6.32a depicts a periodic window of period six, followed by a chaotic band (b, c). Figure 6.33 shows chaos (a), interrupted by periodic motion of period five (b) and a subsequent period doubling bifurcation (c, d), and followed by a transition back to chaos (e–h).

In Fig. 6.34 the projections of the three-dimensional phase trajectories on all three coordinate planes are shown for the period-doubling route to chaos. Unlike in Figs. 6.29–33, the transient behavior is also shown here. The starting point of the trajectories has been chosen close to the unstable homogeneous NDC steady state $(\varepsilon_0, v_0, v_{t_1 0})$. The trajectories asymptotically approach the limit cycles (a–d), after an initial large excursion in phase space. This initial swing represents the rapid ap-proach of the trajectories to the "slow manifold" on which the dielectric-relaxation oscillations around the unstable focus $(\varepsilon_0, v_0, v_{t_1 0})$ occur. After only a few oscillations the trajectories have reached the limit cycle. For ε_0 close to the Hopf bifurcation (Fig. 6.34a) the limit cycle is approximately elliptic, and has a small amplitude. The corresponding oscillations of $\varepsilon(t)$, $v(t)$, $v_{t_1}(t)$ are sinusoidal (harmonic). With in-creasing control parameter ε_0 the limit cycle grows in amplitude, and becomes nonlinearly distorted. The period-doubling sequence is marked by a folding over of the trajectory. In the $v - v_{t_1}$ projection (center column) the shape becomes almost triangular, while the $v_{t_1} - \varepsilon$ projection assumes a long and thin shape stretched in the direction of the first diagonal. The triangular shape in the concentration phase plane $v - v_{t_1}$ can be understood as composed of two pieces of slow motion close to the two surfaces in $(v, v_{t_1}, \varepsilon)$ space

$$\dot{v} = \psi_0(v, v_{t_1}, \varepsilon) = 0 \quad \text{and} \tag{6.3.28}$$

$$\dot{v}_{t_1} = \psi_1(v, v_{t_1}, \varepsilon) = 0 \;, \tag{6.3.29}$$

and a rapid, almost vertical transition between these two sheets. The corresponding slow rise in v, its fast decay, and its slow transit to the next cycle show up clearly in the respective time series (Fig. 6.35, left column). This spiky shape is characteristic of relaxation oscillations. It reflects the time-scale separation between the slow electric field, and the fast relaxing ("slaved") carrier concentrations. For the g-r mechanism (6.2.51) the surfaces (6.3.28) and (6.3.29) are given explicitly by

$$v_{t_1} = \frac{X_1^S + [X_1^*(\varepsilon)N_D^* - X_1^S - T_1^S N_A]v - [T_1^S + X_1^*(\varepsilon)]N_D^* v^2}{X_1^S + [X_1^*(\varepsilon) - X_1(\varepsilon)]N_D^* v} \tag{6.3.30}$$

and

$$v_{t_1} = \frac{T^*(1 - v)}{T^* + X^* + X_1(\varepsilon)N_D^* v} \tag{6.3.31}$$

respectively. The surface (6.3.31) is monotonically decreasing and convex in v and ε, whereas the surface (6.3.30) is S-shaped ("rippled") for large enough ε, such that the intersection of these two surfaces produces the static S-shaped $v(\varepsilon_0)$ characteristic.

In Fig. 6.36 the local field maxima ε_n are plotted versus the control parameter ε_0. The Feigenbaum period-doubling cascade, as well as chaotic bands at $\varepsilon_0 \geq 105.45$ and noise-free windows of period six at $\varepsilon_0 = 105.475$, and of period five at $\varepsilon_0 = 105.575$ can be seen. The lack of very sharp bifurcation points is due to the long non-asymptotic transients which occur in the vicinity of bifurcations [6.62]. The inset shows the Poincaré map of ε_{n+1} versus ε_n, reconstructed from successive maxima of $\varepsilon(t)$ at $\varepsilon_0 = 105.5$. It is strongly reminiscent of the one-dimensional quadratic maps studied in the theory of discrete dynamic systems (cf., Fig. 20 in Ref. [6.59a]). Similar return maps were obtained for other values of ε_0 (Fig. 6.37). Because of the finite resolution of the bifurcation diagram no definite asymptotic value of the Feigenbaum number δ, defined in (6.3.4), could be obtained.

In experiments the chaotic attractor is usually reconstructed from the stroboscopic time series of a single variable, say ε, by the p-tupels $\{\varepsilon(t_n), \varepsilon(t_n + \tau), \varepsilon(t_n + 2\tau) \ldots, \varepsilon(t_n + (p - 1)\tau)\}$ where $t_n = n\Delta t$, Δt is the sampling time, and τ is a suitable delay

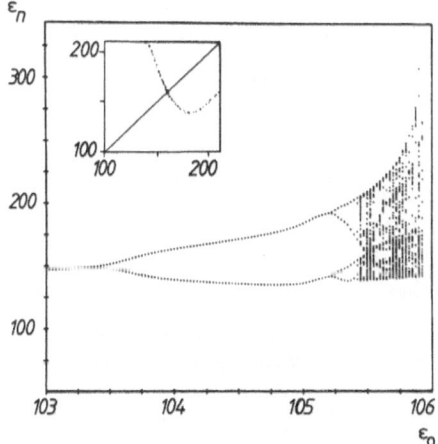

Fig. 6.36. Bifurcation diagram of the field maxima ε_n versus the control parameter ε_0. The inset shows the return map ε_{n+1} versus ε_n, where $\{\varepsilon_1, \ldots, \varepsilon_n, \varepsilon_{n+1}, \ldots\}$ is the set of successive field maxima, for $\varepsilon_0 = 105.5$. Numerical parameters as in Fig. 6.29. Approximately 100 points were plotted for each $\varepsilon_0 \geq 105.45$ (time interval $10^7 \tau_M$)

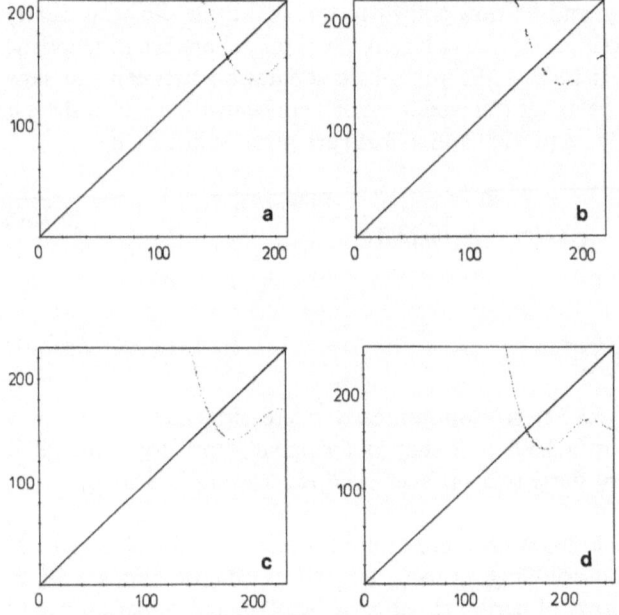

Fig. 6.37a–d. Return maps of ε_{n+1} versus ε_n for the following values of ε_0: (a) 105.5, (b) 105.6, (c) 105.7, (d) 105.8. Numerical parameters as in Fig. 6.29

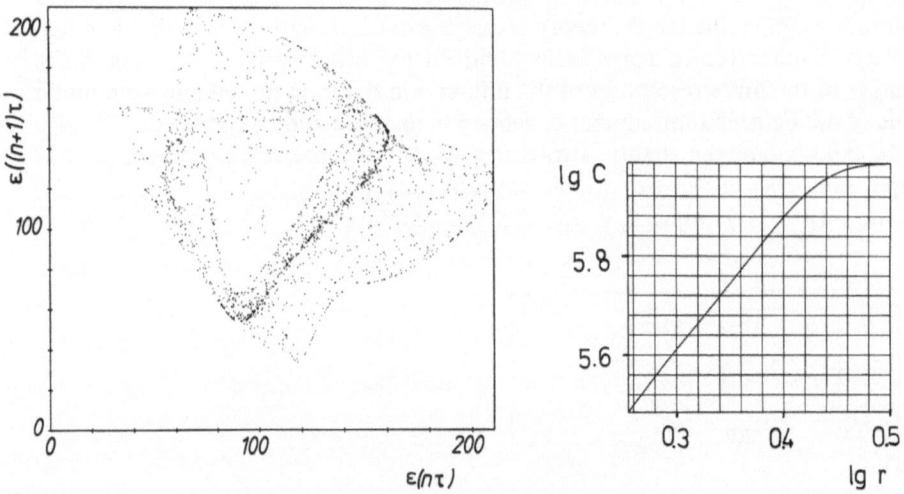

Fig. 6.38. Stroboscopic plot of $\varepsilon((n+1)\tau)$ versus $\varepsilon(n\tau)$ for $\tau = 10^5\tau_M$ and $\varepsilon_0 = 105.5$ with 2000 points plotted, showing the strange attractor in an embedding dimension $p = 2$

Fig. 6.39. Plot of $\lg C$ versus $\lg r$ for the strange attractor at $\varepsilon_0 = 105.5$. $C(r)$ is the number of pairs of points on the attractor in a p-dimensional embedding space whose distance is less than r. A total of 1000 stroboscopic points on the attractor was used, and $p = 15$, $\tau = 10^5\tau_M$ was chosen

time. Figure 6.38 depicts the attractor of Fig. 6.29e with $\tau = \Delta t = 10^5 \approx 0.8 T_0$, where T_0 is the intrinsic period of oscillation, in an embedding space of dimension $p = 2$. The fractal dimension of the chaotic attractor can in principle be determined in such a reconstructed phase space of sufficiently large embedding dimension p by applying the Grassberger-Procaccia [6.69] algorithm, see (6.3.7). However, care should be taken, since the asymptotic scaling region of r-values may not be reached, even if the plot appears to conserve the scaling law $C(r) \sim r^\nu$ perfectly (Fig. 6.39). The slope (e.g., 2.7 for $p = 15$) increases with p, reflecting a dependence upon the total length of signal $(p - 1)\tau$ used to reconstruct the attractor, as was also found for the Lorenz attractor [6.70b, p. 180].

The physical origin of the obtained chaotic dielectric-relaxation oscillations can be understood as follows: injected charge is trapped, which increases the electric field. This enhances impact ionization of the trapped charge, which creates more free carriers and leads to a reduced field due to increased dielectric relaxation. Hereby the trapping rate becomes dominant over the ionization rate, which completes the cycle. While this is similar in the mechanism for *driven* chaos proposed by *Teitsworth* and *Westervelt* (TW), see Sect. 6.3.3, the present model differs from theirs in two essential points: first, there is no periodic driving force; instead, the additional internal dynamic degree of freedom which is necessary to produce chaos in *autonomous* nonlinear systems is furnished by the redistribution of trapped carriers between the ground and the excited state. The mechanism that provides the mixing of orbits on the strange attractor is the *displaced re-injection* of phase trajectories in three-dimensional phase space $(\varepsilon, \nu, \nu_{t_1})$ onto the bistable ("rippled") slow submanifold of dielectric-relaxation oscillations. This represents a novel physical example of spiral-type chaos [6.57]. Second, it suffices to use the simple Shockley formula for the impact-ionization coefficients, while the TW mechanism requires a more elaborate, nonmonotonic dependence of the coefficient upon the field, which is still the subject of investigation [6.106].

The numerical parameters chosen are representative of high-purity materials at low temperatures. Inspection of Figs. 6.29–36 reveals a close similarity with experimentally observed time series, phase portraits, power spectra, and bifurcation diagrams, see Figs. 6.21–26, although an optimum fit of the parameters has not yet been attempted. In the figures presented, all the times, fields, and concentrations can be scaled by varying the temperature, the impurity energy, and the doping density. For p-Ge at 4.2 K [6.84, 88–94], with $\mu_0 = 10^6$ cm^2/Vs, $\epsilon_s = 16$, effective doping $\approx 10^{11}$ cm^{-3}, and an acceptor level at 10 meV, for example, the physical units in Figs. 6.29–39 are $\tau_M \approx 10^{-4}$ µs, $(kTE_t)/(eL_D) \approx 0.2$ V/cm, while for semi-insulating GaAs at 300 K [6.95] with $\mu_0 = 10^4$ cm^2/Vs, $\epsilon_s = 12.5$, $N_D^* = 10^{10}$ cm^{-3}, and a trap level at 700 meV, $\tau_M \approx 10^{-4}$ ms, $(kTE_t)/(eL_D) \approx 10$ V/cm. The oscillation frequency is lower for lower doping and lower mobility.

The condition for an oscillatory instability (6.2.21) requires a delicate balance between the differential mobility, the g-r time constants, and a low, but not too low, carrier density which can be sensitively controlled by the temperature, by optical radiation, or by magnetic freeze-out due to a small transverse magnetic field. This elucidates the role of such additional control parameters in some of the experiments.

The longitudinal oscillatory instability may be coupled with the transverse filamentary instability of SNDC samples. This can result in an oscillatory periodic or chaotic "breathing" of the current filaments (cf., Sect. 6.1.1iv) which shows up as small amplitude current oscillations. Such microampère oscillations superimposed to a *dc* current of some milliampères have indeed been observed in the filamentary SNDC regime, in connection with intermittency, see Fig. 6.24.

An extension of the model which also includes the time-dependent balance equation of the mean energy E per carrier, cf. (1.1.15), has been proposed recently [6.102]. If the timescales of the dielectric field-relaxation and the energy relaxation are well separated, then the dynamics is dominated by the respective slow variable, while the fast variable is slaved. In case of fast energy relaxation, the energy balance assumes a quasi-steady state almost instantaneously, determining E as a function of the electric field ε, which varies slowly according to (6.2.28), and the dielectric relaxation model discussed above is obtained. If, on the other hand, in the post-breakdown regime, or at higher doping levels, the increased number of carriers leads to a shorter dielectric relaxation time, E can now take the role of the slow dynamic variable. *Energy relaxation oscillations* may thus arise from the coupling of impact ionization with energy relaxation of the hot carriers. Here the mean energy E, rather than the field ε, represents the dynamic variable complementary to v: The free carriers are heated up by the applied bias, until their energy is sufficient to impact ionize. This leads to cooling of the free carriers due to the energy loss of the ionizing carriers. This, in turn, decreases the impact ionization rate until trapping prevails, which completes the cycle of oscillating carrier density and mean carrier energy. The mechanism differs from that of dielectric relaxation oscillations [6.100] in that the self-inhibitory step of the cycle is energy loss of hot carriers due to impact ionization, rather than dielectric relaxation of the electric field. In the simplest, spatially homogeneous model, the energy balance equation is given by

$$\dot{E} = ev\varepsilon - (E - E_0)/\tau_e - E_{th}\dot{v}_{ii}/v - E\dot{v}/v \qquad (6.3.32)$$

where τ_e is the energy relaxation time, $E_0 = 3kT_L/2$ is the thermal energy, E_{th} is the average energy lost by each ionizing carrier, and $\dot{v}_{ii}$ is the impact ionization rate. The essential nonlinearity is the energy loss through impact ionization, since the impact ionization coefficient depends strongly upon the mean energy per carrier.

The energy relaxation model can also be applied to experiments showing cross-talk and spatial correlations between two electrically separated parts of a single crystal [6.103]. Both the shift of the static I–V characteristics of the individual subsystems and the dynamic correlations of the oscillatory chaotic behaviour observed in the postbreakdown regime of p-Ge at 4.2 K can be explained by a coupled two-cell model, where each cell (denoted by a subscript 1 or 2, respectively) is described by a set of g-r and energy balance equations. Additionally, the carrier energies E_1 and E_2 are coupled by heat diffusion $\dot{E}_{1,\text{Diff}} := -D(E_1 - E_2)$ due to the rapid exchange of phonons. (Note the excellent heat conductivity of a Ge crystal!) Here D is a heat diffusion constant. The dynamic behaviour of the coupled two-cell system is very complex and allows for periodic as well as chaotic solutions [6.104].

Another possibility for the onset of g-r induced chaos in semiconductors has been pointed out in connection with the discrete nature of the individual g-r events [6.105]. One starts with a Chapman-Kolmogoroff equation for the probability of finding the system at time t in a state with a certain number of carriers as a result of g-r processes. A recurrence relation for the average carrier concentration at successive discrete time intervals is derived by averaging the Chapman-Kolmogoroff equation. The resulting difference equation is similar to the non-invertible one-dimensional maps which have been found to lead to chaos in the theory of discrete dynamic systems.

6.3.5 Two-Carrier Effects

The above examples of chaos were essentially single-carrier effects, associated with impact ionization of shallow impurities. In the following, we discuss chaotic behavior in a two-carrier (electron-hole) plasma, which was recently found in Ge at 77 K in parallel electric and magnetic fields [6.73]. The physical origin of the observed instability is connected with the so-called *oscillistor effect* which we mentioned already in Sect. 6.1.1 [6.40–43]: spontaneous coherent current oscillations of an electron-hole plasma in Ge, Si, and InSb in a static electric field and a parallel time-independent magnetic field are known to be the result of screw-shaped helical density waves [6.44]. *Held* et al. [6.73–76] discovered that when this instability is strongly excited by an increasing static electric field there are quasi-periodic as well as period-doubling transitions to chaos. They used $1 \times 1 \times 10$ mm^3 n-Ge samples with a net donor concentration of $N_D \approx 10^{12}$ cm^{-3} and double injection from an n$^+$ and a p$^+$ contact on opposite 1×1 mm^2 ends producing an electron-hole plasma of density $n \approx p \approx 10^{13}$ cm^{-3}. A 4 kG magnetic field B_0 was applied parallel to the length of the sample.

The voltage V_0 was taken as a control parameter and increased from 0 to 25 V. For $V_0 < 6$ V, the current I was constant in time. At $V_0 = 6$ V, I spontaneously became periodic. Regions of chaotic dynamics occurred in several subsequent voltage intervals. Such an interval with a period-doubling sequence is shown in Figs. 6.40, 41. Starting in Fig. 6.40a, at $V_0 = 10.0$ V, $I(t)$ is oscillating at a fundamental frequency $f_0 \approx 1.2 \times 10^5$ s^{-1}. As V_0 is increased, $I(t)$ shows a period-doubling bifurcation; the power spectrum shown in Fig. 6.40b displays the emergence of a line at $f_0/2$. At larger V_0, another period-doubling bifurcation occurs (Fig. 6.40c) with new spectral components at $f_0/4$, $3f_0/4$. At slightly larger V_0, $I(t)$ becomes nonperiodic (Fig. 6.40d), and the power spectrum enters a region of broad-band noise. For further increases of V_0 there appear noise-free windows of periods 3, 4, 5, ... within this region of broad-band noise; see Fig. 6.40e for period three. This sequence ends at $V_0 = 10.7$ V, with a return to period one oscillations.

Figure 6.41 is a bifurcation diagram, obtained by plotting the local current maxima I_{max} vs. V_0. The period-doubling route to chaos as well as the noise-free windows can be clearly seen.

For several different values of B_0, Held et al. observed a quasi-periodic route to chaos. As the voltage V_0 was increased, the onset of a quasi-periodic state was followed by a transition to turbulence. For instance, at $B_0 = 11.5$ kG and $V_0 = 2.865$

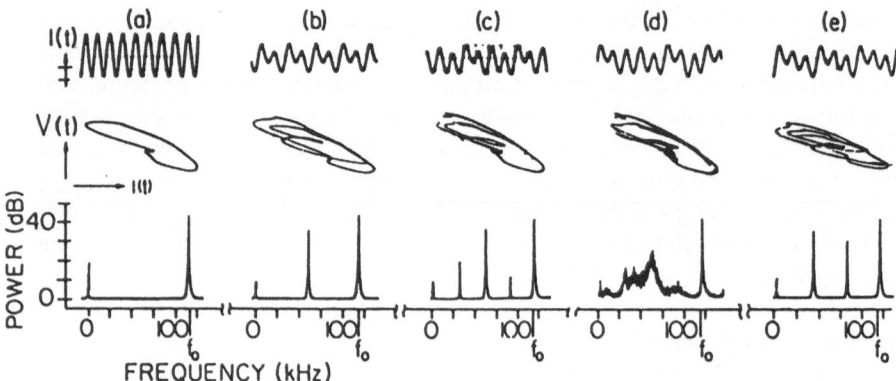

Fig. 6.40. Measured current oscillations $I(t)$; phase portrait of voltage $V(t)$ vs. current $I(t)$; and power spectra $S(\omega)$ for a Ge crystal under double injection and parallel electric and magnetic fields. The applied voltage increases from (a) to (e) as shown in Fig. 6.41 [6.73]

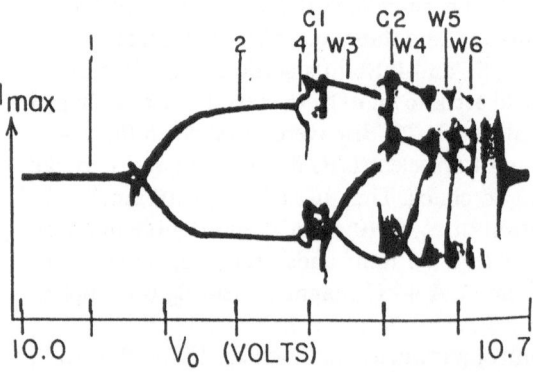

Fig. 6.41. Measured bifurcation diagram of local current maxima, I_{max}, vs. applied voltage, V_0, for the same sample as in Fig. 6.40. Labels 1, 2, 4, C1, and W3 refer to regions displayed in Figs. 6.40a–e, respectively. Labels W4, W5, and W6 refer to noise-free windows of periods four, five, and six, respectively. [6.73]

V, the current oscillated at a fundamental frequency $f_1 = 6.3 \times 10^4$ s^{-1}. At $V_0 =$ 2.907 V, a second spectral component, incommensurate with f_1, appeared at $f_2 = 1.4 \times 10^4$ s^{-1}; thus the system became quasiperiodic. At $V_0 = 2.942$ V, the system was still quasi periodic, however, the two modes were interacting and thus the power spectrum contained sums and differences of the form $mf_1 + nf_2$ (m and n integers). The onset of turbulence was indicated by a slight broadening of the spectral peaks. As V_0 was increased further, the electron-hole plasma exhibited increased turbulence, followed by a return to quasiperiodicity at $V_0 = 3.125$ V and, subsequently, simple periodicity at $V_0 = 3.442$ V.

Two-dimensional Poincaré sections of the flow in a reconstructed phase space are obtained by plotting the local current maxima, I_{n+1} vs. I_n. Periodic motion corresponds to a one-dimensional phase trajectory: a limit cycle. The Poincaré section in this case is simply a point. Similarly, when the system is quasiperiodic (corresponding to motion which fills a whole two-dimensional torus, as shown in Fig. 6.19) the Poincaré section is a set of points which lie dense on a closed curve in (I_n, I_{n+1}) phase space. When the system, however, is turbulent, the points (I_n, I_{n+1}) fill an entire region of this two-dimensional phase space. This region forms a strange

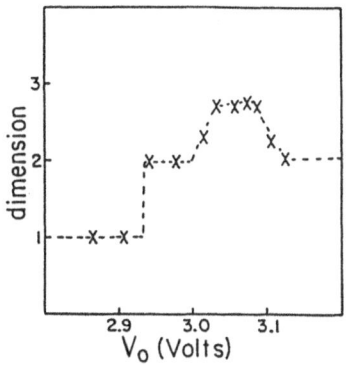

Fig. 6.42. Fractal dimension, d, of the strange attractor vs. applied voltage, V_0, measured for a Ge crystal under double injection and parallel electric and magnetic fields. [6.73]

attractor, the dimension of which cannot be determined by visual inspection of the Poincaré section. For these attractors, Held et al. calculated the fractal dimension from the experimental data. To this end successive values of $I(t)$ were recorded at $T = 200$ µs intervals. From these data the p-dimensional phase spaces $\{I_n, I_{n+1}, \ldots, I_{n+p-1}\}$, where $I_n := I(t_0 + nT)$, were reconstructed. In such a p-dimensional space the number of points $n(r)$ on the attractor which are contained in a p-dimensional hypersphere of radius r should scale as r^d for small r, where d is the fractal dimension of the attractor, provided that p is sufficiently greater than d. This follows from (6.3.6), assuming that the total number of points on the attractor is $n(r)N(r) = $ const. The number $n(r)$ was measured for many r and increasingly larger p, until

$$d = \log n(r)/\log r$$

converged. The result is shown in Fig. 6.42. Within the chaotic regime, the fractal dimension d varies between two and three. At $V_0 = 3.058$ V, for instance, the fractal dimension is $d = 2.7$. If the observed turbulence were due to thermal noise or other stochastic processes, then a measurement of the dimension d would not have converged for relatively small p; the dimension d would have been of the order of 10^{23}.

Oscillations or chaos were not observed on a similar sample with ohmic contacts (which inject only electrons). When this sample, however, was optically pumped (creating both electrons and holes), current oscillations and period-doubling transitions to chaos occurred. This indicates that the observed behavior is in fact a property of the two-carrier plasma. As a theoretical explanation, Held et al. suggested nonlinear couplings between an unstable oscillistor wave and a pair of damped plasma density waves. Coupled wave equations of this type are of a form known to exhibit a period-doubling route to chaos [6.107].

References

Chapter 1

Dielectric Breakdown in Solids

1.1   K.W. Wagner: Trans. American Inst. Electr. Engin. **41**, 288 (1922);
H. Lueder, W. Schottky, E. Spenke: Naturwiss. **24**, 61 (1936)

Early Papers on Nonequilibrium Phase Transitions in Semiconductors

Overheating Instability
1.2   A.F. Volkov, Sh.M. Kogan: Sov. Phys. Usp. **11**, 881 (1969)

Gunn Effect
1.3   E. Pytte, H. Thomas: Phys. Rev. **179**, 431 (1969)

g-r Instability
1.4   P.T. Landsberg, A. Pimpale: J. Phys. C9, 1243 (1976)

Nonequilibrium Phase Transitions

Chemical Reaction Systems
1.5   F. Schlögl: Z. Phys. **253**, 147 (1972)

Laser
1.6   R. Graham, H. Haken: Z. Phys. **213**, 420 (1968); **237**, 31 (1970);
V. DeGiorgio, M.O. Scully: Phys. Rev. A2, 1170 (1970)

Synergetics

1.7   H. Haken, R. Graham: Umschau **6**, 191 (1971)
1.8   H. Haken: *Synergetics, An Introduction*, 3rd ed. (Springer, Berlin, Heidelberg 1983)
1.9   H. Haken: *Advanced Synergetics* (Springer, Berlin, Heidelberg 1983);
see also Springer Series in Synergetics, ed. by H. Haken (Springer, Berlin, Heidelberg)

Textbooks on Applied Semiconductor Physics

1.10   B.G. Streetman: *Solid State Electronic Devices*, 2nd ed. (Prentice-Hall, Englewood Cliffs, NJ 1980)
1.11   S.M. Sze: *Physics of Semiconductor Devices*, 2nd ed. (Wiley, New York 1981)

Dissipative Structures

1.12   G. Nicolis, I. Prigogine: *Self-Organization in Non-Equilibrium Systems* (Wiley, New York 1977)

Bifurcation Theory

1.13   D.H. Sattinger: *Topics in Stability and Bifurcation Theory* (Springer, Berlin 1973)
1.14   O. Gurel, O.E. Rössler (eds.): *Bifurcation Theory and Applications in Scientific Disciplines*. Annals of the New York Academy of Sciences, Vol. 316 (New York Academy of Sciences, New York 1979)
1.15   G. Iooss, D.D. Joseph: *Elementary Stability and Bifurcation Theory* (Springer, Berlin, Heidelberg 1981)
1.16   S.N. Chow, J.K. Hale: *Methods of Bifurcation Theory* (Springer, Berlin, Heidelberg 1982)

288   References

Mathematical Theory of Dynamic Systems

2-Variable Systems

1.17   A.A. Andronov, E.A. Leontovich, I.I. Gordon, A.G. Maier: Vol. 1: *Qualitative Theory of Second-Order Dynamic Systems* (Wiley, New York 1973); Vol. 2: *Theory of Bifurcations of Dynamic Systems on a Plane* (Israel Program for Scientific Translations, Jerusalem 1971)

3-Variable Systems

1.18   J. Guckenheimer, P. Holmes: *Nonlinear Oscillations, Dynamical Systems, and Bifurcations of Vector Fields* (Springer, Berlin, Heidelberg 1983)

Catastrophe Theory

1.19   R. Thom: *Structural Stability and Morphogenesis* (Benjamin, New York 1975) (Original ed. Paris 1972)
1.20   E.C. Zeeman: *Catastrophe Theory. Selected Papers* (Addison-Wesley, Reading, Mass. 1977)
1.21   T. Poston, I. Stewart: *Catastrophe Theory and its Applications* (Pitman, London 1978)
1.22   V.I. Arnold: *Catastrophe Theory* (Springer, Berlin, Heidelberg 1984)

Solitons

1.23   R.K. Bullough, P.J. Caudrey (eds.): *Solitons*, Topics Curr. Phys., Vol. 18 (Springer, Berlin, Heidelberg 1980);
G.L. Lamb: *Elements of Soliton Theory* (Wiley, New York 1980);
G. Eilenberger: *Solitons*, Springer Ser. Solid-State Sci., Vol. 19 (Springer, Berlin, Heidelberg 1981)

Chaos

1.24   H.G. Schuster: *Deterministic Chaos* (Physik-Verlag, Weinheim 1984)

Stochastic Measures in Nonequilibrium Thermodynamics

1.25   F. Schlögl: Physics Reports **62**, 267–380 (1980);
Rostocker Physikalische Manuskripte **8**, 30 (1985)

Overviews of Instabilities in Semiconductors

1.26   A.F. Volkov, Sh.M. Kogan: Sov. Phys. Usp. **11**, 881 (1969)
1.27   H. Hartnagel: *Semiconductor Plasma Instabilities* (Elsevier, New York 1969)
1.28   F. Stöckmann: "Elektrische Instabilitäten in Halb- und Photoleitern", in *Festkörperprobleme* **9**, 138 (Vieweg, Braunschweig 1969)
1.29   J.E. Carroll: *Hot Electron Microwave Generators* (Arnold, London 1970)
1.30   V.L. Bonch-Bruevich, I.P. Zvyagin, A.G. Mironov: *Domain Electrical Instabilities in Semiconductors* (Consultant Bureau, New York 1975)
1.31   M.P. Shaw, H.L. Grubin, P. Solomon: *The Gunn-Hilsum Effect* (Academic Press, New York 1979)
1.32   J. Pozhela: *Plasma and Current Instabilities in Semiconductors* (Pergamon, Oxford 1981)
1.33   M. Asche, Z.S. Gribnikov, V.V. Mitin, O.G. Sarbei: *Hot Electrons in Many-Valley Semiconductors* (in Russian) (Naukova Dumka, Kiev 1982)
1.34   M.P. Shaw, N. Yildirim: Adv. Electr. Electron Phys. **60**, 307 (1983)
1.35   M.P. Shaw, H.L. Grubin, E. Schöll: *The Physics of Instabilities in Solid State Electron Devices* (Plenum, New York), in preparation

Stochastic Methods

1.36   N.G. van Kampen: *Stochastic Processes in Physics and Chemistry* (North-Holland, Amsterdam 1981)
1.37   C.W. Gardiner: *Handbook on Stochastic Methods*, 2nd. ed., Springer Ser. Syn., Vol. 13 (Springer, Berlin, Heidelberg 1986)

1.38   W. Horsthemke, R. Lefever: *Noise-Induced Transitions*, Springer Ser. Syn., Vol. 15 (Springer, Berlin, Heidelberg 1983)
1.39   H. Risken: *The Fokker-Planck Equation*, Springer Ser. Syn., Vol. 18 (Springer, Berlin, Heidelberg 1984)

Combinations of SNDC and NNDC Instabilities (see also [Ref. 1.33, p. 147 (Fig. 33)])

1.40   J. Peinke, A. Mühlbach, R.P. Huebener, J. Parisi: Phys. Lett. **108A**, 407 (1985)
1.41   C.L. Dick, B. Ancker-Johnson: Phys. Rev. **B5**, 526 (1972)
1.42   V.A. Pogrebnyak: Sov. Phys. Semicond. **14**, 1210 (1980)
1.43   V.V. Mitin: Sov. Phys. Semicond. **11**, 727 (1977)

Domains and Filaments in NDC Elements

1.44   B.K. Ridley: Proc. Phys. Soc. **82**, 954 (1963)
        H. Thomas: Lecture Notes: Conf. on Fluctuation Phenomena. Chania, Crete (1969)
1.45   E. Schöll: *Festkörperprobleme* **26**, 309 (Vieweg, Braunschweig 1986)

Violation of Minimum Entropy Production Principle in Current Filaments

1.46   A.F. Volkov, Sh.M. Kogan: Sov. Phys. JETP **25**, 1095 (1967)
1.47   K. Takeyama, K. Kitahara: J. Phys. Soc. Jpn. **39**, 125 (1975)

Validity of Minimum Entropy Production Principle

1.48   R. Landauer: Phys. Rev. **A12**, 636 (1975)
1.49   F. Schlögl: "On the statistical background of the Glansdorff-Prigogine criterion", in *Thermodynamics and regulation of biological processes*, ed. by A.I. Zotin (Nauka, Moscow 1984)

Influence of the Load Line on the Stability of NDC States

1.50   J.L. Jackson, M.P. Shaw: Appl. Phys. Lett. **25**, 666 (1974)
1.51   M. Büttiker, H. Thomas: Z. Phys. **B34**, 301 (1979)

NDC Devices (Original Papers)

Tunnel Diode
1.52   L. Esaki: Phys. Rev. **109**, 603 (1958)
1.53   P.T. Landsberg, M.S. Abrahams: Electron. Lett. **21**, 59 (1985)

Gunn Diode (see also [6.7–15])
1.54   J.B. Gunn: IBM J. Res. Develop. **8**, 141 (1964)

IMPATT Diode
1.55   W. Shockley: Bell Syst. Tech. J. **33**, 799 (1954)
1.56   W.T. Read: Bell Syst. Tech. J. **37**, 401 (1958)

Thyristor
1.57a) W. Fulop: IEEE Trans. ED-10, 120 (1963);
1.57b) M. Stoisiek, R. Sittig: *Festkörperprobleme* **26**, 361 (Vieweg, Braunschweig 1986)

Pin Diode
1.58   R.C. Prim: Bell Syst. Tech. J. **32**, 665 (1953)

Ovonic Switch
1.59   S.R. Ovshinsky: Phys. Rev. Lett. **21**, 1450 (1968)

Reviews on Mechanisms for Drift Instabilities

1.60   J.C. McGroddy, M.I. Nathan, J.E. Smith, jr.: IBM J. Res. Develop. **13**, 543, 554 (1969)
1.61   H. Thomas: In *Synergetics*, ed. by H. Haken (Stuttgart, Teubner 1973), p. 87

Original Papers
1.62   M. Büttiker, H. Thomas: Solid-State Electron. **21**, 95 (1978); Z. Phys. **B33**, 275 (1979); **34**, 301 (1979)

Mechanisms for g-r instabilities (see also [6.16–21])

Field-Assisted Trapping (NNDC)
1.63   B.K. Ridley, T.B. Watkins: J. Phys. Chem. Solids **22**, 155 (1961)
1.64   B.K. Ridley, R.G. Pratt: J. Phys. Chem. Solids **26**, 21 (1965)
1.65   V.L. Bonch-Bruevich: Sov. Phys. Solid State **6**, 1615 (1965)
1.66   V.L. Bonch-Bruevich, S.G. Kalashnikov: Sov. Phys. Solid State **7**, 599 (1965)
1.67   K.W. Böer: IBM J. Res. Develop. **13**, 573 (1969)

Low Temperature Impurity Breakdown (SNDC) (see also [2.3–13, 24–36])
1.68   R.P. Khosla, J.R. Fischer, B.C. Burkey: Phys. Rev. **B7**, 2551 (1973)
1.69   E. Schöll: Z. Phys. **B46**, 23 (1982)

Switching in Amorphous Thin Films (SNDC) (see also [1.59])
1.70   D. Adler, M.S. Shur, M. Silver, S.R. Ovshinsky: J. Appl. Phys. **51**, 3289 (1980)
1.71   P.T. Landsberg, D.J. Robbins, E. Schöll: Phys. Status Solidi (a) **50**, 423 (1978)

Double Injection in pin Diodes (SNDC)
1.72   M.A. Lampert, P. Mark: *Current Injection in Solids* (Academic, New York 1970);
        M.A. Lampert, R.B. Schilling: Semicond. Semimetals **6**, 1 (1970);
        R. Baron, J.W. Mayer: Semicon. Semimetals **6**, 202 (1970)

IMPATT Devices (SNDC)
1.73   W.E. Schroeder, G.I. Haddad: Proc. IEEE **61**, 153 (1973)
1.74   P. Bauhahn, G.I. Haddad: IEEE Trans. **ED-24**, 634 (1977)

Scattering-Induced NDC in 2-Dimensional Systems

1.75   B.K. Ridley: Proc. 17th Int'l. Conf. Physics of Semiconductors (San Francisco 1984), ed. by J.D. Chadi, W.A. Harrison (Springer, New York 1985) p. 401;
        M.A.R. Al-Mudares, B.K. Ridley: Physica **134B**, 526 (1985)

Reviews on Thermal Breakdown

1.76   L. Altcheh, N. Klein: IEEE Trans. **ED-20**, 801 (1973)
1.77   N. Klein: Thin Solid Films **100**, 335 (1983)

Reviews on Transport in Semiconductors

1.78   C.M. Snowden: Rep. Prog. Phys. **48**, 223 (1985)
1.79   C. Jacoboni, L. Reggiani: Rev. Mod. Phys. **55**, 645 (1983)
1.80   H.L. Grubin, D.K. Ferry, G.J. Iafrate, J.R. Barker: VLSI Electronics Microstructure Sci. **3**, 197–299 (Academic, New York 1982)
1.81   J.R. Barker; In *Physics of Non-linear Transport in Semiconductors*, ed. by D.K. Ferry, J.R. Barker, C. Jacoboni. (Plenum, New York 1980) p. 127 and 589
1.82   L. Reggiani (ed.): *Hot Electron Transport in Semiconductors*, Topics Appl. Phys., Vol. 58 (Springer, Berlin, Heidelberg 1985)
1.83   G. Baccarani, M.R. Wordeman: Solid-State Electron. **28**, 407 (1985);
        G. Baccarani, M. Rudan, R. Guerrieri, P. Ciampolini: "Physical Models for Numerical Device Simulation", in *Advances in CAD for VLSI*, Vol. 1, ed. by W. Engl (North Holland, Amsterdam 1986), pp. 107–158

1.84   E.M. Conwell: *High-Field Transport in Semiconductors* (Academic, New York 1967)
1.85   B. Nag: *Theory of Electrical Transport in Semiconductors* (Pergamon, Oxford 1972)
1.86   K.C. Kao, W. Hwang: *Electrical Transport in Solids* (Pergamon, Oxford 1981)

First-Principle Derivation of Macroscopic Transport Equations

1.87   H. Stumpf: *Quantum Processes in Polar Semiconductors and Insulators*, Pts. 1 and 2 (Vieweg, Braunschweig 1983)

Textbooks on Semiconductor Physics

1.88   O. Madelung: *Grundlagen der Halbleiterphysik* (Springer, Berlin, Heidelberg 1970)
1.89   K.H. Seeger: *Semiconductor Physics*, 3rd ed. (Springer, Berlin, Heidelberg 1985)

Semiconductor Junctions and Contacts (see also [4.40–44])

1.90   M.P. Shaw: "Properties of Junctions and Barriers", in *Handbook on Semiconductors*, Vol. 4, ed. by T.S. Moss (North Holland, Amsterdam 1981) Chap. 1
1.91   A. Herlet, E. Spenke: Z. Angew. Phys. **7**, 99, 149, 195 (1955)
1.92   H. Benda, A. Hoffmann, E. Spenke: Solid-St. Electron. **8**, 887 (1965)
1.93   J.A.G. Slatter, J.P. Whelan: Solid-St. Electron. **23**, 1235 (1980)

Coherent Macroscopic Electronic Excitations

1.94   A. Stahl, I. Balslev: *Electrodynamics of the Semiconductor Band Edge* (Springer, Berlin, Heidelberg 1987)

Solution of Quantum-Mechanical Boltzmann Equation by Cumulant Expansion

1.95   B.C. Eu: J. Chem. Phys. **80**, 2123 (1984)

Transport Processes in Gases

1.96   S. Chapman, T.G. Cowling: *The Mathematical Theory of Non-Uniform Gases* (Cambridge Univ. Press, Cambridge 1970)

Critical Behavior

Equilibrium Phase Transitions
1.97   H.E. Stanley: *Phase Transitions and Critical Phenomena* (Clarendon, Oxford 1971)
1.98   S.K. Ma: *Modern Theory of Critical Phenomena* (Benjamin, New York 1976)
1.99   V. Dohm, R. Folk: *Festkörperprobleme* **22**, 1 (Vieweg, Braunschweig 1982)
1.100   V. Dohm: Z. Phys. **B60**, 61 (1985); **61**, 193 (1985)
1.101   M.N. Barber: In *Phase Transitions and Critical Phenomena*, Vol. 8, ed. by C. Domb, J.L. Lebowitz (Academic, London 1983)
1.102   H.W. Diehl, S. Dietrich: *Festkörperprobleme* **25**, 39 (Vieweg, Braunschweig 1985)
1.103   E. Eisenriegler: Z. Phys. **B61**, 299 (1985);
        Th.W. Burkhardt, E. Eisenriegler: J. Phys. **A18**, L83 (1985)
1.104   F. Schlögl, E. Schöll: Z. Phys. **B51**, 61 (1983)

Nonequilibrium Phase Transitions
1.105   A. Nitzan, P. Ortoleva, J. Deutch, J. Ross: J. Chem. Phys. **61**, 1056 (1974)

Bit Number Cumulants as Characteristic Critical Quantities (see also [1.25])
1.106   F. Schlögl: Z. Phys. **267**, 77 (1974); **B20**, 177 (1975); **B22**, 301 (1975)
1.107   F. Schlögl: Z. Phys. **B52**, 51 (1983)

Tricritical Exponents
1.108   R.B. Griffiths: Phys. Rev. B7, 545 (1973)

Tricritical Behavior in Laser Phase Transitions
1.109   D. Walgraef, P. Borckmans, G. Dewel: Z. Physik **B30**, 437 (1978)

Examples of Various Synergetic Systems (see also [1.2–9, 12])

Tunnel Diode
1.110   R. Landauer: J. Appl. Phys. **33**, 2209 (1962)

Chemical Reaction Systems (see also [1.5, 12, 105])
1.111   W. Ebeling: *Physik der Selbstorganisation und Evolution* (Akademie-Verlag, Berlin 1982)

Ballast Resistor and Superconducting Hot Spots
1.112   D. Bedeaux, P. Mazur: Physica **105A**, 1 (1981);
        L. Freytag, R.P. Huebener: J. Low Temp. Phys. **60**, 377 (1985)

Semiconductors with g-r Instabilities (see also [1.4, 45])
1.113   E. Schöll, P.T. Landsberg: Proc. R. Soc. (London) Ser. **A365**, 495 (1979);
        E. Schöll: Z. Phys. **B46**, 23 (1982); **48**, 153 (1982); **52**, 321 (1983)
1.114   P.T. Landsberg: Eur. J. Phys. **1**, 31 (1980) (review of the early work)

Semiconductor Lasers
1.115   E. Schöll, P.T. Landsberg: J. Opt. Soc. Am. **73**, 1197–1206 (1983);
        E. Schöll, D. Bimberg, H. Schumacher, P.T. Landsberg: IEEE J. **QE-20**, 394–399 (1984);
        D. Bimberg, K. Ketterer, E. Schöll, H.P. Vollmer: Electron. Lett. **20**, 640–641 (1984); Physica
        129B, 469 (1985);
        D. Bimberg, K. Ketterer, E.H. Böttcher, E. Schöll: Int. J. Electronics **60**, 23 (1986) (review)

Traffic Flow on Motorways
1.116   R. Kühne: Physik in unserer Zeit **15**, 84 (1984)

Recent Conference Proceedings
        see Springer Series in Synergetics, ed. by H. Haken, Vols. 2–6, 8, 9, 11, 12, 17, 21–24, 26–32
        G. Benedek, H. Bilz, R. Zeyher (eds.): *Statics and Dynamics of Nonlinear Systems* Springer Ser.
        Solid-State Sci., Vol. 46 (Springer, Berlin, Heidelberg 1983)

Large Bibliography up to 1981
1.117   P.C. Hohenberg, J.S. Langer: J. Stat. Phys. **28**, 193 (1982)

Hot Electron Phenomena (see also [1.79–84])

1.118   E. Gornik (ed.): Proc. 4th Int'l. Conf. on Hot Electrons in Semiconductors (Innsbruck 1985),
        Physica **134B** (1985)
1.119   D. Bimberg, H. Münzel, A. Steckenborn: Physica **117** and **118B**, 214 (1983)
1.120   T. Grave, E. Schöll, H. Wurz: J. Phys. **C16**, 1693 (1983)

Nonequilibrium Recombination Statistics (Reviews and Monographs)

1.121   P.T. Landsberg: *Festkörperprobleme* **6**, 174 (Vieweg, Braunschweig 1966)
1.122   D.A. Evans: In *Solid State Theory*, ed. by P.T. Landsberg (Wiley, New York 1969)
1.123   P.T. Landsberg: "Semiconductor Statistics", in *Handbook on Semiconductors*, ed. by T.S. Moss,
        Vol. I, Chap. 8 (North Holland, Amsterdam 1982)
1.124   J.S. Blakemore: *Semiconductor Statistics* (Pergamon, Oxford 1962)
1.125   P.T. Landsberg, M.J. Adams: J. Lum. **7**, 3 (1973)
1.126   W. Schulz: *Festkörperprobleme* **5**, 165 (Vieweg, Braunschweig 1966)
1.127   H.J. Hoffmann, F. Stöckmann: *Festkörperprobleme* **19**, 271 (Vieweg, Braunschweig 1979)

Three-Particle Auger Processes

1.128   P.T. Landsberg: J. Phys. **C9**, L 111 (1976);
        A. Pimpale: J. Phys. **C11**, 1085 (1978)

Reviews on Impact Ionization

1.129   D.J. Robbins: Phys. Status Solidi (b) **97**, 9, 387; **98**, 11 (1980)
1.130   F. Capasso: "Physics of avalanche photodiodes", in *Semicond. Semimetals* **22D**, 1–172 (Academic, New York 1985)

Band-Band Impact Ionization

1.131   W. Shockley: Solid-State Electron. **2**, 35 (1961)
1.132   B.K. Ridley: J. Phys. C**16**, 3373; 4733 (1983);
         M.G. Burt: J. Phys. C**18**, L477 (1985);
         S. McKenzie, M.G. Burt: J. Phys. C**19**, 1959 (1986)

Impact Ionization from Impurities (see also [1.95])

1.133   A. Zylbersztejn: Phys. Rev. **127**, 744 (1962)
1.134   M.E. Cohen, P.T. Landsberg: Phys. Rev. **154**, 683 (1967); Phys. Status Solidi (b) **64**, 39 (1974)
1.135   D.J. Robbins, P.T. Landsberg: J. Phys. C**13**, 2425 (1980)
1.136   V.F. Bannaya, L.I. Velesova, E.M. Gershenzon, V.A. Chuenkov: Sov. Phys. Semicond. **7**, 1315 (1974)
1.137   R.M. Westervelt, S.W. Teitsworth: J. Appl. Phys. **57**, 5457 (1985)
1.138   V.V. Mitin: Appl. Phys. A**39**, 123 (1986)

Nonlinear Oscillations

1.139   J.J. Stoker: *Nonlinear Vibrations* (Wiley, New York 1950)
1.140   N. Minorsky: *Nonlinear Oscillations* (Van Nostrand, Toronto 1962)
1.141   A.A. Andronov, A. Witt, S.E. Khaikin: *Theory of Oscillators* (Pergamon, London 1966)
1.142   P. Hagedorn: *Nonlinear Oscillations* (Oxford Univ. Press, Oxford 1981)

Global Bifurcation of Limit Cycles in Chemical Reaction Systems

1.143   C. Escher: Z. Phys. B**35**, 351 (1979); Ber. Bunsenges. Phys. Chem. **84**, 387 (1980)

Early Investigation of Solitons

1.144   H. Seeger, H. Donth, A. Kochendörfer: Z. Phys. **134**, 173 (1953)

Chapter 2

Review and Classification of g-r Processes

2.1   E. Schöll: Proc. Roy. Soc. A**365**, 511–21 (1979)
2.2   P.T. Landsberg, M.J. Adams: J.Lum. **7**, 3 (1973)

One-Carrier Models for SNDC

2.3   R.S. Crandall: J. Phys. Chem. Solids **31**, 2069 (1970); Phys. Rev. B**1**, 730 (1970)
2.4   A.A. Kastalskij: Phys. Status Solidi (a) **15**, 599 (1973)
2.5   A.G. Zabrodskij, I.S. Shlimak: Sov. Phys. Solid State **16**, 1528 (1975)
2.6   W. Pickin: Solid State Electr. **21**, 309, 1299 (1978)
2.7   E. Schöll, P.T. Landsberg: Proc. 14th Int. Conf. Physics of Semiconductors (Edinburgh 1978), ed. by B.L.H. Wilson, *Inst. Phys. Conf. Ser.* **43**, 461 (Institute of Physics, Bristol 1979)
2.8   E. Schöll, P.T. Landsberg: Proc. R. Soc. (London) Ser. A**365**, 495 (1979)
2.9   E. Schöll: Proc. 3rd Int'l. Conf. on Hot Carriers in Semiconductors (Montpellier 1981), J. Physique C**7**, 57 (1981)

294   References

2.10   E. Schöll: Z. Phys. **B46**, 23 (1982)
2.11   E. Schöll, W. Heisel, W. Prettl: Z. Phys. **B47**, 285–291 (1982)
2.12   W.G. Proctor, P. Lawaetz, Y. Marfaing, R. Triboulet: Phys. Status Solidi (b) **110**, 637 (1982)
2.13   E. Schöll: Z. Phys. **B52**, 321 (1983)

Impurity Breakdown in Semiconductors

2.14   S.H. Koenig, R.D. Brown, W. Schillinger: Phys. Rev. **128**, 1668 (1962)
2.15   M.E. Cohen, P.T. Landsberg: Phys. Rev. **154**, 683 (1972)
2.16   A.E. McCombs, A.G. Milnes: Int. J. Electron. **32**, 361 (1972)

g-r Kinetics of F Centers in Alkali Halogenides

2.17   F. Lüty: Halbleiterprobleme **6**, 238 (1961)
2.18   H.J. Hoffmann: Phys. Status Solidi (b) **57**, 123 (1973)
2.19   H. Stumpf: *Quantum Processes in Polar Semiconductors and Insulators* (Vieweg, Braunschweig 1983)
2.20   R. Swank, F. Brown: Phys. Rev. **130**, 34 (1963)
2.21   F. De Martini, U.M. Grassano, F. Simoni: Opt. Commun. **11**, 8 (1974)

Nonexistence Theorem for Limit Cycles

2.22   P. Hanusse: C.R. Acad. Sci. C**274**, 1245 (1972);
2.23   J.J. Tyson, L.C. Light: J. Chem. Phys. **59**, 4164 (1973)

Experiments on SNDC Induced by Impurity Impact Ionization

Ge
2.24   N. Sclar, E. Burstein: J. Phys. Chem. Solids **2**, 1 (1957)
2.25   A.L. McWhorter, R.H. Rediker: Proc. Inst. Radio Engrs. **47**, 1207 (1959)
2.26   I. Melngailis, A.G. Milnes: J. Appl. Phys. **33**, 995 (1962)
2.27   A. Zylbersztejn: J. Phys. Chem. Solids **23**, 297 (1962)
2.28   F. Brown, D. Parker, J. Heyman, N. Newbury: Appl. Phys. Lett. **49**, 1548 (1986)

GaAs
2.29   P.J. Oliver: Phys. Rev. **127**, 1045 (1962)
2.30   R.A. Reynolds: Solid State El. **11**, 385 (1968)
2.31   T.O. Poehler: Phys. Rev. **B4**, 1223 (1971)
2.32   G.E. Stillman, C.M. Wolfe, J.O. Dimmock: *Semiconductors and Semimetals* **12**, 169 (Academic, New York 1977)
2.33   W. Heisel, W. Böhm, W. Prettl: Int. J. Infrared Millim. **2**, 829 (1981)

CdSe
2.34   R.P. Khosla, J.R. Fischer, B.C. Burkey: Phys. Rev. **B7**, 2551 (1973)

CdTe
2.35   N.V. Agrinskaya, M.V. Alekseenko, O.A. Matveev: Sov. Phys. Semicond. **9**, 341, 1286 (1976)

InSb
2.36   E.H. Putley: *Semiconductors and Semimetals* **1**, 289 (Academic, New York 1966)

Space Charge Limited Currents

2.37   M.A. Lampert, P. Mark: *Current Injection in Solids* (Academic, New York 1970)

Tricritical Behaviour in Equilibrium

2.38   R.B. Griffiths: Phys. Rev. Lett. **24**, 715 (1970)
2.39   R.B. Griffiths: Phys. Rev. **B7**, 545 (1973)

2.40   E.G.D. Cohen: In *Fundamental Problems in Statistical Mechanics*, Vol. III, ed. by E.G.D. Cohen (North Holland, Amsterdam 1975) p. 47

Critical Behaviour in Equilibrium (see also [1.104])

2.41   R.B. Griffiths, J.C. Wheeler: Phys. Rev. A2, 1047 (1970)

Critical Behaviour of Bit-Number Variance (see also [1.106])

2.42   F. Schlögl: Z. Phys. B52, 51 (1983)
2.43   E. Schöll, unpublished

Nonlinear FIR Magneto-Photoconductivity and -Absorption in n-GaAs

2.44   C.R. Pidgeon, A. Vass, G.R. Allan, W. Prettl, L. Eaves: Phys. Rev. Lett. 50, 1309 (1983)
2.45   W. Prettl, A. Vass, G.R. Allan, C.R. Pidgeon: Int. J. Infrared Millim. Waves 4, 561 (1983)

Cyclotron Resonance Induced Nonequilibrium Phase Transitions in n-GaAs

2.46   R. Obermaier, W. Böhm, W. Prettl, P. Dirnhofer: Phys. Lett. 105A, 149 (1984)
2.47   M. Weispfenning, I. Hoeser, W. Böhm, W. Prettl, E. Schöll: Phys. Rev. Lett. 55, 754 (1985)

Mobility of n-GaAs Under Cyclotron Resonance Absorption

2.48   H.J.A. Bluyssen, I.C. Maan, T.B. Tan, P. Wyder: Phys. Rev. B22, 749 (1980)

Two-Carrier Model for Second-Order Phase Transitions

2.49   P.T. Landsberg, A. Pimpale: J. Phys. C9, 1243 (1976)
2.50   A. Pimpale, P.T. Landsberg: J. Phys. C10, 1447 (1977)

Two-Carrier Models for NDC and Threshold Switching

2.51   D. Adler, M.S. Shur, M. Silver, S.R. Ovshinsky: J. Appl. Phys. 51, 3289 (1980)
2.52   N. Klein, P. Solomon: J. Appl. Phys. 47, 4364 (1976)
2.53   I. Kashat, N. Klein: J. Appl. Phys. 48, 5217 (1977)
2.54   N. Klein: J. Appl. Phys. 53, 5828 (1982)

Two-Carrier Models for SNDC and Threshold Switching (see also [2.7, 8])

2.55   P.T. Landsberg, D.J. Robbins, E. Schöll: Phys. Status Solidi (a) 50, 423–6 (1978)
2.56   D.J. Robbins, P.T. Landsberg, E. Schöll: Phys. Status Solidi (a) 65, 353–64 (1981)

Experimental Data on Trapping Cross Sections

2.57   V.L. Bonch-Bruevich, E.G. Landsberg: Phys. Status Solidi 29, 9 (1968)

Reviews on Threshold Switching (in particular, in amorphous thin films)

2.58   D. Adler: *Amorphous Semiconductors* (CRC Press, Cleveland 1971), Chap. VII
2.59   H. Fritzsche: In *Amorphous and Liquid Semiconductors*, ed. by J. Tauc (Plenum, New York 1974) Chap. 6
2.60   D. Adler, H.K. Henisch, N.F. Mott: Rev. Mod. Phys. 50, 209 (1978)
2.61   R. Landauer, J.W.F. Woo: Comments on Sol. State Phys. 4, 139 (1972) (thermal switching)
2.62   M.P. Shaw, N. Yildirim: Adv. Electr. Electron Phys. 60, 307 (1983)

Threshold Switching in Crystalline Solids, Organic Films, Liquids (see also [2.5, 35])

2.63   A. Szymanski, D.C. Larson, M.M. Labes: Appl. Phys. Lett. 14, 88 (1969)

2.64   V.K. Zaitsev, O.A. Golikova, M.M. Kazanin, V.M. Orlov, E.N. Tkalenko: Soviet Phys. –
       Semicond. **9**, 1372 (1976)
2.65   A.I. Popov, I.K. Geller, V.K. Shemetova: Phys. Status Solidi (a) **44**, K71 (1977)

Threshold Switching in Amorphous Semiconductors (mainly experimental) (see also [2.51])

2.66   S.R. Ovshinsky: Phys. Rev. Lett. **21**, 1450 (1968)
2.67   H.J. Stocker, C.A. Barlow, jr., D.F. Weirauch: J. Non-Cryst. Solids **4**, 523 (1970)
2.68   M.P. Shaw, S.H. Holmberg, S.A. Kostylev: Phys. Rev. Lett. **31**, 542 (1973)
2.69   M.P. Shaw, S.C. Moss, S.A. Kostylev, L.H. Slack: Appl. Phys. Lett. **22**, 114 (1973)
2.70   D. Adler: J. Vacuum Sci. Technol. **10**, 728 (1973)
2.71   S.M. Rivkin, I.S. Shlimak: In *Amorphous and Liquid Semiconductors*, ed. by J. Stuke and
       W. Brenig (Taylor Francis, London 1974) p. 1155
2.72   P.J. Walsh, G.C. Vezzoli: In *Amorphous and Liquid Semiconductors*, ed. by J. Stuke and W. Brenig
       (Taylor Francis, London 1974) p. 1391
2.73   K.E. Petersen, D. Adler: J. Appl. Phys. **47**, 256 (1976)
2.74   D. Adler: Proc. 7th Int'l. Conf. on Amorphous and Liquid Semiconductors (Edinburgh 1977)
       p. 695
2.75   D.K. Reinhard: Appl. Phys. Lett. **31**, 52 (1977)
2.76   M.P. Shaw: IEEE Trans. ED-**26**, 1766 (1979)
2.77   Ch. Chiang: Phys. Status Solidi (a) **54**, 735 (1979)
2.78   M.P. Shaw, K.F. Subhani: Solid State Electr. **24**, 233 (1981)
2.79   J. Kotz, M.P. Shaw: J. Appl. Phys. **55**, 427 (1984)

Reviews on Breakdown in Insulators (in particular SiO$_2$) (see also [1.1])

2.80   N. Klein: Adv. Electron. Electron Phys. **26**, 309 (1969)
2.81   J.J. O'Dwyer: *The Theory of Electrical Conduction and Breakdown in Solid Dielectrics* (Clarendon,
       Oxford 1973)
2.82   L. Altcheh, N. Klein: IEEE Trans. ED-**20**, 801 (1973)
2.83   P. Solomon: J. Vac. Sci. Technol. **14**, 1122 (1977)
2.84   N. Klein: Thin Solid Films **50**, 223 (1978); **100**, 335 (1983)

Switching and Delay Times (experimental)

2.85   H.K. Charles jr., C. Feldman: J. Appl. Phys. **46**, 819 (1975);
       N.P. Kalmykova, B.T. Kolomiets, E.A. Smorgonskaya, V.Kh. Shpunt: Sov. Phys. Semicond. **14**,
       1280 (1980);
       G.C. Vezzoli, L.W. Doremus, S. Levy, G.K. Gaulé, B. Lalevic, M. Shoga: J. Appl. Phys. **52**, 833
       (1981)

Delay Time Statistics in Superfluorescence

2.86   F. Haake, J.W. Haus, H. King, G. Schröder, R. Glauber: Phys. Rev. Lett. **45**, 558 (1980); Phys.
       Rev. A**23**, 1322 (1981)
2.87   F. Haake, J.W. Haus, R. Glauber: Phys. Rev. A**23**, 3255 (1981)

Auger-Recombination Induced Tristability

2.88   E. Schöll: Verh. DPG **3**, 364 (1980);
       Lecture Notes VI. Sitges Conference on Systems Far From Equilibrium, Sitges/Spain, June 1980
       (unpublished)

Reviews on Excitons and Electron-Hole Droplet Condensation

2.89   H. Haken, S. Nikitine (eds.): *Excitons at High Density*, Springer Tracts Mod. Phys., Vol. 73
       (Springer, Berlin, Heidelberg 1975)

2.90   E. Hanamura, H. Haug: Phys. Repts. C33, 209 (1977)
2.91   T.M. Rice: Sol. State Phys. 32, 1 (1977)
2.92   J.C. Hensel, T.G. Phillips, G.A. Thomas: Sol. State Phys. 32, 87 (1977)
2.93   E.O. Göbel, G. Mahler: Festkörperprobleme 19, 105 (Vieweg, Braunschweig 1979)
2.94   K. Cho (ed.): *Excitons* (Springer, Berlin, Heidelberg 1979)
2.95   C.D. Jeffries, L.V. Keldysh (eds.): *Electron-Hole Droplets in Semiconductors* (North Holland, Amsterdam 1983)

Exciton g-r Kinetics

2.96   E.L. Nolle: Sov. Phys. – Solid State 9, 90 (1967)
2.97   C. Benoit à la Guillaume, J.M. Debever, F. Salvan: Phys. Rev. 177, 567 (1969)
2.98   H.B. Bebb, E.W. Williams: *Semicond. and Semimetals* 8, 181 (Academic, New York 1972)
2.99   T.K. Lo: Ph.D. Thesis (University of California, 1974)
2.100   H. Haug, P. Mengel: J. Lum. 12/13, 629 (1976)
2.101   H. Sternheim, E. Cohen: Solid State Electr. 21, 1343 (1978)
2.102   G.C. Osbourn, S.A. Lyon, K.R. Elliott, D.L. Smith, T.C. McGill: Solid State Electr. 21, 1339 (1978)
2.103   W. Klingenstein, W. Schmid: Phys. Rev. B20, 3285 (1979)
2.104   K. Aoki, T. Kobayashi, K. Yamamoto: Phys. Lett. 79A, 445 (1980) (GaAs)
2.105   M. Ganser, M. Seelmann-Eggebert, R.P. Huebener: Phys. Status Solidi (b) 111, 131 (1982)
2.106   Ch. Nöldeke, W. Metzger, R.P. Huebener, H. Schneider: Phys. Status Solidi (b) 129, 224 (1985) (Ge)

Excitonic Mechanism for Limit-Cycle Oscillations

2.107   A. Pimpale, P.T. Landsberg, L.L. Bonilla, M.G. Velarde: J. Phys. Chem. Solids 42, 873 (1981)

Excitonic Mechanisms for Bistability

2.108   E. Schöll, unpublished

Theory of Electron-Hole Droplet Formation

2.109   S.W. Koch: *Dynamics of First-Order Phase Transitions in Equilibrium and Nonequilibrium Systems*, Lecture Notes Phys., Vol. 207 (Springer, Berlin, Heidelberg 1984)

Reviews on Optical Bistability in Semiconductors

2.110   D.A.B. Miller, S.D. Smith, C.T. Seaton: IEEE J. QE-17, 312 (1981)
2.111   E. Abraham, S.D. Smith: J. Phys. E15, 33 (1982)
2.112   H. Haug: *Festkörperprobleme* 22, 149 (Vieweg, Braunschweig 1982)
2.113   M.H. Pilkuhn (ed.): *High Excitation and Short Pulse Phenomena* (North Holland, Amsterdam 1985);
        H.M. Gibbs, P. Mandel, N. Peyghambarian, S.D. Smith (eds.): *Optical Bistability III*, Springer Proc. in Physics, Vol. 8 (Springer, Berlin, Heidelberg 1986)

Optical Bistability Involving Excitons

2.114   H.M. Gibbs, A.C. Gossard, S.L. McCall, A. Passner, W. Wiegmann, T.N.C. Venkatesan: Sol. State Commun. 30, 271 (1979) (GaAs)
2.115   J. Goll, H. Haken: Phys. Status Solidi (b) 101, 489 (1980)
2.116   N. Peyghambarian, H.M. Gibbs, M.C. Rushford, D.A. Weinberger: Phys. Rev. Lett. 51, 1692 (1983) (CuCl)
2.117   M. Dagenais, H.G. Winful: Appl. Phys. Lett. 44, 574 (1984) (Cd)

Optical Bistability in InSb

2.118   D.A.B. Miller, S.D. Smith: Opt.Commun. 31, 101 (1979)
2.119   A.K. Kar, J.G.H. Mathew, S.D. Smith, B. Davis, W. Prettl: Appl. Phys. Lett. 42, 334 (1983)

Spatial Effects (Kinks, Phase Coexistence) in Optical Bistability

2.120   S.W. Koch, H.E. Schmidt, H. Haug: J. Lum. **30**, 232 (1985)
2.121   H.M. Gibbs, G.R. Olbright, N. Peyghambarian, H.E. Schmidt, S.W. Koch, H. Haug: Phys. Rev.
        A**32**, 692 (1985)
2.122   H. Haug, S.W. Koch: IEEE J. QE-**21**, 1385 (1985)

Chaos in Optical Bistability

2.123   H.M. Gibbs, F.A. Hopf, D.L. Kaplan, R.L. Shoemaker: Phys. Rev. Lett. **46**, 474 (1981)
2.124   R. Neumann, S.W. Koch, H.E. Schmidt, H. Haug: Z. Phys. B**55**, 155 (1984)
2.125   J.R. Ackerhalt, P.W. Milonni, M.L. Shik: Phys. Repts. **128**, 205 (1985);
        H. Haug: In *Lasers and Synergetics*, ed. by R. Graham, Springer Proc. in Phys. (Springer, Berlin,
        Heidelberg 1987)

Optically Induced Avalanche

2.126   T.W. Nee, C.D. Cantrell, J.F. Scott, M. Scully: Phys. Rev. B**17**, 3936 (1978)
2.127   T. Grave, E. Schöll, H. Wurz: J. Phys. C**16**, 1693–1711 (1983)

Chapter 3

General Analysis of Linear Modes in NDC Elements (for specific examples see, e.g. [1.2, 3, 62])

3.1   B.K. Ridley: Proc. Phys. Soc. **82**, 954 (1963)
3.2   H. Thomas: Lecture Notes, Conf. on Fluctuation Phenomena, Chania, Crete (1969)
3.3   H. Thomas: In *Synergetics*, ed. by H. Haken (Teubner, Stuttgart 1973) p. 87
3.4   H. Thomas: "Current Instabilities", in *Cooperative Effects*, ed. by H. Haken (North Holland,
       Amsterdam 1974) p. 171
3.5   V.L. Bonch-Bruevich, I.P. Zvyagin, A.G. Mironov: *Domain Electrical Instabilities in Semiconduc-
       tors* (Consultant Bureau, New York 1975)

Linear Modes of One-Carrier g-r Mechanisms

3.6   E. Schöll: *Festkörperprobleme* **26**, 309 (Vieweg, Braunschweig 1986)
3.7   E. Schöll: Z. Phys. B**48**, 153 (1982)

Generalized Shockley-Read-Hall-Kinetics

3.8   D.A. Evans: In *Solid State Theory*, ed. by P.T. Landsberg (Wiley, London 1969)

Anomalous Tilted SNDC Characteristics

Poole-Frenkel Tunneling
3.9   A.A. Sukhanov: Sov. Phys. Semiconductors **5**, 1160 (1972)

Sasaki Effect in Many-Valley Semiconductors
3.10   Z.S. Gribnikov, V.A. Kochelap, V.V. Mitin: Sov. Phys. JETP **32**, 991 (1971); see [1.33, §26, Fig. 33]

Nerve-Axon Excitations in Neurobiology
3.11   S. Thiesen: Thesis RWTH Aachen (1983); submitted to Physica D

Bessel Functions

3.12   E. Jahnke, F. Emde, F. Lösch: *Tables of Higher Functions* (Teubner, Stuttgart 1960)

Continuous Bifurcation in Reaction-Diffusion Systems

3.13   K. Kirchgässner: In *Synergetics*, A Workshop, ed. by H. Haken, Springer Ser. Syn., Vol. 2
       (Springer, Berlin, Heidelberg 1977) p. 34
3.14   M. Büttiker, H. Thomas: Phys. Rev. A**24**, 2635 (1981)

Bifurcation of Current Filaments in Thermally Induced SNDC

3.15   H. Lueder, W. Schottky, E. Spenke: Naturwiss. **24**, 61 (1936)

Moving Gunn Domains in NNDC Elements

3.16   M.P. Shaw, P.R. Solomon, H.L. Grubin: IBM J. Res. Develop. **13**, 587 (1969)
3.17   P.R. Solomon, M.P. Shaw, H.L. Grubin, R. Kaul: IEEE Trans. ED-**22**, 127 (1975)

Competing Interaction of Filaments and Domains

3.18   C.L. Dick, B. Ancker-Johnson: Phys. Rev. B5, 526 (1972)

Linear Mode Analysis of SNDC Elements

3.19   M.P. Shaw, H.L. Grubin, I.J. Gastman: IEEE Trans. ED-**20**, 169 (1973)
3.20   F.G. Bass, Yu.G. Gurevich, S.A. Kostylev, N.A. Terent'eva: Sov. Phys. Semicond. **17**, 808 (1983)

Oscillatory Instability in a 2-Level g-r Mechanism

3.21   E. Schöll: Physica **134B**, 271 (1985)

Relaxation Semiconductors

3.22   W. van Roosbroeck: Phys. Rev. **119**, 636 (1960)

Bifurcation of Travelling and Standing Waves in Two-Component Reaction-Diffusion Systems

3.23   S. Thiesen, H. Thomas: Z. Phys. B **65**, 397 (1987)

Chapter 4

Filamentation in a One-Carrier g-r Mechanism

4.1   E. Schöll: J. Physique C7, 57 (1981)
4.2   E. Schöll: Z. Phys. **B48**, 153 (1982)
4.3   E. Schöll: In *Dynamical Systems and Chaos*, ed. by L. Garrido, Lecture Notes Phys., Vol. 179 (Springer, Berlin, Heidelberg 1983) p. 204
4.4   E. Schöll: Proc. 17th Int'l. Conf. Physics of Semiconductors (San Francisco 1984), ed. by J.D. Chadi, W. Harrison (Springer, New York 1985) p. 1353
4.5   E. Schöll: Solid-State Electron. **29**, 687 (1986)
4.6   E. Schöll: unpublished

Phase-Portrait Analysis of NNDC Instabilities

4.7   K.W. Böer, P.L. Quinn: Phys. Stat. Sol. **17**, 307 (1966)
4.8   K.W. Böer, G. Döhler: Phys. Rev. **186**, 793 (1969)
4.9   K.W. Böer: IBM J. Res. Develop. **13**, 573 (1969)
4.10   J.B. Gunn: IBM J. Res. Develop. **13**, 591 (1969)
4.11   E.M. Conwell: IEEE Trans. ED-**17**, 262 (1970)
4.12   V.L. Bonch-Bruevich, I.P. Zvyagin, A.G. Mironov: *Domain Electrical Instabilities in Semiconductors* (Consultant Bureau, New York 1975)
4.13   H. Tateno, S. Kataoka, K. Tomizawa: Solid State and El. Dev. **3**, 145 (1979)

Equal-Areas Rules (see also [4.2, 5])

4.14   F. Schlögl: Z. Phys. **253**, 147 (1972)
4.15   D. Adler, M.S. Shur, M. Silver, S.R. Ovshinsky: J. Appl. Phys. **51**, 3289 (1980)
4.16   A.F. Volkov, Sh. M. Kogan: Sov. Phys. Usp. **11**, 881 (1969)

4.17   P.N. Butcher: Phys. Lett. **19**, 546 (1965);
        M.P. Shaw, H.L. Grubin, P. Solomon: *The Gunn-Hilsum Effect* (Academic, New York 1979)

Universal Unfolding, Catastrophe Theory

4.18   R. Thom: *Structural Stability and Morphogenesis* (Benjamin, New York, 1975)

Measured Filamentary SNDC Characteristics

4.19   G.E. Stillman, C.M. Wolfe, J.O. Dimmock: *Semicond. and Semimetals* **12**, 169 (Academic, New York 1977)

Reviews on Current Filaments in SNDC Elements (see also [4.12, 16])

4.20   A.M. Barnett: *Semicond. and Semimetals* **6**, 141 (Academic New York 1970);
        K.C. Kao, W. Hwang: *Electrical Transport in Solids* (Pergamon, Oxford 1981)
4.21   E. Schöll: *Festkörperprobleme* **26**, 309 (Vieweg, Braunschweig 1986)

Filamentation in Electron Overheating Mechanisms

4.22   A.F. Volkov, Sh.M. Kogan: Sov. Phys. JETP **25**, 1095 (1967);
        Sh.M. Kogan: Sov. Phys. JETP **27**, 656 (1968);
        F.G. Bass, V.S. Bochkov, Yu.G. Gurevich: Sov. Phys. JETP **31**, 972 (1970)
4.23   B.S. Kerner, V.V. Osipov: Sov. Phys. JETP **44**, 807 (1976), **47**, 874 (1978); Sov. Phys. Solid State **21**, 1348 (1979); Sov. Phys. Semicond. **13**, 424, 523 (1979); Sov. Phys. JETP **52**, 1122 (1980)

Filamentation in pin-Diodes

4.24   A.M. Barnett: IBM J. Res. Develop. **13**, 522 (1969)
4.25   V.V. Osipov, V.A. Kholodnov: Sov. Phys. Semicond. **4**, 1033 (1971)
4.26a) H. Juling, D. Jäger: BEDO **15**, 143 (1982);
        Th. Pioch, H. Baumann, D. Jäger: BEDO **18**, 133 (1985)
4.26b) H. Baumann, T. Pioch, H. Dahmen, D. Jäger: Scanning Electron Microscopy (SEM Inc., AMF O'Hare, Chicago 1986/II) p. 441
4.26c) D. Jäger, H. Baumann, R. Symanczyk: Phys. Lett. A**117**, 141 (1986)

Filamentation in pnpn-Diodes (see also [1.57b])

4.27   I.V. Varlamov, V.V. Osipov: Sov. Phys. Semicond. **3**, 803 (1970);
        I.V. Varlamov, V.V. Osipov, E.A. Poltoratskii: Sov. Phys. Semicond. **3**, 978 (1970);
        G. Wachutka: Verh. DPG **1** (1987) HL-22.2

Filamentation in Amorphous Chalcogenide Films

4.28   J.R. Bosnell, C.B. Thomas: Solid State Electr. **15**, 1261 (1972)
4.29   K.E. Petersen, D. Adler: J. Appl. Phys. **47**, 256 (1976)

Filamentation in Low Temperature Impurity Breakdown

4.30   K.M. Mayer, R. Gross, J. Parisi, J. Peinke, R.P. Huebener: Solid State Commun. **63**, 55 (1987)

Observation of Current Filaments see [4.20, 24, 26, 28, 29, 30]

Submicron Structures

4.31   H.L. Grubin, D.K. Ferry, G.J. Iafrate, J.R. Barker: "The Numerical Physics of Micron-Length and Submicron-Length Semiconductor Devices" in *VLSI Electronics: Microstructure Science* **3**, Chap. 6 (Academic, New York 1982)

4.32   Nato Summer School on the Physics of Submicron Structures. San Miniato, Italy, July 1983;
Proc. 2nd Int'l. Conf. on Superlattices, Microstructures and Microdevices, Gothenburg, Sweden,
August 1986;
F. Nizzoli, K.-H. Rieder, R.F. Willis (eds.): *Dynamical Phenomena at Surfaces, Interfaces and
Superlattices*, Springer Ser. Surf. Sci., Vol. 3 (Springer, Berlin, Heidelberg 1985)

Influence of Boundaries in NNDC Elements (see also [4.8, 9, 11, 13])

4.33   H. Kroemer: IEEE Trans. ED-11, 819 (1968)
4.34   T.E. Hasty, R. Stratton, E.L. Jones: Phys. Rev. 39, 4623 (1968)
4.35   M.P. Shaw, P.R. Solomon, H.L. Grubin: IBM J. Res. Develop. 13, 587 (1969);
P.R. Solomon, M.P. Shaw, H.L. Grubin, R. Kaul: IEEE Trans. ED-22, 127 (1975)

Surface Defects

4.36   J. Pollmann: *Festkörperprobleme* 20, 117 (Vieweg, Braunschweig 1980)

Surface g-r Kinetics and Recombination Statistics

4.37   P.T. Landsberg, C.M. Klimpke: Solid-State Electron. 23, 1139 (1980)
4.38   P.T. Landsberg, IEEE Trans. ED-29, 1284 (1982)

Boundary Conditions at Surfaces and Interfaces

4.39   S.M. Sze: *Physics of Semiconductor Devices* (Wiley, New York 1981)
4.40   P.T. Landsberg, M.S. Abrahams: Proc. 16th Photovoltaic Specialist Conf. San Diego (1982)
4.41   Y.H. Chen, F.A. Lindholm: J. Appl. Phys. 55, 964 (1984)
4.42   O.v. Roos, F.A. Lindholm: J. Appl. Phys. 57, 415 (1985)
4.43   A.H. Marshak, K.M. van Vliet: Solid-State Electron. 23, 1223 (1980)
4.44   P.T. Landsberg, M.S. Abrahams: J. Appl. Phys. 55, 4284 (1984)
4.45   E. Schöll: J. Appl. Phys. 60, 1434 (1986)

Influence of Boundaries in General Dissipative Systems (see also [1.112])

4.46   E. Schöll: Z. Phys. B62, 245 (1986)
4.47   M. Hanson: J. Chem. Phys. 66, 5551 (1977)
4.48   G. Jetschke: Doctoral Thesis, Jena (1978);
G. Jetschke: Phys. Lett. 72A, 265 (1979)
4.49   J. Ibañez, M. Velarde: Phys. Rev. A18, 750 (1978)
4.50   P. Hemmer, M. Velarde: Z. Phys. B31, 111 (1978)
4.51   V. Fairén, M.G. Velarde: Repts. Math. Phys. 16, 421 (1979)
4.52   M. Velarde: In *Stability of Thermodynamic Systems*, eds. by J. Casas-Vásquez, G. Lebon (Springer
Berlin, Heidelberg 1982) p. 248
4.53   M.A. Livshits, G.T. Gurija, B.N. Belintsev, M.V. Volkenstein: J. Math. Biology 11, 295 (1981)
4.54   L. Kramer: Physica 13D, 357 (1984)

Optical Threshold Switches

4.55   E. Schöll: unpublished
4.56   R. Arrathoon, M.H. Hassoun: Opt. Lett. 10, 143 (1984);
D.A.B. Miller, D.S. Chemla, T.C. Damen, A.C. Gossard, W. Wiegmann, T.H. Wood, C.A. Burrus:
Appl. Phys. Lett. 45, 13 (1984)

Recombination Limited Ambipolar Currents

4.57   M.A. Lampert, P. Mark: *Current Injection in Solids* (Academic, New York 1970)

Recombination Statistics

4.58   D.A. Evans: In *Solid State Theory*, ed. by P.T. Landsberg (Wiley, New York 1969) Chap. VII

Multiple Filaments (see also [4.23] (theory), [4.26c] (experiment))

4.59   B.S. Kerner, V.F. Sinkevich: Sov. Phys. JETP Lett. **36**, 437 (1982)
4.60   Ch. Radehaus, K. Kardell, H. Baumann, D. Jäger, H.G. Purwins: Z. Phys. B **65**, 515 (1987)

Pattern Formation in Activator-Inhibitor Systems

4.61   P.C. Fife: J. Math. Biol. **4**, 353 (1977)

Tristability in Chemical Reaction Systems

4.62   G. Czajkowski: Z. Phys. **270**, 25 (1974)

Quantum Mechanical Treatment of Current Filamentation

4.63   H. Haken: Physica Scripta **32**, 274 (1985)

Chapter 5

Stability of Dissipative Structures in Infinite Systems

g-r Instabilities
5.1   E. Schöll: Z. Phys. B**52**, 321 (1983)

Chemical Reaction – Diffusion Systems
5.2   L.A. Turski, D. dinh Thanh: J. Phys. C**11**, L823 (1978)
5.3   E. Magyari: J. Phys. A**15**, L139 (1982)
5.4   F. Schlögl, C. Escher, R.S. Berry: Phys. Rev. A**27**, 2689 (1983);
        F. Schlögl, R.S. Berry: Phys. Rev. A**21**, 2078 (1980)
5.5   M. Büttiker, H. Thomas: Phys. Rev. A**24**, 2635 (1981)

Flame Fronts in Combustion
5.6   G.I. Barenblatt, Ya.B. Zel'dovich: Prikl. Mat. Mekh. **21**, 856 (1957)

Electron Overheating Instabilities in Semiconductors
5.7   A.F. Volkov, Sh.M. Kogan: Sov. Phys. JETP **25**, 1095 (1967)
5.8   Sh.M. Kogan: Sov. Phys. JETP **27**, 656 (1968)

Gunn Domains, Recombination Domains
5.9   B.W. Knight, G.A. Peterson: Phys. Rev. **155**, 393 (1967)
5.10   R. Enderlein: Phys. Status Solidi **28**, 519 (1968)

Semiconductor Instabilities (Reviews)
5.11   A.F. Volkov, Sh.M. Kogan: Sov. Phys. Usp. **11**, 881 (1969)
5.12   V.L. Bonch-Bruevich, I.P. Zvyagin, A.G. Mironov: *Domain Electrical Instabilities in Semiconductors* (Consultant Bureau, New York 1975)

Bragg Scattering Induced Field Domains
5.13   M. Büttiker, H. Thomas: Z. Phys. B**34**, 301 (1979)

Domain Walls in Ferromagnets
5.14   E. Magyari, H. Thomas: Phys. Rev. Lett. **51**, 54 (1983)
5.15   E. Magyari, H. Thomas: Z. Phys. B**57**, 141 (1984)

Defective Degeneracy of Goldstone Mode

5.16   E. Magyari, H. Thomas: Phys. Rev. Lett. **53**, 1866 (1984)

Properties of the Schrödinger Equation and Related Differential Equations

5.17   A. Messiah: *Quantum Mechanics*, Vol. I (North Holland, Amsterdam 1972)
5.18   E.L. Ince: *Ordinary differential equations* (Dover, New York 1956)
5.19   E. Kamke: *Differentialgleichungen, Lösungsmethoden und Lösungen*, Vol. I (Akadem. Verlagsgesellschaft, Leipzig 1967)
5.20   S. Flügge: *Practical Quantum Mechanics*, Vol. 1 (Springer, Berlin, Heidelberg 1971)
5.21   see [Ref. 5.17, Chap. IX, § 4]
5.22   E.A. Coddington, N. Levinson: *Theory of ordinary differential equations* (Mc Graw-Hill, New York 1955) Chap. 8, § 1

Stability of Dissipative Structures with Finite Boundaries

g-r Instability in Semiconductors
5.23   E. Schöll: Proc. 17th Int'l. Conf. Physics Semiconductors (San Francisco 1984), ed. by J.D. Chadi, W.A. Harrison (Springer, New York 1985) p. 1353

Electron Overheating Instability in Semiconductors
5.24   F.G. Bass, V.S. Bochkov, Yu.G. Gurevich: Sov. Phys. JETP **31**, 972 (1970)

Electrothermal Instability in Semiconductors
5.25   J.L. Jackson, M.P. Shaw: Appl. Phys. Lett. **25**, 666 (1974)

Ballast Resistor
5.26   D. Bedeaux, P. Mazur: Physica **105A**, 1 (1981)

Chemical Reaction – Diffusion Systems
5.27   M. Hanson: J. Chem. Phys. **66**, 5551 (1977)
5.28   G. Jetschke: Ph.D. Thesis, Jena (1978); Phys. Lett. **72A**, 265 (1979)
5.29   M. Velarde: In *Stability of Thermodynamic Systems*, ed. by J. Casas-Vázques, G. Lebon (Springer, Berlin, Heidelberg 1982) p. 248
5.30   L.L. Bonilla, M.G. Velarde: J. Math. Phys. **21**, 2586 (1980)
5.31   J.L. Ibáñez, M.G. Velarde: Phys. Rev. **A18**, 750 (1978)
5.32   V. Fairén, M.G. Velarde: Repts. Math. Phys. **16**, 421 (1979)
5.33   N. Chafee: J. Diff. Eq. **18**, 111 (1975)
5.34   H. Malchow, W. Ebeling: Z. Phys. Chemie Leipzig **265**, 49 (1984)
5.35   E. Schöll: Z. Phys. **B62**, 245 (1986)

Population Dynamics
5.36   E. Yanagida: J. Math. Biol. **15**, 37 (1982)
5.37   A. Schiaffino, A. Tesei: J. Math. Biol. **15**, 93 (1982)

Dynamics of Phase Transitions

5.38   S.W. Koch: *Dynamics of First-Order Phase Transitions in Equilibrium and Nonequilibrium Systems*, Lect. Notes Phys. Vol. 207 (Springer, Berlin, Heidelberg 1984)

Spinodal Decomposition

5.39   J.W. Cahn, J.E. Hilliard: J. Chem. Phys. **28**, 258 (1958); **31**, 668 (1959); J.W. Cahn: Trans. Metall. Soc. AIME **242**, 166 (1968)
5.40   J.S. Langer: Ann. Phys. (N.Y.) **65**, 53 (1971)
5.41   J.S. Langer, M. Baron, H.D. Miller: Phys. Rev. **A11**, 1417 (1975)
5.42   e.g., D.W. Heermann: Z. Phys. **B55**, 309 (1984); Phys. Rev. Lett. **52**, 1126 (1984)

Nucleation Theory

5.43   R. Becker, W. Döring: Ann. Phys. (Leipzig) **24**, 719 (1935)
5.44   K. Binder, D. Stauffer: Adv. Phys. **25**, 342 (1976)
5.45   J.D. Gunton, M. San Miguel, P.S. Sahni: In *Phase Transitions and Critical Phenomena* **8**, 267
        (Academic, New York 1983);
        M. San Miguel, J.D. Gunton, G. Dee, P.S. Sahni: Phys. Rev. **B23**, 2334 (1981)
        (tricritical models)

Relative and Absolute Stability in Bistable Systems

5.46   L. Kramer: Z. Phys. **B41**, 357 (1981); **45**, 167 (1981)

Noise-Induced Phase Transitions

5.47   W. Horsthemke, R. Lefever: *Noise-Induced Transitions, Theory and Applications in Physics,
        Chemistry, and Biology*, Springer Ser. Syn., Vol. 15 (Springer, Berlin, Heidelberg 1984)
5.48   W. Horsthemke, D.K. Kondepudi (eds.): *Fluctuations and Sensitivity in Nonequilibrium Systems*,
        Springer Proc. Phys., Vol. 1 (Springer, Berlin, Heidelberg 1984)

Solitary Waves

5.49   K. Parliński, P. Zieliński: Z. Phys. **B44**, 317 (1981)
5.50   F. Falk: Z. Phys. **B54**, 159 (1984)
5.51   A.S. Mikhailov: Z. Phys. **B41**, 277 (1981)
5.52   F. Schlögl: Z. Phys. **253**, 147 (1972)
5.53   A. Nitzan, P. Ortoleva, J. Ross: Faraday Symp. Chem. Soc. **9**, 241 (1974)
5.54   P. Ortoleva, J. Ross: J. Chem. Phys. **60**, 5090 (1974); **63**, 3398 (1975)
5.55   T. Dehrmann: J. Phys. **A15**, L649 (1982)
5.56   M.A. Livshits: Z. Phys. **B53**, 83 (1983)
5.57   J. Rinzel, J.B. Keller: Biophys. J. **13**, 1313 (1973);
        J. Rinzel: J. Math. Biol. **2**, 205 (1975)
5.58   K.P. Hadeler, F. Rothe: J. Math. Biol. **2**, 251 (1975)
5.59   P.C. Fife, J.B. McLeod: Arch. Rat. Mech. Anal. **65**, 335 (1977)
5.60   P.C. Fife: *Mathematical Aspects of Reacting and Diffusing Systems* (Springer, Berlin, Heidelberg
        1979)
5.61   K. Kirchgässner: J. Diff. Eqs. **45**, 113 (1982)
5.62   P. Grassberger: Z. Phys. **B47**, 365 (1982)

Chapter 6

Circuit-Induced Oscillations

6.1   B. Van der Pol: London, Edinburgh and Dublin Phil. Mag. **3**, 65 (1927)
6.2   J.J. Stoker: *Nonlinear Vibrations* (Wiley, New York 1950)
6.3   N. Minorsky: *Nonlinear Oscillations* (Van Nostrand, Princeton 1962)

in NNDC Systems
6.4   P.R. Solomon, M.P. Shaw, H.L. Grubin: J. Appl. Phys. **43**, 159 (1972)

in SNDC Systems
6.5   M.P. Shaw, I.J. Gastman: Appl. Phys. Lett. **19**, 243 (1971); J. Non-Crystalline Solids **8–10**, 999
        (1972)
6.6   M.P. Shaw, H.L. Grubin, I.J. Gastman: IEEE Trans. ED-**20**, 169 (1973)

Gunn Domain Oscillations

6.7   B.K. Ridley, T.B. Watkins: Proc. Phys. Soc. (London) **78**, 293 (1961);
        B.K. Ridley: Proc. Phys. Soc. (London) **82**, 954 (1963)
6.8   C. Hilsum: Proc. IRE **50**, 185 (1962)
6.9   J.B. Gunn: Solid State Commun. **1**, 88 (1963);
        J.B. Gunn: Proc. 7th Int'l. Conf. Phys. Semicond. (Dunod, Paris 1964);
        J.B. Gunn: IBM J. Res. Dev. **10**, 300 (1966)
6.10   H. Kroemer: Proc. IEEE **52**, 1736 (1964)
6.11   D.E. McCumber, A.G. Chynoweth: IEEE Trans. ED-**13**, 4 (1966)
6.12   P.N. Butcher: Phys. Lett. **19**, 546 (1965); Rep. Prog. Phys. **30**, 97 (1967)
6.13   B.W. Knight, G.A. Peterson: Phys. Rev. **155**, 393 (1967)
6.14   H.L. Grubin, M.P. Shaw, P.R. Solomon: IEEE Trans. ED-**20**, 63 (1973)
6.15   M.P. Shaw, H.L. Grubin, P.R. Solomon: *The Gunn-Hilsum-Effect* (Academic, New York 1979);
        M.P. Shaw, H.L. Grubin, E. Schöll: *The Physics of Instabilities in Solid State Electron Devices*
        (Plenum, New York) in preparation

Recombination Domain Oscillations (see also reviews [1.26, 30])

6.16   V.L. Bonch-Bruevich, S.G. Kalashnikov: Sov. Phys.-Solid State **7**, 599 (1965);
        V.L. Bonch-Bruevich, Sh.M. Kogan: Sov. Phys.-Solid State **7**, 15 (1965);
        V.L. Bonch-Bruevich: Sov. Phys.-Solid-State **8**, 1397 (1966)
6.17   K.W. Böer: J. Phys. Chem. Solids **22**, 123 (1961)
6.18   V.A. Sablikov: Sov. Phys.-Semicond. **13**, 785 (1979);
        P.V. Bogun, I.V. Karpova, V.A. Sablikov: Sov. Phys.-Semicond. **14**, 132 (1980)
6.19   D.C. Northrop, P.R. Thornton, K.E. Tresize: Solid-State Electron. **7**, 17 (1964)
6.20   B.K. Ridley, P.H. Wisbey: Brit. J. Appl. Phys. **18**, 761 (1967);
        B.K. Ridley, R.G. Pratt: J. Phys. Chem. Sol. **26**, 21 (1965)
6.21   M. Kaminska, J.M. Parsey, J. Lagowski, H.C. Gatos: Appl. Phys. Lett. **41**, 989 (1982)
6.22   I.D. Vagner, I.V. Ioffe: Sov. Phys.-Sol. State **13**, 2926 (1972); Sov. Phys.-Semicond. **6**, 1092
        (1973);
        P. Rosenau, I.D. Vagner: Phys. Lett. **108A**, 50 (1985)

Oscillations in pin Diodes

6.23   N. Holonyak, Jr., S.F. Bevacqua: Appl. Phys. Lett. **2**, 71 (1963)
6.24   J.S. Moore: Appl. Phys. Lett. **10**, 58 (1967);
        B.G. Streetman, M.M. Blouke, N. Holonyak, Jr.: Appl. Phys. Lett. **11**, 200 (1967);
        J.S. Moore, N. Holonyak, Jr., M.D. Sirkis: Solid-State Electron. **10**, 823 (1967)
6.25   B.G. Streetman, N. Holonyak, Jr.: IBM J. Res. Develop. **13**, 529 (1969)
6.26   W.H. Weber, G.W. Ford: Solid-State Electron. **15**, 1277 (1972)
6.27   R. Kassing, I. Dudeck: Phys. status Solidi (a) **31**, 431 (1975)

Bulk g-r oscillations

6.28   V.V. Vladimirov, V.N. Gorshkov: Sov. Phys.-Semicond. **14**, 247 (1980); V.V. Vladimirov, V.N.
        Gorshkov, P.M. Golovinskij: Sov. Phys.-Semicond. **15**, 894 (1981)
6.29   M.A. Rizakhanov, E.M. Zobov: Sov. Phys. Semicond. **13**, 1185 (1979)

Temperature – g-r Oscillations

6.30   V.L. Vinetskii, I.D. Konozenko, S.I. Shakhovtsova: Sov. Phys.-Sol. State **5**, 1971 (1964)
6.31   H. Gobrecht, G. Gschwind, D.H. Haberland, N. Nelkowski, A. Stais: Phys. Status Solidi (a) **4**,
        K19 (1970)
6.32   N.N. Degtyarenko, V.F. Elesin, V.A. Furmanov: Sov. Phys.-Semicond. **7**, 1147 (1974)
6.33   V.L. Bonch-Bruevich, Le Vu Ky: Phys. Status Solidi (b) **124**, 111 (1984)

306   References

Nonlinear Transient Response of Extrinsic Photoconductors

6.34   R.M. Westervelt, S.W. Teitsworth: J. Appl. Phys. **57**, 5457 (1985)

Exciton-Induced Limit Cycle Oscillations

6.35   A. Pimpale, P.T. Landsberg, L.L. Bonilla, M.G. Velarde: J. Phys. Chem. Sol. **42**, 873 (1981)
6.36   E. Schöll, unpublished
6.37   E.F. Gross, S.A. Permogorov, V.V. Travnikov, A.V. Selin: Sov. Phys. Solid-State **14**, 1193 (1972)

Breathing of Filaments

6.38   K.M. Mayer, R. Gross, J. Parisi, J. Peinke, R.P. Huebener: Solid State Commun. **63**, 55 (1987)
6.39   E. Schöll, unpublished

Helical Waves

6.40   I.L. Ivanov, S.M. Ryvkin: Sov. Phys. Techn. Phys. **3**, 722 (1958)
6.41   R.D. Larrabee, M.C. Steele: J. Appl. Phys. **31**, 1519 (1960)
6.42   M. Glicksman: Phys. Rev. **124**, 1655 (1961)
6.43   T. Misawa, T. Yamada: Jap. J. Appl. Phys. **2**, 19 (1963)
6.44   C.E. Hurwitz, A.L. McWhorter: Phys. Rev. **134**, A1033 (1964)

Oscillations in Parallel, Transverse, and Vanishing Magnetic Fields

6.45   M. Cardona, W. Ruppel: J. Appl. Phys. **31**, 1826 (1967)

Generalised Equal Areas Rules

6.46   E. Schöll, P.T. Landsberg: Generalised Equal Areas Rules, to be published

Circuit-Induced Bifurcations

6.47   E. Schöll, unpublished

Introductions to Chaos

6.48   H. Haken: *Synergetics, An Introduction*, 3rd ed., Springer Ser. Syn., Vol. 1 (Springer, Berlin, Heidelberg 1983)
6.49   H. Haken: *Advanced Synergetics*, Springer Ser. Syn., Vol. 20 (Springer, Berlin, Heidelberg 1983)
6.50   H.G. Schuster: *Deterministic Chaos* (Physik-Verlag, Weinheim 1984)
6.51   H. Bai-Lin: *Chaos* (World Scientific, Singapore 1985) (Introduction, and reprints of original key papers)
6.52   A. Kunick, W.H. Steeb: *Chaos in dynamischen Systemen* (Bibliographisches Institut, Mannheim 1986)
6.53   P. Collet, J.P. Eckmann: *Iterated Maps on the Interval as Dynamical Systems* (Birkhäuser, Basel 1980)
6.54   J. Guckenheimer, P. Holmes: *Nonlinear Oscillations, Dynamical Systems, and Bifurcations of Vector Fields* (Springer, New York 1983)
6.55   G.M. Zaslavsky: *Chaos in Dynamic Systems* (G + B/harwood, Chur 1985)
6.56   A.V. Holden (ed.): *Chaos* (Manchester Univ. Press, Manchester 1986)

Characterisation of Strange Attractors

6.57   O.E. Rössler: Phys. Lett. **57A**, 397 (1976)
6.58   O.E. Rössler: In *Synergetics, A Workshop*, ed. by H. Haken, Springer Ser. Syn., Vol. 2 (Springer, Berlin, Heidelberg 1977) p. 184

6.59a) E. Ott: Rev. Mod. Phys. **53**, 655 (1981);
6.59b) R. Shaw: Z. Naturforsch. **36a**, 80 (1981)
6.60   J.P. Eckmann, D. Ruelle: Rev. Mod. Phys. **57**, 617 (1985)
6.61   F. Schlögl: J. Statist. Phys. **46**, 135 (1987) (Variance of Information loss)
6.62   W.F. Wolff, B.A. Huberman: Z. Phys. **B63**, 397 (1986) (transients)

Routes to Chaos

6.63   M.J. Feigenbaum: J. Stat. Phys. **19**, 25 (1978)
6.64   S. Newhouse, D. Ruelle, F. Takens: Commun. Math. Phys. **64**, 35 (1978)
6.65   Y. Pomeau, P. Manneville: Commun. Math. Phys. **77**, 189 (1980)
6.66   J.P. Eckmann: Rev. Mod. Phys. **53**, 643 (1981)

Fractal Dimension

6.67   J.D. Farmer, E. Ott, J.A. Yorke: Physica **7D**, 153 (1983)
6.68   P. Grassberger: J. Stat. Phys. **26**, 173 (1981)
6.69   P. Grassberger, I. Procaccia: Phys. Rev. Lett. **50**, 346 (1983); Physica **9D**, 189 (1983);
        P. Grassberger: In *Chaos*, ed. by A.V. Holden (Manchester University Press, Manchester 1986)
        p. 291
6.70a) B. Mandelbrot: *The Fractal Geometry of Nature* (Freeman, San Francisco 1982);
6.70b) G. Mayer-Kress (ed.): *Dimensions and Entropies in Chaotic Systems*, Springer Ser. in Syn.,
        Vol. 32 (Springer, Berlin, Heidelberg 1986)

Chaos in Van der Pol Oscillators

6.71   J. Guckenheimer: Physica **1D**, 227 (1980)

Chaos in Semiconductors

Induced by Domains in Gunn Diodes
6.72   K. Nakamura: Progr. Theoret. Phys. **57**, 1874 (1977)

Induced by Helical Plasma Waves
6.73   G.A. Held, C. Jeffries, E.E. Haller: Phys. Lett. **52**, 1037 (1984); *Proc. 17th Int. Conf. Phys. Semicond.*
        (San Francisco 1984) ed. by J.D. Chadi and W.A. Harrison (Springer, New York 1985) p. 1289
6.74   G.A. Held, C. Jeffries: Phys. Rev. Lett. **55**, 887 (1985); **56**, 1183 (1986)
6.75   C. Jeffries: Physica Scripta **T9**, 11 (1985) (review)
6.76   G.A. Held, C. Jeffries: In Springer Ser. in Syn., Vol. 32, ed. by G. Mayer-Kress (Springer, Berlin,
        Heidelberg 1986) p. 158 (review)

Induced by g-r Processes
6.77   K. Aoki, T. Kobayashi, K. Yamamoto: J. Physique **C7**, 51 (1981)
6.78   K. Aoki, T. Kobayashi, K. Yamamoto: J. Phys. Soc. Jap. **51**, 2373 (1982)
6.79   K. Aoki, K. Miyamae, T. Kobayashi, K. Yamamoto: Physica **117 & 118B**, 570 (1983)
6.80   K. Aoki, K. Yamamoto: Phys. Lett. **98A**, 72 (1983)
6.81   K. Aoki, O. Ikezawa, K. Yamamoto: Phys. Lett. **98A**, 217 (1983); **106A**, 343 (1984); K. Aoki,
        N. Mugibayashi: Phys. Lett. **114A**, 425 (1986) (theoretical)
6.82   K. Aoki, O. Ikezawa, N. Mugibayashi, K. Yamamoto: Proc. 4th Int. Conf. on Hot Electrons in
        Semicond., Innsbruck, Physica **134B**, 288 (1985)
6.83   K. Aoki, N. Mugibayashi, K. Yamamoto: Physica Scripta, Vol. **T14** (1986); Proc. 18th Int.
        Conf. Phys. Semicond. (Stockholm 1986), ed. by O. Engström (World Scientific, Singapore 1987)
        p. 1543. (theoretical)
6.84   S.W. Teitsworth, R.M. Westervelt, E.E. Haller: Phys. Rev. Lett. **51**, 825 (1983)
6.85   E.G. Gwinn, R.M. Westervelt: Phys. Rev. Lett. **57**, 1060 (1986)
6.86   R.M. Westervelt, S.W. Teitsworth, E.G. Gwinn: In *Perspectives in Nonlinear Dynamics*, ed. by
        M.F. Shlesinger (World Scientific, Singapore 1986) (review)

308   References

6.87   D.G. Seiler, C.L. Littler, R.J. Justice, P.W. Milonni: Phys. Lett. **108A**, 462 (1985); Proc. 17th Int. Conf. Phys. Semicond. (San Francisco 1984), ed. by J.D. Chadi and W.A. Harrison (Springer, New York 1985) p. 1385
6.88   J. Peinke, A. Mühlbach, R.P. Huebener, J. Parisi: Phys. Lett. **108A**, 407 (1985)
6.89   J. Peinke, B. Röhricht, A. Mühlbach, J. Parisi, Ch. Nöldeke, R. P. Huebener, O.E. Rössler: Z. Naturforsch. **40a**, 562 (1985)
6.90   B. Röhricht, B. Wessely, J. Parisi, J. Peinke: Appl. Phys. Lett. **48**, 233 (1986)
6.91   B. Röhricht, B. Wessely, J. Peinke, A. Mühlbach, J. Parisi, R.P. Huebener: Proc. 4th Int. Conf. on Hot Electrons in Semiconductors, Innsbruck, Physica **134B**, 281 (1985)
6.92   J. Peinke, A. Mühlbach, B. Röhricht, B. Wessely, J. Mannhart, J. Parisi, R.P. Huebener: Physica **23D**, 176. (1986)
6.93   J. Parisi, J. Peinke, B. Röhricht, R.P. Huebener: Proc. 18th Int. Conf. Phys. Semicond. (Stockholm 1986), ed. by O. Engström (World Scientific, Singapore 1987) p. 1571
6.94   B. Röhricht, J. Parisi, J. Peinke, O.E. Rössler: Z. Phys. **B65**, 259 (1986)
6.95   G.N. Maracas, W. Porod, D.A. Johnson, D.K. Ferry, H. Goronkin: Proc. 4th Int. Conf. on Hot Electrons in Semiconductors, Innsbruck, Physica **134B**, 276 (1985)
6.96a) S.B. Bumelienė, Yu.K. Pozhela, K.A. Pyragas, A.V. Tamaševičius: Proc. 4th Int. Conf. on Hot Electrons in Semiconductors, Innsbruck, Physica **134B**, 293 (1985)
6.96b) S.B. Bumelienė, Yu.K. Pozhela, K.A. Pyragas, A.V. Tamaševičius: Proc. 18th Int. Conf. Phys. Semicond. (Stockholm 1986), ed. by O. Engström (World Scientific, Singapore 1987) p. 1563; S.B. Bumelienė, Yu.K. Pozhela, A.V. Tamaševičius: Phys. Status Solidi (b) **134**, K71 (1986)
6.97   A. Brandl, T. Geisel, W. Prettl: Europhys. Lett. **3**, 401 (1987)

Models for g-r Induced Chaos in Semiconductors

Driven Chaos
6.98   S.W. Teitsworth, R.M. Westervelt: Phys. Rev. Lett. **53**, 2587 (1984); Phys. Rev. Lett. **56**, 516 (1986) (Experimental)

Field-Enhanced Trapping
6.99   K.A. Pyragas: Sov. Phys.-Semicond. **17**, 652 (1983) [Fiz. Tekh. Poluprov. **17**, 1035 (1983)]

Impact-Ionization Induced Dielectric-Relaxation Oscillations (see also [6.83])
6.100   E. Schöll: Proc. 4th Int. Conf. on Hot Electrons in Semiconductors, Innsbruck 1985, Physica **134B**, 271 (1985); Phys. Rev. **B34**, 1395 (1986)
6.101   E. Schöll, unpublished

Impact Ionization Induced Energy Relaxation Oscillations
6.102   E. Schöll: Proc. 18th Int. Conf. Phys. Semicond. (Stockholm 1986), ed. by O. Engström (World Scientific, Singapore 1987) p. 1555
6.103   E. Schöll, J. Parisi, B. Röhricht, J. Peinke, R.P. Huebener: Phys. Lett. A **119**, 419 (1987)
6.104   E. Schöll: Nonlinear energy relaxation oscillations in semiconductors. To be publ.

Discrete g-r Processes (1-Dimensional Map)
6.105   P.T. Landsberg, E. Schöll, P. Shukla: Physica D (to be publ.)

Theories of Nonmonotonic Impurity Impact Ionization Coefficients

6.106   B.C. Eu: J. Chem. Phys. **80**, 2123 (1984); V.V. Mitin: Appl. Phys. **A39**, 123 (1986)

Model for Helical-Wave Induced Chaos

6.107   J.M. Wersinger, J.M. Finn, E. Ott: Phys. Fluids **23**, 1142 (1980)

Subject Index